WILEY-PRAXIS SERIES IN ATMOSPHERIC PHYSICS
Series Editor: **John Mason, B.Sc., Ph.D.**
Consultant Editor: **David Sloggett, M.Sc., Ph.D.**

This series reflects the major developments which are occurring in studies of the physical and dynamic processes which take place within the Earth's atmosphere. With increasing concern over important environmental issues such as ozone depletion, global warming and dispersal of pollutants, this is a vibrant research area.

Written by international authors chosen for their expertise and high reputations in all aspects of atmospheric physics, the titles in this series provide a forum for the presentation and dissemination of new ideas and scientific results in this important field.

These books will appeal to: professional meteorologists; atmospheric physicists; civil, aeronautical, mechanical and chemical engineers; applied mathematicians; aerobiologists; architects, environmental scientists and ecologists; together with post-graduate and undergraduate students in these fields and, in some cases, high-flying 'A'-level students and non-scientists with a keen interest in environmental issues.

PHYSICS OF THE UPPER POLAR ATMOSPHERE
Asgeir Brekke, Professor in Physics, The Auroral Observatory, Institute of Mathematical and Physical Science, University of Tromsø, Tromsø, Norway

APPLICATIONS OF WEATHER RADAR SYSTEMS: A Guide to Uses of Radar Data in Meteorology and Hydrology Second edition
Christopher G. Collier, Professor of Environmental Remote Sensing, Telford Research Institute, University of Salford

Forthcoming titles
PHYSICS OF THE ENVIRONMENT AND CLIMATE
Gérard Guyot, Research Director, Laboratory of Bioclimatology of INRA, Avignon, France

OPTICS OF LIGHT SCATTERING MEDIA: Problems and Solutions
Alex A. Kokhanovsky, Institute of Physics, Academy of Sciences, Minsk, Belarus

DYNAMICS OF METEOROLOGY AND CLIMATE
Richard Scorer, Emeritus Professor and Senior Research Fellow in Mathematics and Environmental Technology, Imperial College of Science, Technology and Medicine, London

PHYSICS OF THE
UPPER POLAR ATMOSPHERE

PHYSICS OF THE UPPER POLAR ATMOSPHERE

Asgeir Brekke
The Auroral Observatory
Institute of Mathematical and Physical Science
University of Tromsø, Tromsø, Norway

JOHN WILEY & SONS
Chichester • New York • Brisbane • Toronto • Singapore

Published in association with
PRAXIS PUBLISHING
Chichester

Copyright © 1997 Praxis Publishing Ltd
The White House,
Eastergate, Chichester,
West Sussex, PO20 6UR, England

Published in 1997 by
John Wiley & Sons Ltd
in association with Praxis Publishing Ltd

All rights reserved

No part of this book may be reproduced by any means,
or transmitted, or translated into a machine language
without the written permission of the publisher

Wiley Editorial Offices

John Wiley & Sons Ltd, Baffins Lane,
Chichester, West Sussex PO19 1UD, England

John Wiley & Sons, Inc., 605 Third Avenue,
New York, NY 10158-0012, USA

Jacaranda Wiley Ltd, G.P.O. Box 859, Brisbane,
Queensland 4001, Australia

John Wiley & Sons (Canada) Ltd, 22 Worcester Road,
Rexdale, Ontario M9W 1L1, Canada

John Wiley & Sons (Asia) Pte Ltd, 2 Clementi Loop #02-01,
Jin Xing Distripark, Singapore 129809

Library of Congress Cataloging-in-Publication Data
Brekke, Asgeir.
  Physics of the upper polar atmosphere / Asgeir Brekke.
  p. cm. -- (Wiley-Praxis series in atmospheric physics)
  Includes bibliographical references and index.
  ISBN 0-471-96018-7 (Hardcover : alk. paper)
  1. Atmosphere, Upper--Polar regions. 2. Geophysics--Polar regions. I. Title. II Series.
QC879.232.P64B74 1995
551.5'14'0911--dc20   95-18695
                                                                                                   CIP

A catalogue record for this book is available from the British Library

ISBN 0-471-96018-7

Printed and bound in Great Britain by Hartnolls Ltd, Bodmin

Table of contents

Preface xi

1 The Sun as a radiation source 1
   1.1   General about the Sun 1
   1.2   The solar atmosphere 3
   1.3   The electromagnetic radiation from the Sun 7
   1.4   The sunspots – solar cycle 18
   1.5   The sunspots 25
   1.6   The electromagnetic radiation from the disturbed Sun 28
   1.7   References 31
   1.8   Exercises 31

2 The solar wind and the interplanetary magnetic field 33
   2.1   Particle emissions from the quiet Sun 33
   2.2   Laval's nozzle 36
   2.3   The solar wind equation 38
   2.4   Indications of magnetic fields on the Sun 42
   2.5   The frozen-in field concept 44
   2.6   The electric field in the solar wind 49
   2.7   The garden hose effect 50
   2.8   Hydromagnetic waves in interplanetary space 56
   2.9   References 63
   2.10 Exercises 63

3 The atmosphere of the Earth 65
   3.1   Nomenclature 65
   3.2   The greenhouse effect 68
   3.3   The temperature structure of the atmosphere 72
   3.4   Atmospheric drag on satellites 76
   3.5   The atmosphere as an ideal gas 80
   3.6   The exosphere 86

- 3.7 Height-dependent temperature 88
- 3.8 The adiabatic lapse rate 89
- 3.9 Thermal structure of the atmosphere 91
- 3.10 Diffusion 96
- 3.11 Chemistry of an oxygen atmosphere 99
- 3.12 The ozone layer 103
- 3.13 The destruction of ozone 111
- 3.14 The height profile of ozone 115
- 3.15 The role of the CFC gases 118
- 3.16 Possible pollution effects in the mesosphere 121
- 3.17 References 123
- 3.18 Exercises 124

4 The Earth's magnetic field 127
- 4.1 An historical introduction 127
- 4.2 Description of the Earth's magnetic field 130
- 4.3 Mathematical representation of the Earth's magnetic system 136
- 4.4 E-field mapping along conducting magnetic field lines 142
- 4.5 Secular variations in the Earth's magnetic field 145
- 4.6 The source of the magnetic field of the Earth 152
- 4.7 The unipolar inductor 157
- 4.8 Motion of a charged particle in a magnetic field 158
  - 4.8.1 General introduction of the guiding centre 158
  - 4.8.2 Motion in a static and uniform magnetic field 160
  - 4.8.3 Zeroth-order drift 163
  - 4.8.4 First-order drift motions 168
  - 4.8.5 Curvature drift 170
  - 4.8.6 Second-order drift 172
  - 4.8.7 The first adiabatic invariant – the magnetic moment 173
  - 4.8.8 The magnetic mirror 175
  - 4.8.9 The second adiabatic invariant 178
- 4.9 Størmer's calculation of particle motions in magnetic fields 178
- 4.10 The radiation belt 186
- 4.11 References 189
- 4.12 Exercises 189

5 The ionosphere 191
- 5.1 The production of ionization by solar radiation 191
- 5.2 The ionization profile of the upper atmosphere 201
- 5.3 The Chapman ionization profile 209
- 5.4 The recombination process 213
- 5.5 The O^+ dominant ionosphere 216
- 5.6 Ambipolar diffusion 221
- 5.7 Multicomponent topside ionosphere 223
- 5.8 Diffusion in the presence of a magnetic field 225
- 5.9 The E-layer ionization and recombination 228

5.10	The time constant of the recombination process	231
5.11	The D-region ionization and recombination	233
5.12	The plasmasphere	241
5.13	Ferraro's theorem	243
5.14	The magnetospheric convection close to the Earth	244
5.15	Equatorial fountain effect	248
5.16	References	254
5.17	Exercises	255

6 Dynamics of the neutral atmosphere ... 257

6.1	The general equations	257
6.2	The vertical motion	259
6.3	The equation of motion of the neutral gas	262
6.4	The geostrophic and thermal winds	264
6.5	The equation of motion in polar coordinates	266
6.6	The wind systems at lower altitudes	268
6.7	The wind systems of the upper atmosphere	271
6.8	Observations of the neutral wind	274
6.9	Collisions between particles	275
6.10	Collisions in gases with different temperatures	278
6.11	Drag effects	279
6.12	Thermospheric neutral winds	282
6.13	The E-region winds	290
6.14	Observations of E-region neutral winds	292
6.15	Tidal oscillations in the neutral atmosphere	297
6.16	The Brunt–Väisälä frequency	303
6.17	Internal waves in the atmosphere	306
6.18	References	311
6.19	Exercises	311

7 Currents in the ionosphere ... 313

7.1	The steady-state approach	313
7.2	Rotation of the ion velocity by height in the ionosphere	320
7.3	The current density in the ionosphere	324
7.4	Currents due to gravity and diffusion	327
7.5	Height-dependent currents and heating rates	329
7.6	Heating due to collisions	335
7.7	Heating of an oscillating electric field	337
7.8	Height-integrated currents	341
7.9	Height-integrated currents and magnetic fluctuations	344
7.10	Equivalent current systems	347
7.11	Currents at the Harang discontinuity	352
7.12	Equivalent currents at different latitudes	354
7.13	The S_q current system	360
7.14	The Kamide–Richmond–Matsushita (KRM) method	365
7.15	Polar cap conductance and current distribution	376

7.16	Mapping of E-fields in the ionosphere	378
7.17	Polarization fields around an auroral arc	382
7.18	References	385
7.19	Exercises	386

8 The magnetosphere ... 389

8.1	The magnetic field away from the Earth	389
8.2	The magnetic tail	399
8.3	Magnetic field merging	402
8.4	Some magnetohydrodynamic concepts	407
8.5	The energy flux into the magnetosphere	410
8.6	Some aspects of the energy balance	412
8.7	Currents in a collisionless plasma	418
8.8	Space charges in the magnetosphere	422
8.9	Currents related to an auroral arc	424
8.10	High-latitude convection patterns	426
8.11	High-latitude convection and field-aligned currents	430
8.12	References	433
8.13	Exercises	434

9 The aurora ... 435

9.1	An historical introduction	435
9.2	The height of the aurora	438
9.3	The occurrence frequency of the aurora	440
9.4	The global distribution of the aurora	443
9.5	The auroral appearance	448
9.6	Auroral particles	452
9.7	Precipitation patterns of auroral particles	457
9.8	The energy deposition profiles of auroral particles	457
9.9	Deriving energy spectra from electron density profiles	463
9.10	Excitation processes in the aurora	466
9.11	The quenching process	469
9.12	The proton aurora	471
9.13	The auroral substorm	477
	9.13.1 The quiet phase	477
	9.13.2 The growth phase	477
	9.13.3 The expansion phase	479
	9.13.4 The recovery phase	479
9.14	References	480
9.15	Exercises	481

Symbols ... 483

Index ... 485

To the students of UNIS

Preface

The upper polar atmosphere sets the scene for one of the nature's most beautiful celestial phenomena, the aurora borealis or the northern lights. The colourful, dynamical and airy forms are the end product and the most spectacular of a long chain of plasma processes initiated by particle eruptions on the Sun. Such plasma processes are thought to be of fundamental importance all over the universe. Therefore the polar atmosphere is a natural laboratory in which we can study physical processes that give us insight into the understanding of similar light phenomena at other planets and celestial bodies. In fact, the polar atmosphere is the nearest laboratory in space from which we can expand our knowledge into the most remote places in our environment.

This book tries to follow the chain of processes which take place when the stream of particles (the solar wind) leaves the Sun, travels through interplanetary space and ends up as energetic particle beams producing the spectacular auroral forms in the polar sky.

This book has been made possible through encouragements from colleagues and students, and I will in particular thank Professor Leroy C. Cogger and Professor Nobuo Matuura who made it possible for me to spend extended visits at the universities of Calgary and Nagoya during which I was able to work on this book. I am also grateful to valuable discussions with Professor Yoshuke Kamide, Dr. Satonori Nozawa, and my Norwegian colleagues Professor Alv Egeland, Dr. Chris Hall, and Dr. Jøran Moen. Special thanks go to my students Trygve Sparr and Mårten Blixt. My secretary Liv Larssen who has typed the manuscript over and over again and who has supplied many of the illustrations in the book is greatly acknowledged for her patience.

Without the support of my wife who took all the extra loads during my many trips away from home this book would never have been a reality.

Tromsø, August 1996

Asgeir Brekke

1
The Sun as a radiation source

1.1 GENERAL ABOUT THE SUN

The mean distance from the Sun to the Earth is called 1 AU – one astronomical unit and is about 1.5×10^{11} m. As the solar radius $R_\odot$ or the radius of the visible disc (Fig. 1.1) is close to 7×10^8 m, the distance to the Sun as seen from the Earth is about 215 $R_\odot$. Seen from the Earth the Sun covers approximately 1920 arc seconds or about half a degree. Since the speed of light (c) is close to 3×10^8 m/s, it will take the light about 500 seconds or close to 8 minutes to pass from the Sun to the Earth. This is the shortest warning we might have that something is happening on the Sun. If an optical event occurs at the Sun which also emits X-rays, these will be observed first almost simultaneously with the light, and other effects such as particle streams can be expected to follow after some time, depending on their speeds through the interplanetary medium.

The mass of the Sun ($M_\odot$) is close to 2×10^{30} kg, and for the given radius of the Sun it leaves the medium mass density (ρ) at about 1.4×10^3 kg/m^3 or 1.4 times as dense as water. The acceleration ($g_\odot$) due to gravity at the Solar surface is about 2.7×10^2 m/s^2 and the escape velocity ($v_{\odot c}$) is 6.2×10^5 m/s.

The temperature at the solar centre is assumed to be as high as 1.5×10^7 K. At this high temperature protons will be converted to helium nuclei by thermonuclear reactions such as the proton–proton and carbon-cycle chains. The proton–proton chain can be illustrated as follows:

$$
\begin{aligned}
^1\text{H} + {}^1\text{H} &\rightarrow {}^2\text{H} + e^+ + \nu + 0.42 \text{ MeV} &\text{(a)} \\
^1\text{H} + {}^2\text{H} &\rightarrow {}^3\text{He} + \gamma + 5.5 \text{ MeV} &\text{(b)} \\
^3\text{He} + {}^3\text{He} &\rightarrow {}^4\text{He} + 2\,{}^1\text{H} + 12.8 \text{ MeV} &\text{(c)}
\end{aligned}
\qquad (1.1)
$$

where e^+, ν and γ represent a positron, a neutrino and a gamma ray quantum, respectively. We notice that 4 protons are converted into one helium nucleus while an energy of 24.64 MeV ($2 \times 0.42 + 2 \times 5.5 + 12.8$) MeV is released. In addition, energy will be released as the positron is annihilated by an electron, and the kinetic energy of the neutrino must also be included.

In the central portions of the Sun it is expected that the temperature can be high enough that the carbon cycle can take place. This will result in a helium nucleus according to the following scheme:

$$
\begin{aligned}
^{12}\text{C} + ^{1}\text{H} &\rightarrow\ ^{13}\text{N} + \gamma \text{ (unstable)} & \text{(a)} \\
^{13}\text{N} &\rightarrow\ ^{13}\text{C} + e^{+} + \nu & \text{(b)} \\
^{13}\text{C} + ^{1}\text{H} &\rightarrow\ ^{14}\text{N} + \gamma & \text{(c)} \\
^{14}\text{N} + ^{1}\text{H} &\rightarrow\ ^{15}\text{O} + \gamma \text{ (unstable)} & \text{(d)} \\
^{15}\text{O} &\rightarrow\ ^{15}\text{N} + e^{+} + \nu & \text{(e)} \\
^{15}\text{N} + ^{1}\text{H} &\rightarrow\ ^{12}\text{C} + ^{4}\text{He} & \text{(f)}
\end{aligned}
\tag{1.2}
$$

A helium nucleus has been created and the rest product is ^{12}C which can start the process all over again. The kinetic energy which is released in this manner will rapidly be converted to local thermal energy which will be transported out of the central regions.

The Sun rotates with a synodic angular velocity (relative to the Earth) ω_s which depends on the heliocentric latitude θ_s. At the Solar equator this is about 13.4° per day while at 75° latitude it is only 11° per day. Observations of sunspots have lead to the following empirical formula for this differential rotation:

$$\omega_s = 13.4 - 2.7 \cdot \sin^2 \theta_s \text{ (degrees per day)} \tag{1.3}$$

and the corresponding synodic rotation period will be:

$$T_s = 26.9 + 5.4 \cdot \sin^2 \theta_s \text{ days} \tag{1.4}$$

The sidereal angular velocity ω_{si} as measured with respect to a fixed star will be

$$\omega_{si} = 14.15 - 4.19 \sin^2 \theta_s \text{ (degrees per day)} \tag{1.5}$$

The solar rotations are assigned with numbers according to a rotation period of 27 days corresponding to a latitude of 8° ($\sin^2 \theta_s = 0.02$). In this system the rotation starting on February 8, 1832 was given the number unity according to Bartels. At the present time the rotation numbers are close to 2150. Bartels also noticed that some magnetic disturbances on the Earth have a tendency to repeat themselves by a period of close to 27 days. He related this to so-called M-regions which today are believed to be associated with coronal holes.

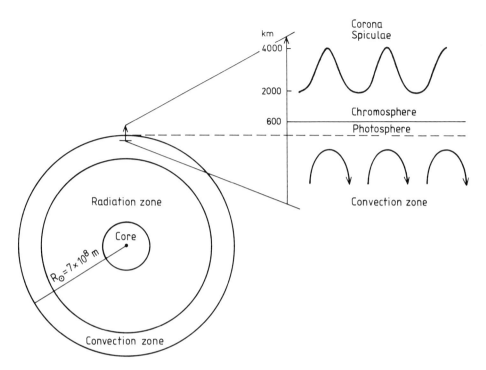

Fig. 1.1. A schematic diagram showing the different zones of the Sun: the core, the radiation zone and the convection zone. Also indicated is the structure of the solar atmosphere above the convection zone. The innermost region reaching out to about 600 km above the top of the convection zone is *the photosphere*, between 600 and 2000 km is *the chromosphere* and outside this *the corona*. So-called spiculae can reach out to 3–4000 km above the base of the photosphere.

1.2 THE SOLAR ATMOSPHERE

The heat located in the interior of the Sun is transported outwards in the solar atmosphere by radiation and convection. Photons and energetic particles emitted from the interior will be absorbed in layers immediately above. These layers will emit new photons, and by absorption and emission at successive layers the radiation energy will be transported outwards in the solar atmosphere.

We discriminate between 3 regions in the solar atmosphere without any sharp demarcation lines (Fig. 1.1). The innermost region where most of the visible light is coming from is called *the photosphere*. This is supposed to be about 500–600 km thick. Negative hydrogen ions, which strongly absorb visible radiation, are present in the photospheric atmosphere, thus visible radiation emitted from a layer deeper than 400 km below the photospheric surface cannot be observed. In the photosphere the temperature (T) as well as the density (ρ) decreases. In the innermost layer (Fig. 1.2), at present within reach for observations, the temperature is about 7500 K and it decreases to about 4200 K at the top of the photosphere. The ionization ratio, the number of free electrons divided by the total number of particles, varies around 10^{-4} in the photosphere (see Table 1.1).

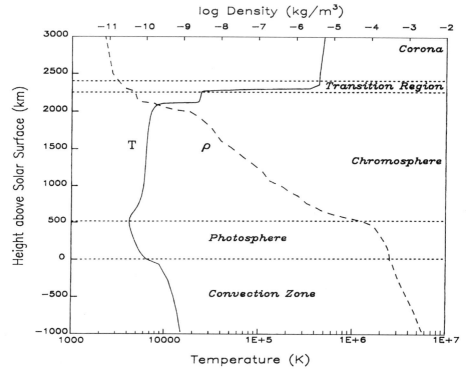

Fig. 1.2. Temperature and density profiles in the solar atmosphere showing the steep temperature increase in the transition region. The zero level refers to about one solar radius (1 $R_\odot = 7 \times 10^8$ km) from the centre of the solar disc. (From Lean, 1991.)

Situated above the photosphere is *the chromosphere* which is transparent to the visible part of the spectrum except for the strong absorption lines which are formed here. The density decreases continuously by height in the chromosphere, but the temperature which reaches a minimum at the top of the photosphere increases slowly and reaches a value of 9000 K at 2000 km or so. Within 500 km above the top of the chromosphere, in the so-called transition region, the temperature increases tremendously and reaches a value of 4×10^5 K at 2500 km above the base of the photosphere. In the chromosphere the ionization ratio increases from close to 10^{-3} at the bottom to 3×10^{-1} at 2000 km above the base of the photosphere. In the transition region of the chromosphere where the temperature gradient is so strong, the ionization ratio increases rapidly to 1.

Table 1.1. Model solar atmosphere. After Allen (1973) and Tohmatsu (1990)

Region	Altitude above base of photosphere (km)	Temperature (K)	Number density (m^{-3})	Electron density (m^{-3})	Ionization ratio
Photosphere	−56	8100	1.6×10^{23}	8.7×10^{20}	5.4×10^{-3}
	0	6430	1.5×10^{23}	6.5×10^{19}	4.4×10^{-4}
	136	5140	6.6×10^{22}	5.1×10^{18}	7.7×10^{-5}
	278	4640	2.0×10^{22}	1.3×10^{18}	6.5×10^{-5}
	420	4370	5.8×10^{21}	3.9×10^{17}	6.7×10^{-5}
	560	4180	1.6×10^{21}	1.1×10^{17}	6.9×10^{-5}
Chromosphere	840	5280	1.1×10^{20}	7.8×10^{16}	7.1×10^{-4}
	1004	5750	2.5×10^{19}	8.9×10^{16}	3.6×10^{-3}
	1580	7150	4.1×10^{17}	6.5×10^{16}	1.6×10^{-1}
	2000	9000	1.0×10^{17}	3.0×10^{16}	3.3×10^{-1}
Corona	7000 $= 0.01 \times R_\odot{}^*$	10^6	3.0×10^{14}	3.0×10^{14}	1.0
	70 000 $= 0.1 \times R_\odot$	10^6	9.1×10^{13}	9.1×10^{13}	1.0
	280 000 $= 0.4 \times R_\odot$	10^6	1.5×10^{13}	1.5×10^{13}	1.0
Interplanetary space	$3\,R_\odot$	10^6	2.8×10^{11}	2.8×10^{11}	1.0
	$20\,R_\odot$	–	2.0×10^{9}	2.0×10^{9}	1.0
	$215\,R_\odot =$ 1 AU	2×10^5	5.0×10^{5}	5.0×10^{5}	1.0

${}^*R_\odot = 7 \times 10^8$ m.

Outside the transition region we find the hottest part of the solar atmosphere, *the corona*, where the temperature is fairly uniform and close to 10^6 K. In this region the solar atmosphere is fully ionized. This is also true in the interplanetary space at the distance of the Earth (1 AU). Here the temperature, however, is decreased to a few times 10^5 K (see Table 1.1).

During solar eclipses the corona can be photographed (Fig. 1.3) at distances out to 4–5 $R_\odot$ while some extreme rays can be seen to reach out to 15 $R_\odot$. From scintillation studies of radio stars the corona can in fact be traced as far out as 100 $R_\odot$ which is almost half the distance to the Earth.

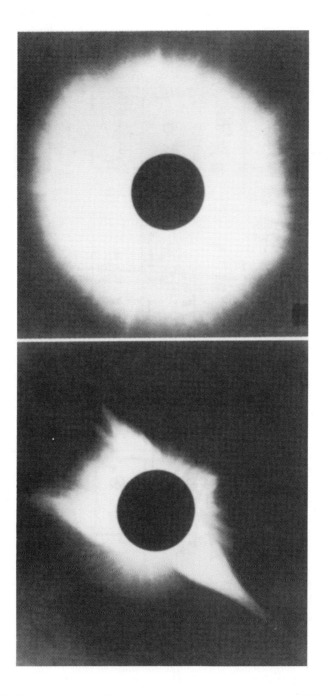

Fig. 1.3. Photographs of the solar corona at solar maximum (1947, upper panel) and solar minimum (1952, lower panel). (From Tohmatsu, 1990.)

The corona is by no means uniform as it has a very complicated fine structure, with rays, plumes, fans and arcs. These structures are closely connected to the shape of the magnetic field in the corona which changes in a complicated but partly systematic way in the course of a solar cycle.

The thermal conductivity in the corona is very high, where hydrogen as well as helium are almost totally ionized. Therefore is the corona more or less isothermal at a temperature of close to 10^6 K.

1.3 THE ELECTROMAGNETIC RADIATION FROM THE SUN

The observed radiation spectrum from the Sun as measured from the Earth's surface in the wavelength region 0.2–3.2 μm (2000–32 000 Å) is shown in Fig. 1.4. The black areas in the spectrum represent the absorption due to water vapour (H_2O), carbon dioxide (CO_2), oxygen (O_2) and ozone (O_3) in the Earth's atmosphere. The solar spectrum, as it is to be observed outside the Earth's atmosphere, is also illustrated together with the spectrum from a black body at a temperature of 6000 K. The peak intensity of the solar spectrum is close to 0.15 W/m^2 Å.

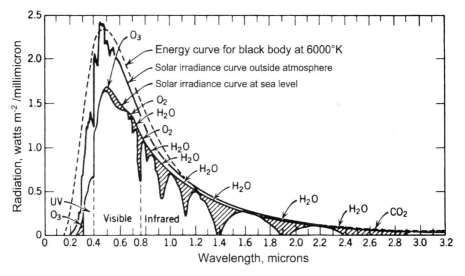

Fig. 1.4. The energy distribution of solar radiation observed at sea level compared with at blackbody spectrum at 6000 K and the spectrum observed outside the Earth's atmosphere. The black areas indicate absorption windows due to H_2O, CO_2, O_2, O_3 in the Earth's atmosphere. (From Pettit, 1951.)

The ultraviolet (UV) and X-ray radiation shorter than 0.2 μm are strongly absorbed by air. The UV radiation region is usually divided into two parts: (1) the far ultraviolet region between 0.1 and 0.2 μm, and (2) the extreme ultraviolet region (EUV) between 0.01 and 0.1 μm. The term Schumann region is sometimes used for the region 0.120 to 0.185 μm.

In Fig. 1.5 the solar ultraviolet and X-ray radiation spectrum between 1 and 2000 Å (between 0.1 nm and 0.2 μm) is shown. The continuum spectrum in the far ultraviolet region originates from the uppermost layer of the photosphere, therefore the effective blackbody temperature for this spectral region decreases to about 4500 K as compared with 6000 K in the visible regime. The solar radiation in this wavelength region is completely absorbed by molecular oxygen at altitudes between 80 and 120 km in the terrestrial atmosphere. It is therefore an important heat source of the upper atmosphere (*mesosphere* and *thermosphere*) and also for the dissociation of molecular oxygen in these regions.

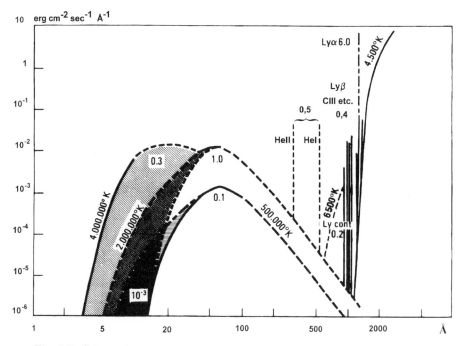

Fig. 1.5. Schematic representation of the solar UV spectrum. The temperatures are indicating those for the emitting regions of thermal radiations. The cross-hatched areas give an idea of the range of X-rays variability. (From Giraud and Petit, 1978.)

In the EUV regime there is a number of strong emission lines such as the hydrogen Balmer lines at 121.6 nm (Ly$_\alpha$), 102.5 nm (Ly$_\beta$) etc., the helium line at 58.4 nm and the oxygen line at 130.4 nm. Another group of emission lines are the ion lines such as the He$^+$ line at 30.4 nm and the emission lines from multi-ionized Si, Fe, Mg, N and O.

The EUV radiation originates in the chromosphere which is the region extending above the photosphere in the solar atmosphere. It is almost transparent in the visible spectral region since the absorption due to negative hydrogen ions

Sec. 1.3] The electromagnetic radiation from the Sun 9

is negligibly small. Therefore the solar chromosphere can never be seen with the naked eye because of the strong continuum emission from the photosphere below, unless during total solar eclipses when the bright solar photosphere is shielded by the moon.

The strong atomic emission lines are all originating in the low temperature region of the chromosphere while the ionic lines are emitted from the high

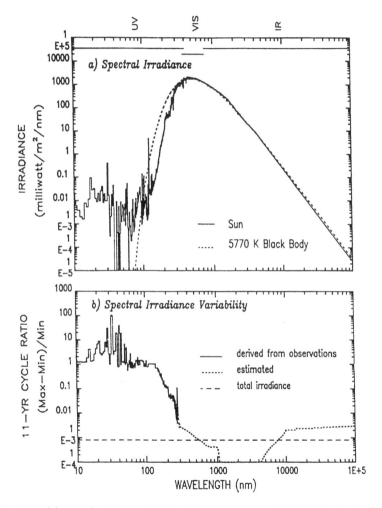

Fig. 1.6. (a) The Sun's spectral irradiance or energy distribution typical for a solar minimum condition compared with a blackbody spectrum at 5770 K. The spectral bands indicated are the ultraviolet (UV), visible (VIS) and infrared (IR). (b) The approximate amplitude of the Sun's spectral irradiance variation from the maximum to the minimum of the 11-year solar cycle with respect to the intensity at minimum. The variability for wavelengths shorter than 300 nm is outstanding. (From Lean, 1991.)

temperature region near the coronal base. Since the temperature in the chromosphere varies considerably from approximately 4500 K at the bottom to some hundred thousand K at the top, the emission spectrum from different layers in the chromosphere shows different features corresponding to the degree of ionization, temperature and density there.

In order to understand the behaviour of the upper atmosphere of the Earth and its ionized part, *the ionosphere,* it is important to have a precise knowledge of the solar EUV radiation flux since it is this radiation which is responsible for the formation of the ionosphere through the photoionization process.

While the solar spectrum between 0.2 and 3.2 μm is rather constant during a solar cycle, the UV lines and the X-ray emissions are more variable and often enhanced during so-called solar flare events. The emissions in the EUV and X-ray regimes (< 50 nm) are often connected to sunspots and other visible phenomena on the solar surface related to the solar activity. The variability in these emissions is illustrated in Fig. 1.5, and estimated emission curves for temperatures varying between 5×10^5 K and 4×10^6 K are indicated. The estimated peak intensity is about 10^{-2} ergs/cm^2 sec Å or 10^{-5} W/m^2 Å, much less than the peak in the visible part of the spectrum. This variability of the EUV and X-ray radiation intensity is also illustrated in Fig. 1.6 where the Sun's spectral irradiance typical of solar minimum conditions are presented in the upper panel. The approximate amplitude of the Sun's spectral irradiance variation from the maximum to the minimum phase relative to the amplitude at minimum for an 11-year solar cycle is presented in the lower panel of Fig. 1.6. Clearly there is a much larger variability in the solar radiation below 300 nm than in any other part of the spectrum.

Let us assume that the Sun is radiating its energy approximately as a black body with a temperature T, then the spectral brightness per frequency band, B_ν, according to the Planck's radiation law will be:

$$B_\nu = \frac{2h\nu^3}{c^2} \frac{1}{\exp\left(\frac{h\nu}{\kappa T}\right) - 1} \tag{1.6}$$

where ν is the radiation frequency, c is the speed of light ($= 3 \times 10^8$ m/s), h is Planck's constant ($= 6.63 \times 10^{-34}$ Js), κ is the Boltzmann constant ($= 1.38 \times 10^{-23}$ J/K), and T is the temperature in K. B_ν is then given in W Hz^{-1} sr^{-1} m^{-2} in the frequency range $\nu + d\nu$ to ν. We notice that B_ν only depends on the temperature and not on the geometry of the body. We can also express B_ν per unit wavelength in the wavelength range $\lambda + d\lambda$ to λ in the following manner when noting that $d\nu = -(c/\lambda^2)d\lambda$ and $B_\nu d\nu = B_\lambda d\lambda$:

$$B_\lambda = \frac{2hc^2}{\lambda^5} \frac{1}{\exp\left(\frac{hc}{\kappa \lambda T}\right) - 1} \tag{1.7}$$

We find that the maximum in B_λ occurs at the wavelength λ_m which satisfies the following transcendental equation:

$$5(1 - \exp(-x_m)) = x_m \tag{1.8}$$

and where
$$x_m = \frac{hc}{\kappa \lambda_m T} \tag{1.9}$$

Equation (1.8) has the solution
$$x_m = 4.9651 \tag{1.10}$$

And finally we find:
$$\lambda_m \cdot T = \frac{h \cdot c}{x_m \cdot \kappa} = 2.898 \times 10^{-3} \text{ m K} \tag{1.11}$$

which is the Wien displacement law, expressing that the wavelength at maximum radiation for a black body is inversely proportional to the temperature. From the solar spectrum we notice that there is a maximum close to 500.0 nm which corresponds to a temperature of

$$T = 5780 \text{ K}$$

or very close to 6000 K which is the theoretical spectrum introduced in Fig. 1.4.

For a temperature of the order of $2 \cdot 10^6$ K we find a radiation maximum at λ_m given by

$$\lambda_m = \frac{2.890 \times 10^{-3} \text{ m K}}{2 \cdot 10^6 \text{ K}} = 1.45 \text{ nm} \tag{1.12}$$

which is in the X-ray regime as indicated in Fig. 1.5.

According to the Stephan–Boltzmann's law which is the integrated Planck's equation over all wavelengths, we have the total radiated power given by:

$$Q_\odot = \sigma \cdot T_\odot^4 \cdot S_\odot \tag{1.13}$$

where $S_\odot$ is the area of the radiating Sun, $T_\odot$ the average solar temperature, and σ the Stephan–Boltzmann constant given by:

$$\sigma = \frac{2\pi^5 \cdot \kappa^4}{15 c^2 h^3} = 5.67 \times 10^{-8} \text{ W m}^{-2} \text{ K}^{-4} \tag{1.14}$$

In the visible part of the spectrum the radiated power per unit area, when $T_\odot = 5780$ K, will be:

$$E_\odot = \sigma \cdot T_\odot^4 = 6.3 \times 10^7 \text{ W/m}^2 \tag{1.15}$$

Since the solar radius is $R_\odot = 7 \times 10^8$ m, the total radiated power from the Sun will be:

$$Q_\odot = 4\pi R_\odot^2 E_\odot = 3.9 \times 10^{26} \text{ W} \tag{1.16}$$

This energy will radiate in all directions. At a distance from the Sun corresponding to the average position of the Earth ($r = 1$ AU) we find from the conservation of energy:

$$4\pi R_\odot^2 E_\odot = 4\pi r^2 E_e \tag{1.17}$$

E_e is the radiation per unit area hitting the Earth given by:

$$E_e = E_\odot \left(\frac{R_\odot}{r}\right)^2 = 1380 \text{ W/m}^2 \qquad (1.18)$$

Of course, the Earth does not move around the Sun at a constant distance, since aphelion (r_a) is 1.46×10^{11} m and perihelion (r_p) is 1.52×10^{11} m. Therefore, although the radiation density from the Sun may be constant at a fixed distance, it will vary at the position of the Earth by:

$$\left(\frac{r_p}{r_a}\right)^2 = 1.08$$

or 8% during one year.

The radiation from the Sun in the visible part of the spectrum is, however, rather constant and changes only by a fraction of a per cent in the course of time. Therefore E_e is often called the solar constant. The short wavelength part of the spectrum of the EUV and the X-ray regime is, as we have seen, highly variable, and so also is the long wavelength region, the radio emissions from the Sun.

The radio emissions from the quiet Sun has been studied over a wide range of wavelengths from the mm- to the m-region. For radio frequencies corresponding to this wavelength regime (10^4–10^2 MHz) we notice that the energy related to them, is:

$$h\nu = 6.6 \times 10^{-34} \text{ J} \cdot \text{s} \cdot 10^3 \cdot 10^6 \text{ s}^{-1} = 6.6 \times 10^{-25} \text{ J} \qquad (1.19)$$

This is much less than the thermal energy ($\kappa T_\odot$) for a reasonable solar temperature ($T_\odot = 6000$ K) which corresponds to

$$\kappa T_\odot = 8.4 \times 10^{-20} \text{ J} \qquad (1.20)$$

Therefore, in the radio wavelength regime we can use the Rayleigh–Jeans approximation of Planck's radiation law ($h\nu \ll \kappa T$):

$$B_\nu = \frac{2h\nu^3}{c^2} \frac{1}{1 + h\frac{\nu}{\kappa T} + \cdots - 1} = \frac{2\kappa T \nu^2}{c^2} = \frac{2\kappa T}{\lambda^2} \qquad (1.21)$$

From Fig. 1.7 we notice that the intensity of the radio emissions for a quiet Sun decreases by more than two orders of magnitude in the wavelength regime 1 cm to 10 m. Strictly speaking, the effective temperature T will depend on the radio wavelength such that for cm-waves it is about 10^4 K, for dm-waves about 10^5 K and for m-waves it is close to 10^6 K.

The solar irradiance at the Earth's distance can be found from the spectral brightness, B_ν, by calculating the radiation flux passing through a unit area of the Earth perpendicular to the solar rays during unit time.

In Fig. 1.8 we notice that the radiation energy emitted from an infinitesimal area $d\sigma$ perpendicular to the Sun–Earth line at point P on the Sun and hitting the surface of the Earth is:

$$dI_\nu = B_\nu d\sigma \cdot \Omega = B_\nu d\sigma \frac{S}{r_1^2} \qquad (1.22)$$

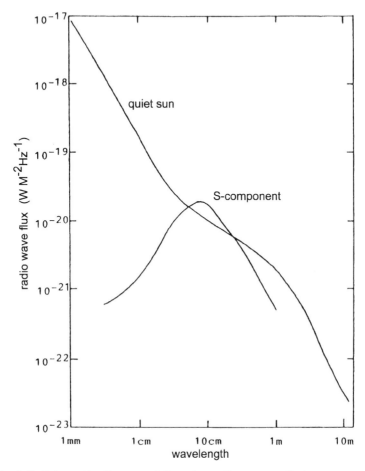

Fig. 1.7. Schematic diagram of the solar radio wave radiation spectrum for the quiet Sun as well as for the slowly varying component (the S-component). (From Tohmatsu, 1990.)

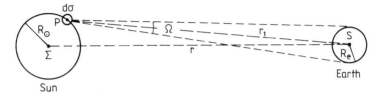

Fig. 1.8. The Sun–Earth geometry where S is the projection of the Earth's surface to a plane perpendicular to the Sun–Earth line. r is equal to 1 AU, and $d\sigma$ is the projection of an infinitesimal area on to a plane perpendicular to the line from a point P on the Sun to the Earth's centre. Since $r \gg R_\odot$, where $R_\odot$ is the solar radius, r_1 can be considered equal to r.

where S is the projection of the surface of the Earth facing the Sun, on a plane perpendicular to line from P to the Earth's centre, r_1 is the distance from P to the Earth ($r_1 \gg R_e$) and Ω is the solid angle corresponding to the surface S as seen from P. Assuming now that the brightness, B_ν, is isotropic all over the solar surface, and that $r \gg R_\odot$ so that $r = r_1$, then the total radiation flux per unit area reaching the Earth is:

$$\phi_\nu = \frac{I_\nu}{S} = \int_\Sigma B_\nu \frac{d\sigma}{r^2} = \pi \left(\frac{R_\odot}{r}\right)^2 B_\nu \qquad (1.23)$$

where the integration is made over the surface Σ of the complete Sun. Introducing B_ν from (1.21) we have:

$$\phi_\nu = \pi \left(\frac{R_\odot}{r}\right)^2 \cdot \frac{2\kappa}{c^2} \nu^2 T \qquad (1.24)$$

and applying the proper constants we find:

$$\phi_\nu = 2.09 \times 10^{-44} \nu^2 T \ [\text{W m}^{-2} \text{ Hz}^{-1}] \qquad (1.25)$$

Superposed on these thermal background solar radio wave emissions there is a component varying with phenomena on the Sun such as sunspots and plages. It is predominant in the cm- and dm-regions and its flux varies with a time scale of a few days, therefore it is often referred to as the slowly varying component or the S component (see Fig. 1.7). It is rather well correlated with the fluxes of the EUV and X-ray radiation and appears to be emitted from areas on the solar surface with extremely high temperature. The flux of this component is frequently referred to in the context of ionospheric physics due to its high correlation with the EUV radiation (Fig. 1.9) and because it can be monitored continuously. The 10.7 cm (2.8 GHz) radio wave flux measured at Dominion Astronomical Observatory, Ottawa has been commonly used and represented as $F_{10.7}$ or S_a in units of 10^{-22} W m^{-2} Hz^{-1} reduced to 1 AU.

For the 10.7 cm radio flux we find from (1.25) that the radiation flux is given by ($F_\lambda = \phi_\nu$):

$$F_{10.7} = 16.4 \times 10^{-28} \cdot T \ \text{W m}^{-2} \text{ Hz}^{-1} \qquad (1.26)$$

and the index is therefore a direct indicator of the effective temperature. For typical values of $F_{10.7}$ between 50 and 200 we notice that T must be between 3 and 12×10^6 K, much higher than the average temperature of the corona.

Before we discuss the observations of the radio bursts from solar flares we will introduce the electron plasma frequency given by:

$$f_{pe} = \frac{1}{2\pi} \left(\frac{e^2 n_e}{\varepsilon_0 m_e}\right)^{1/2} = K n_e^{1/2} \qquad (1.27)$$

where e is the electronic charge ($= 1.6 \times 10^{-19}$ C), n_e is the electron density (m^{-3}), m_e is the electron mass ($= 9.1 \times 10^{-31}$ kg), and ε_0 is the permittivity of a vacuum ($= 8.854 \times 10^{-12}$ F/m).

Sec. 1.3]   The electromagnetic radiation from the Sun   15

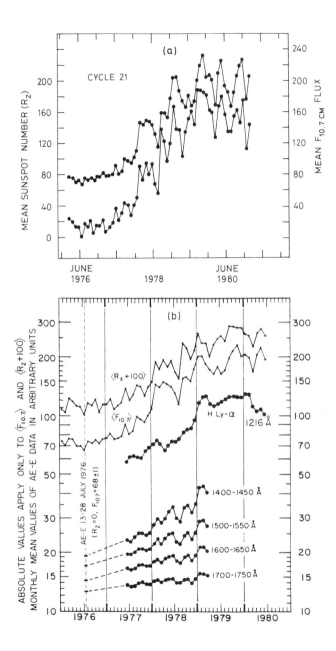

Fig. 1.9. (a) Variations in the monthly mean sunspot numbers (R_z) compared to the mean radio wave flux $F_{10.7}$ during part of solar cycle 21. (After Jursa, 1985.) (b) Variations in the irradiance at wavelengths from 121.6 nm to 175.0 nm (1216 Å – 1750 Å) compared to the variations in the radio wave flux $F_{10.7}$ during the solar cycle 21. (From Hinteregger, 1980.)

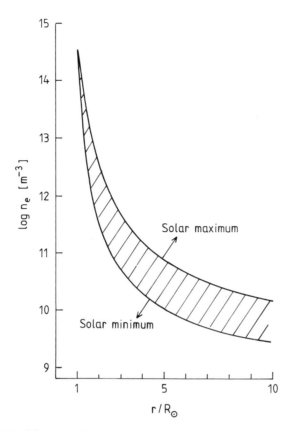

Fig. 1.10. Schematic illustration of the variation in the electron density in the solar atmosphere from the base of the photosphere to a distance of 10 $R_\odot$ from the centre of the solar disc.

From a given level of the solar atmosphere characterized by a specific electron density of n_e only radiation with a frequency $\nu > f_{pe}$ will escape if the influence of the magnetic field is neglected. Most of the radiation will actually originate from layers where f_{pe} is close to ν. In Fig. 1.10 a schematic illustration of the variation on the electron density is presented as function of distance from the solar centre. As can be noticed, the electron density decreases strongly from close to 10^{15} m^{-3} at 1 $R_\odot$ to close to 10^{10} m^{-3} at 10 $R_\odot$. The electron density is on the average lower during solar minimum conditions.

According to Figs. 1.7 and 1.10 and our understanding of the plasma frequency we now realize that by observing radio emissions from the Sun at increasing frequencies (or decreasing wavelengths) we are probing deeper and deeper into the solar atmosphere. It is found that while m-waves are formed far out in the corona, cm-waves are generated deep in the solar atmosphere and mm-waves penetrate almost all the way from the photosphere.

We also find evidence in these measurements for an increasing temperature by distance from the Sun and that the temperature in the corona actually is close to 10^6 K.

Another interesting aspect of the radio wave emissions from the Sun is the apparent magnitude of the solar disc when viewed at different wavelengths. If we compare the size of the solar disc as observed in the radio wavelength regime with the size of the visual Sun, as is done in Fig. 1.11, we notice that at a wavelength of 25 cm the Sun appears as a sharp ring at the rim of the visual Sun. At 5 m wavelength, however, the Sun has a strong emission in the centre while it gets weaker toward the rim. It can, however, be seen at twice the distance from the solar centre compared to the visual Sun. In the 5 m wavelength region the Sun appears as a very large blurred Sun with a hot spot in the centre.

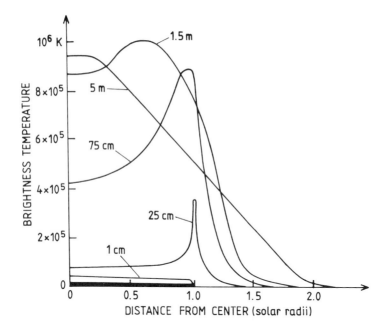

Fig. 1.11. The variation of brightness temperature with distance from the centre of the solar disc at various wavelengths of observation. The optical disc is indicated by the heavy black bar. (From Akasofu and Chapman, 1972.)

Although the spectrum of the electromagnetic radiation from the Sun extends from X-rays with wavelengths of 10^{-8} m or less to radio waves of wavelength 100 m or more, 99% of the energy is concentrated in the range 276 to 4960 nm and 99.9% in the range 217 and 10940 nm. In other words, all but 0.1% of the energy radiating from the Sun is found in the visible, infrared and ultraviolet portion of the spectrum.

1.4 THE SUNSPOTS – SOLAR CYCLE

One of the most fascinating facets of the Sun is the sunspots which have been observed by the naked eye for thousands of years (Fig. 1.12). They have often caused fear and been observed as omens of famine, war, fires etc. But they have also been deeply admired and inspired people to speculate on their cause. It is only the very strongest of the sunspots which under certain favourable circumstances can be observed by the naked eye. They were first observed by a telescope in the beginning of the 17th century by Galilei (in 1616).

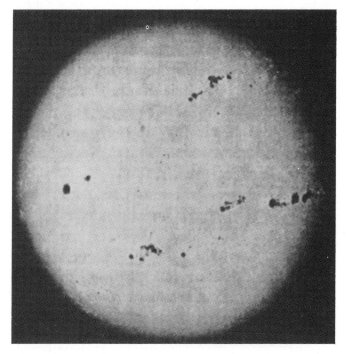

Fig. 1.12. A photograph of the Sun near maximum sunspot activity (Dec. 1957). The sunspot zones are clearly outlined. (From Zirin, 1966.)

The fact that the sunspots come and go in cycles was not appreciated until 1843, when Schwabe (1844) published a short paper based on his observations between 1826 and 1843. Schwabe found the cycle length to be about 10 years. Wolf devised a quantitative definition for a sunspot number and studied sunspot data for the years 1700 to 1848 and indeed confirmed Schwabe's discovery of a cycle, but Wolf found a period of 11.1 years.

In its present form the Wolf sunspot number R is defined as:

$$R = k(10g + f) \tag{1.28}$$

Sec. 1.4] The sunspots – solar cycle 19

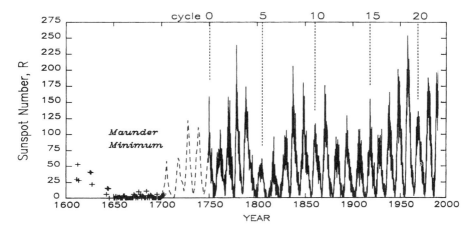

Fig. 1.13. The sunspot number represented by monthly means since 1750 and yearly means from 1610 to 1750. The numbers indicated above represent the sunspot number starting from zero at the cycle maximizing in 1750. Cycle 1 is therefore the one starting at 1755. (From Lean, 1991.)

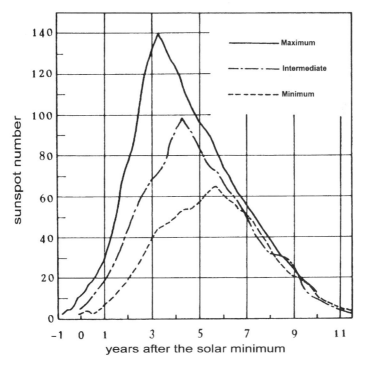

Fig. 1.14. Differences in amplitude and phase of a solar cycle showing that the stronger cycles reach maximum faster than the weaker cycles. (From Waldmeier, 1955).

where f is the total number of sunpots regardless of size, g is the number of sunspot groups, and k is a normalization constant accounting for the different observatories. A typical plot of the yearly mean sunspot number is given in Fig. 1.13. The most prominent feature is an apparent periodicity of the order of 11 years, but longer periods may also be discerned. The distance between two successive minima can vary between 8.5 and 14 years and it may take from 7.3 to 17 years between successive maxima. There is a general tendency for a more active cycle to have its maximum at an earlier phase of the cycle and for a less active cycle to have its maximum at a later phase as shown in Fig. 1.14. It is also noticeable that there in general is a rapid increase from a minimum to the following maximum and a slower decrease from the maximum to the following minimum.

From spectral analysis of the variations of the sunspot numbers it is found that the predominant period is often depending on the length of the time series analysed provided that the time series is long enough to enclose a reasonable number of complete cycles. The time series is therefore not stationary as the different frequency components probably change phase in the course of time. At the moment, 6 periods appear to stand out from this analysis. These are listed as follows (in years) (from Herman and Goldberg, 1978):

$$\begin{array}{ccc} & 5.5 & \\ 8.1 & 9.7 & 11.2 \\ & 100 & \\ & 180 & \end{array}$$

The 3 components 8.1, 9.7 and 11.2 years are believed to be a triplet giving rise to the commonly known 11 years solar cycle.

It has also been indicated that some of the periods actually result from a beating between some of the more fundamental ones.

If T_1 and T_2 are two fundamental periods, we know from the coupling between waves that two other periods (T_3 and T_4) may occur as a result of beating, and these are given by the following relations:

$$\frac{1}{T_3} = \frac{1}{T_1} + \frac{1}{T_2} \quad \text{(a)}$$
$$\frac{1}{T_4} = \frac{1}{T_1} - \frac{1}{T_2} \quad \text{(b)}$$
(1.29)

For example, if $T_1 = 9.7$ years and $T_2 = 11.2$ years, T_3 will be 5.2 years which is close to 5.5 years, but of course, 5.5 years is close to the half of 11.5 years and may therefore merely be the first harmonic of T_2.

The reason for the origin of the solar cycle has been the subject to much research and speculations. One interesting theory relating to this is the planetary tidal theory. There are in the solar system a number of periods which are very close to 11 years – some of those are found among the inner planets. These are listed in Table 1.2 (after Herman and Goldberg, 1978).

Furthermore, among the outer planets there are periods which are very close to 180 years. Some of these are shown in Table 1.3 (after Herman and Goldberg, 1978).

Table 1.2

Period	Earth years
46 sidereal revolutions of Mercury	11.079
18 sidereal revolutions of Venus	11.074
11 sidereal revolutions of Earth	11.000
6 sidereal revolutions of Mars	11.286
(137 synodic revolutions of Moon)	11.077

Table 1.3

Period	Earth years
6 sidereal revolutions of Saturn	176.746
15 sidereal revolutions of Jupiter	177.933
9 synodic periods, Jupiter–Saturn	178.734
14 synodic periods, Jupiter–Neptune	178.923
13 synodic periods, Jupiter–Uranus	179.562
5 synodic periods, Saturn–Neptune	179.385
4 synodic periods, Saturn–Uranus	181.455

Because these periods are so similar, it has been thought that resonances occur in the solar system which give rise to the 11-year and the 180-year cycles. The physical explanation for why these resonances will be more dominant than similar coincidence of periods among the planets is not clear.

A more plausible physical explanation for the periods observed in the sunspot variations can be found in the fact that the Sun itself rotates around the centre of mass of the solar system with a period of 179.2 years or very close to the 180 years as found in the existing data. If we also consider the motion of the planets around the Sun and take into account the relative masses of the planets and their relative distances to the Sun, we can show that the gravitational vector force exerted by the planets on a fixed spot on the Sun varies cyclically with a period of 11.1 years.

If we take these two fundamental periods in the solar system and let $T_1 = 11.1$ years and $T_2 = 179.2$ years, we find the following beat periods:

$$T_3 = 10.4 \text{ years} \quad \text{(a)}$$
$$T_4 = 11.8 \text{ years} \quad \text{(b)} \quad (1.30)$$

If we then take into account these two beating periods and the first harmonic periods of T_1 and T_2 which are 5.5 and 89.6 years respectively, we obtain the following set of periods:

                    5.5
        10.4   11.1   11.8
                89.6
                179.2

These are reasonably close to the periods deduced from the time series of the sunspot number to be within the accuracy of the observations.

We should remember that the sunspot number is an empirical quantity which gives us no direct link to the mechanism causing the sunspots, and therefore we should not expect complete matching between the theoretical values and the deduced periods from the sunspot series.

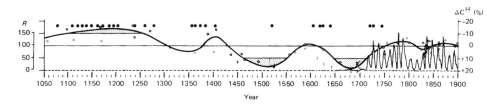

Fig. 1.15. Historical sunspot variability. The heavy solid line indicates the variation in C^{14} concentration in tree rings. The thin curve refers to the Wolf sunspot numbers (R_z). Shaded portions of the C^{14} curve are the Spörer (1460–1550) and Maunder (1645–1715) periods. Circles are from historical Oriental sunspot records and tree ring analyses. (From Eddy, 1976.)

Many attempts have been made to study the solar cycle before 1700, and in Fig. 1.15 a well-known graph by Eddy (1976) is reproduced. The graph indicates that the 11-year solar cycle started abruptly at about 1700 which is an artifact of the analysis method and due to lack of high quality data before this time. According to Eddy (1976) the sunspot numbers (Wolf numbers) from 1818 to 1847 are good, from 1749 to 1847 are questionable, from 1700 to 1748 are poor and before 1700 they of course are worse than poor. It is noticeable, however, that a long periodic variation appears to be interfering with the 11-year solar cycle leading to a strong reduction in the sunspot numbers in the so-called Spörer minimum (1460–1550) and the Maunder minimum (1645–1715). These data do to a large extent lean on studies of C^{14} concentration in tree rings. The exact relationship between the behaviour in the C^{14} concentration and the sunspots is not well known. This is also clear for the period between 1800 and 1820 where the sunspot data are good and showing two very weak cycles while the C^{14} results indicate a minor decrease and a minimum about 10 years delayed with respect to the sunspot numbers.

It has become common practice to refer to each 11-year solar cycle measured from minimum to minimum by a number. The period from 1755 to 1766 has been assigned cycle 1. Thus we are at present in cycle 22 which reached a maximum in 1990 (Fig. 1.16).

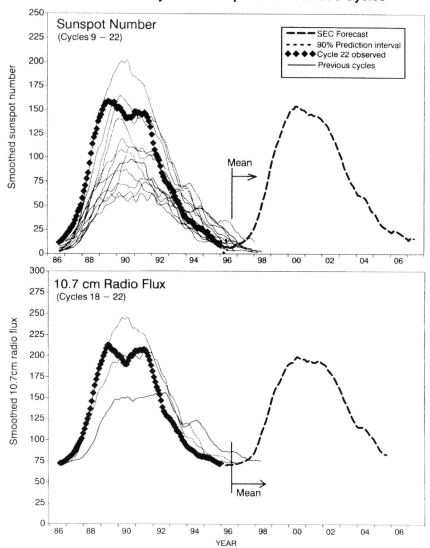

Fig. 1.16. Upper panel: The smoothed sunspot numbers (R) for the present cycle (cycle 22 beginning in Sept. 1986) compared with the same for the preceding cycle. Also indicated are the mean sunspot numbers for cycles 9–22 as well as predictions for the coming cycle. Lower panel: The smoothed variations of the 10.7 cm radio flux ($F_{10.7}$) for the present solar cycle compared with the same for cycles 18–22. (From Joint NOAA-USAF Space Weather Operations, 1996.)

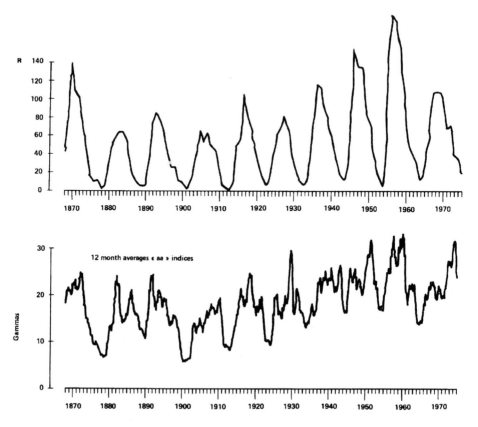

Fig. 1.17. A comparison between the solar sunspot numbers between 1850 and 1975 and the aa index for the same period. The aa index is the average of the amplitude in γ over three hour intervals of the irregular variations of the horizontal components of the Earth's magnetic field measured at two antipodal sites (Europe and Australia) at about 50 degrees latitude. (From Giraud and Petit, 1978.)

The reason there is such a strong interest in the sunspots is that conditions on the Earth appear to vary in concordance with the sunspot cycle. Many attempts have been made to show correlations between the solar cycle and different variable phenomena on the Earth, especially the weather. So far no conclusive results appear to have been obtained, and the discussion will continue with respect to this fundamental problem for life on Earth in the years to come. For scientists working with problems in the Earth's upper atmosphere, the ionosphere, the solar cycle variations stand out very clearly. For example, as shown in Fig. 1.17, a close correlation exists between the variations in the sunspot numbers and in the variations of the Earth's geomagnetic field as evidenced by the so-called aa index. These variations in the Earth's field are due to currents in the ionosphere, which are modulated by radiation from the Sun. We will return to these currents in Chapter 8.

1.5 THE SUNSPOTS

The spots first appear as a small dark pore which gradually grows bigger and develops after a day or so to the characteristic structure with an inner dark umbra surrounded by the penumbra (Fig. 1.18). Around 500 nm the umbra intensity is only 3–5% of the photospheric intensity, while in the penumbra it is about 80%. The effective temperature in the umbra of a large spot is of the order of 4500 K. That is to say that the spot appears cooler than its surroundings, which are closer to 6000 K. The lifetime of the spots is highly variable, but on the average there is a positive correlation with the maximum area that a spot covers. Half of all spots have lifetime less than 2 days while only 10% survive for more than 11 days. Only exceptionally has a spot been observed for more than 5 solar rotations.

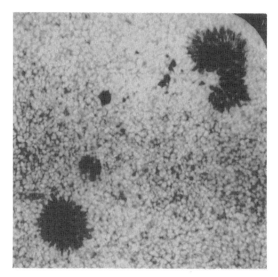

Fig. 1.18. A photograph of a regular sunspot and a complex sunspot group. (From Danielson, 1961.)

Strong magnetic fields are the most characteristic of sunspots. In order for a pore to develop into a spot a minimum field strength is required. As the newly born sunspot develops in size, the magnetic field increases asymptotically towards a value of about 3600 gauss or 0.36 tesla. The topography of the magnetic field may be approximated by

$$B = B_0 \left(1 + \left(\frac{r}{R_S}\right)^2\right)^{-1} \tag{1.31}$$

Here B_0 is the field in the umbra centre, R_S is the radius of the spot and r is the distance from the spot centre.

Spots often appear in groups where the spots of one magnetic polarity cluster in the preceding part, while a group with opposite polarity clusters in the trailing part. Such a set of spots is called a bipolar sunspot group. Of all spots more than 90% appear in this category. Unipolar groups only account for 9% of the spots.

Complex groups with many small spots and no apparent clear distribution of magnetic polarity represent less than 0.5% of the total number of spots. This group is very important geophysically, however, as the probability of flare and disturbance activity increases markedly with the complexity of the group.

The distribution of spots over the solar surface changes with phase in the solar cycle, as illustrated by a "butterfly diagram" in Fig. 1.19. Shortly before a solar minimum is reached new spots appear at heliocentric latitudes around ±30°. As the number of spots increases, the spots appear at lower latitudes and at solar maximum they are at an average latitude of ±12%. The average latitude decreases further and reaches about ±5% at sunspot minimum.

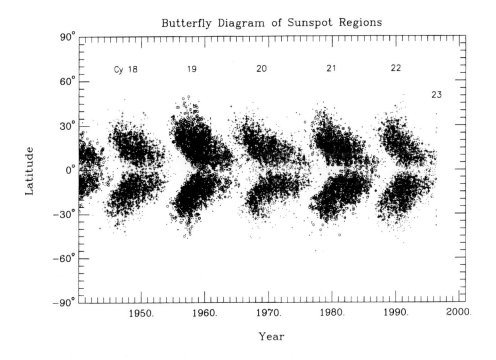

Fig. 1.19. Butterfly diagrams from 1940 until 1996 showing the latitudinal positions of sunspots marked for the solar cycles 18–23. This illustrates the equatorward movement of the active latitude band over each solar cycle. (From Joint NOAA–USAF Space Weather Operations, 1996.)

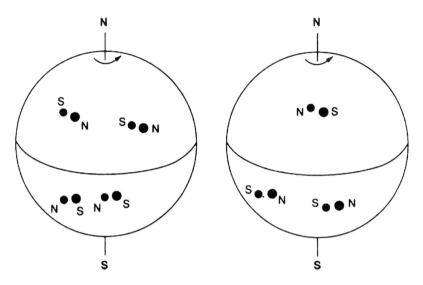

Fig. 1.20. A schematic diagram showing the relationship between magnetic polarity and the sunspots for consecutive solar cycles. (From Egeland et al., 1990.)

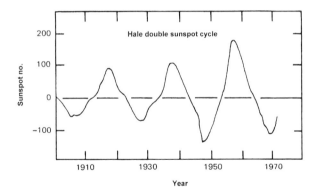

Fig. 1.21. The double sunspot cycle with alternate maxima plotted with opposite signs to indicate the 22-year period related to the reversal of magnetic polarity from one cycle to the next. (From Roberts, 1975.)

The magnetic polarity of the spots changes from cycle to cycle. It appears that within the same cycle the leading spots have the same polarity in the same hemisphere but opposite in the other hemisphere (Fig. 1.20). At minimum the polarity changes, and sometimes new spots at high latitudes with a changed polarity may be seen together with spots from the old cycle at low latitude.

Because of this change in polarity the solar cycle has a characteristic double period close to 22 years as shown in Fig. 1.21.

1.6 THE ELECTROMAGNETIC RADIATION FROM THE DISTURBED SUN

The electromagnetic radiation from the quiet Sun is, as we have indicated, to be explained as thermal radiation from the corona and the solar atmosphere. The variations in these background emissions from X-rays to radio waves are rather smooth and follows the 27-day recurrence tendency corresponding to the rotation period of the Sun as seen from Earth. Fig. 1.22 illustrates this behaviour rather well where the emission at X-rays between 8 and 12 Å ($E(8,12)$) are compared with radio wave emissions at 10.7 cm (2800 MHz). Also to be noticed from Fig. 1.22 is a close correlation with the solar sunspot number R_Z as well as a plage index ($\Sigma A \times I$) with these emissions. Superimposed on those smooth variations, however, are considerable fluctuations with short time scales especially in the X-ray emissions. Similar rapid fluctuations are also found in the UV emission from the Sun.

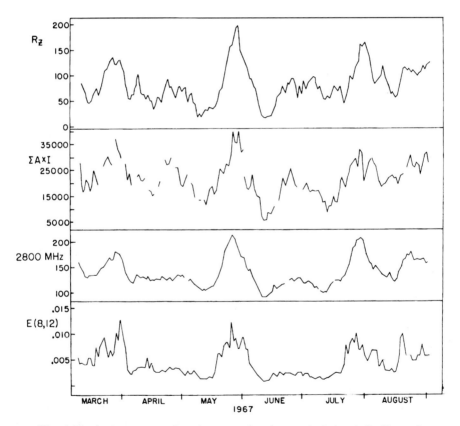

Fig. 1.22. An intercomparison between the time variations of the X-ray flux between 0.8 and 1.2 nm ($E(8,12)$), the 10.7 cm flux density, a plage index ($\Sigma A \times I$) and the solar sunspot number R_z showing the 27-day recurrence period. (From Teske, 1969.)

Sec. 1.6] The electromagnetic radiation from the disturbed Sun 29

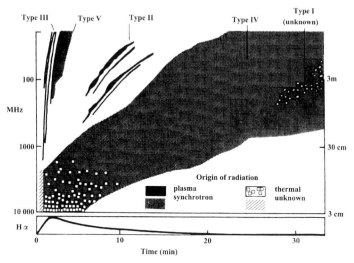

Fig. 1.23. Idealized sketch showing the dynamic spectrum of solar radio wave bursts. (From Wild et al., 1963.)

The most short-lived phenomenon observed from Earth on the solar surface is the flare events which usually occur within a region encompassed by a large magnetically bipolar sunspot group. It lasts usually from a few minutes to a few hours. Its occurrence is defined by a sudden brightening of the H Lyα line, and it is usually accompanied by an enhancement in the X-ray emission (Fig. 1.23).

The X-ray and UV energy from flares arrive at the Earth in about 8 min and is known to produce intense ionization in the upper atmosphere which is manifested in radio wave fade-out and sudden cosmic noise absorption.

The occurrence frequency of small flares is positively correlated with the 11-year solar cycle, and although the frequency of major flares also tends to peak at sunspot maxima, they can occur at any time. There is some evidence for a double peak in major flare occurrence, the second one appearing a few years after sunspot maximum.

The radio emissions from solar flares have little energy compared to X-ray and EUV emissions, but as indicators of fast particles streaming out of the solar atmosphere they are of great importance.

In a time frequency diagram where the intensity is included as a third dimension as is done in Fig. 1.23, so-called radio bursts from the Sun are illustrated and classified as type from I to V. Since the electron density decreases outward in the solar atmosphere, and so does the plasma frequency, we may conclude that if a burst is due to a plasma wave decreasing in frequency with time, the source of the burst moves outward through the corona. The velocity of the wave can then actually be derived from the rate of change of the observed burst frequency.

Let us assume by inspecting Fig. 1.10 and Table 1.1 that to a first approximation the electron density is decaying exponentially by distance from the solar surface, then the electron density $n_e(z)$ at a distance z from the solar surface is given by:

$$n_e(z) = n_{e0} \exp\left(-\frac{z - z_0}{H}\right)$$

where n_{e0} is the electron density at a reference height z_0, and H is the e-folding scale height. From the expression of the plasma frequency (equation (1.27)) we find that

$$\frac{df_{pe}}{dt} = K \cdot \frac{1}{2}[n_e(z)]^{-1/2} \frac{dn_e(z)}{dz} \frac{dz}{dt} = -\frac{f_{pe}}{2H}\frac{dz}{dt}$$

and the propagation velocity v_{RS} of the radio source emitting a radio frequency f equal to the instantaneous plasma frequency f_{pe} is then given by:

$$v_{RS} = \frac{dz}{dt} = -2H \cdot \frac{d}{dt}(\ln f)$$

By observing the rate of change of the received frequency by time and for a given model of the electron density profile in the solar atmosphere, the speed of the radio emission source can be found. While the scale height increases from close to 50 km at the base of the photosphere to more than 500 km at the top of the chromosphere, it is of the order of one solar radius or more in the corona and interplanetary space.

During the first minutes of development after the flare flash has occurred, groups of very narrow bursts appear that show a fast drift from high to low frequency. These bursts called Type-III bursts are supposed to be carried by relativistic electrons moving upward in the corona by speeds between 0.2 and 0.9 times the speed of light. These bursts, assumed to be due to synchrotron radiation, probably originate between 50 and 100 $R_\odot$ from the Sun according to their original frequencies (100–300 MHz). A continuum appearing in the frequency-time diagram as a diffuse extended region called Type-V bursts is sometimes observed 1–3 min after the Type-III burst.

Another type of burst showing much slower drift velocities than the Type-III bursts is called Type-II bursts which usually start between 75 and 100 MHz. They can last from 5 to 10 minutes and are delayed by typically 5 minutes behind the Type-III bursts. The velocity in the source region through the corona for the Type-II bursts is between 500 and 1500 km/s. Since the speed of sound in this part of the corona is typically 175 km/s, these waves are supersonic and usually interpreted as electromagnetic radiation caused by shock waves.

The last type of bursts accompanying flares is the Type-IV bursts. They have a broad bandwidth (continuum) and last from 10 min up to hours after the flare. It is generally assumed that this radiation is due to relativistic particles trapped in the magnetic field of the active region.

Type-I bursts may last for hours and days and are not related to flares. They consist of an enhanced background continuum with short-lived (0.5 s) bursts superimposed. These noise bursts (Type-I) have their origin close to large sunspot groups. It is found that most of the Type-I events are preceded by flare events, however, there is no clear connection between storm characteristics and flare importance. The way Type-I events are created is uncertain.

1.7 REFERENCES

Akasofu, S.-I. and Chapman, S. (1972) *Solar–Terrestrial Physics,* Oxford at the Clarendon Press.
Allen, C. W. (1973) *Astrophysical Quantities,* p. 161, The Athlone Press.
Danielson, R. E. (1961) *Ap. J.,* **134**, 275.
Eddy, J. A. (1976) *Science,* **192**, 1189–1202.
Egeland, A., Henriksen, T. and Kanestrøm, I. (1990) *Drivhuseffekten. Jordens Atmosfære og Magnetfelt,* Universitetet i Oslo.
Giraud, A. and Petit, M. (1978) *Ionospheric Techniques and Phenomena,* D. Reidel, Dordrecht, The Netherlands.
Herman, J. R., and Goldberg, R. A. (1978) *Sun, Weather and Climate,* NASA, Washington, D.C.
Hinteregger, H. E. (1980) in *Proceedings of the Workshop on Solar UV Irradiance Monitors,* NOAA Environmental Research Laboratories (ERL), Boulder.
Joint NOAA–USAF Space Weather Operations (1996) *Solar Geophysical Data,* SWO PRF 1087.
Jursa, A. S. (Ed.) (1985) *Handbook of Geophysics and the Space Environment,* AFGL, USA.
Lean, J. (1991) *Rev. Geophys.,* **29**, 505–536.
Pettit, E. (1951) in *Astrophysics,* Hynek, J. A. (Ed.), p. 259, McGraw–Hill, New York.
Roberts, W. O. (1975) *Relationship between solar activity and climate change,* Bandeen, W. E. and Maran, S. (Eds.), p. 13, Goddard Space Flight Center Spes. Rep. NASA SP-366.
Schwabe, H. (1844) *Astron. Nachr.,* **21**, 233.
Teske, R. G. (1969) *Solar Physics,* **6**, 193.
Tohmatsu, T. (1990) *Compendium of Aeronomy,* Terra Sci. Publ. Comp., Tokyo.
Waldmeier, M. (1955) *Ergebnisse und Problem der Sonnenforschung,* Akademische Verlagsgesellschaft, Leipzig.
Wild, J. P., Smerd, S. F. and Weiss, A. A. (1963) *Ann. Rev. Astron. Astrophys.,* **1**, 291.
Zirin, H. (1966) *The Solar Atmosphere,* Ginn Blaisdell, Waltham, Mass.

1.8 EXERCISES

1. Show that

$$\int_0^\infty B_\lambda d\lambda = \sigma T^4$$

where B_λ is given by (1.7) and determine σ.
Hint: $\int_0^\infty [x^3/(\exp(x)-1)]dx = \pi^4/15$.

2. Fig. 1.24 illustrates the portion of the Earth's (E) orbit around the Sun (S) covered during one "day". When the point A has reached A' one *sidereal day* has passed, while when A has reached A'' one *solar day* has been completed. Show that the difference between one solar day and one sidereal day is approximately equal to 240 s.

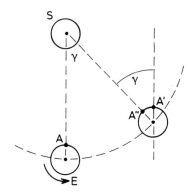

Figure 1.24.

3. Assume that the electron density in the solar atmosphere and the corona can be expressed as function of distance from the solar centre by:

$$n_c(r) = n_\odot \left(\frac{R_\odot}{r}\right)^\gamma$$

where $n_\odot$ is the electron density at the reference distance 1 $R_\odot$ from the solar centre.

We have observed a Type-III radio burst associated with a flare and measured a dispersion of the frequency (df/dt). Let the burst be due to a plasma wave which source is moving upwards in the atmosphere.

(a) Find an expression for the velocity of this source expressed by df/dt.

Assume that the electron density at 1 $R_\odot$ is equal to 10^{16} m^{-3} and at 10 $R_\odot$ it is equal to 10^9 m^{-3}. Let the burst first appear at 90 MHz and assume a frequency dispersion of 8 MHz/s.

(b) At what height did the burst start?
(c) What was the speed of the source at the onset of the burst?

2

The solar wind and the interplanetary magnetic field

2.1 PARTICLE EMISSIONS FROM THE QUIET SUN

The quiet Sun does not only radiate electromagnetic waves in the frequency range from a few nanometres to a few hundred metres, but also particles. A solar wind is always blowing. This wind is not a steady wind, but rather gusty as it varies quite strongly in velocity as well as density at a distance from the Sun corresponding to the Earth's orbit.

During quiet conditions the wind has a medium velocity of 400 km/s and can, however, vary between 200 and 700 km/s (Fig. 2.1). Note that with a speed of 435 km/s the solar wind takes 4 days to reach the Earth from the Sun. The particle density is about 5×10^6 m^{-3} consisting of mainly protons and electrons in the same amount. This density can vary between 10^6 and 2×10^7 m^{-3}. The average energy of the protons is of the order of 1 keV while the electrons have energies of the order of 1 eV. The average particle flux from the Sun can then be estimated to be:

$$\phi = n \cdot v \approx 2 \times 10^{12} \text{ m}^{-2} \text{ s}^{-1} \tag{2.1}$$

The total average particle loss from the Sun is therefore

$$\dot{N} = 4\pi R_\odot^2 \cdot \phi \approx 1.32 \times 10^{31} \text{ s}^{-1} \tag{2.2}$$

where $R_\odot$ is the solar radius. The proton temperature in the solar wind is $10^4 - 2 \times 10^5$ K while the electron temperature is a factor of 3–4 larger during quiet average conditions. During disturbed conditions the proton and electron temperatures are rather similar. The magnetic field strength associated with the solar wind is varying between 1 and 15 γ (1 $\gamma = 10^{-9}$ tesla) (Fig. 2.2). The temperature is always observed higher along the magnetic field $(T_\parallel)$ than perpendicular to it $(T_\perp)$. On the average $T_\parallel \approx 2T_\perp$.

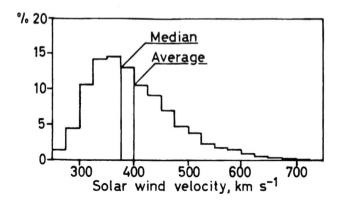
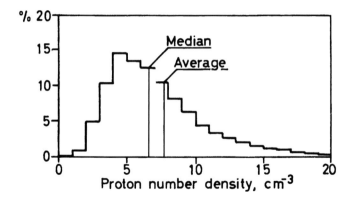
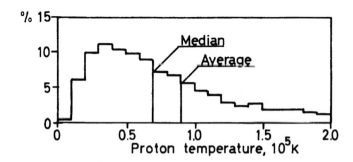

Fig. 2.1. Histograms of occurrence frequency for the values of the solar wind velocity, proton number density and proton temperature in interplanetary space. (From Hundhausen et al., 1970.)

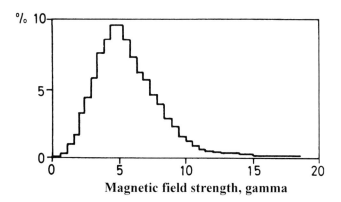

Fig. 2.2. Histogram of occurrence frequency for magnetic field strength values in interplanetary space. (From Ness, 1969; e.g. Fälthammar, 1973.)

For the characteristic parameters we have given for the solar wind so far, we notice that the speed of sound of the solar wind gas is approximately:

$$c_s = \sqrt{\gamma \frac{\kappa T_\odot}{m_p}} = 1.17 \times 10^4 \text{ m/s} \qquad (2.3)$$

where γ is the adiabatic constant equal to $\frac{5}{3}$ for a monoatomic gas and $T_\odot$ is set equal to 10^4 K, $m_p = 1.672 \times 10^{-27}$ kg is the proton mass and κ is Boltzmann's constant.

Since the solar wind speed is about 40 times higher on the average, the solar wind is supersonic.

This high flow wind is produced by the pressure difference between the hot dense gas of the solar atmosphere and the cold tenuous gas in the background interstellar medium. This pressure difference easily overcomes the solar gravitational pull on the plasma. Also the high electrical conductivity of the plasma prevents motions across the magnetic field lines and these lines are therefore frozen into the plasma as a "glue" and forces the plasma to act collectively as a hydrodynamic fluid. Since these field lines are anchored in the solar surface while the wind plasma moves radially outward, the lines of force will form Archimedean spirals making an angle of about 45° with the radius vector to the Sun at the Earth's orbit. We will come back to the structure of this field later.

Much like flow through a rocket engine nozzle, the gas starts with a low outward velocity at the coronal base, accelerates to sonic speed about 2 to 6 $R_\odot$ from the Sun's centre, and becomes supersonic thereafter.

The density of the solar wind gas as it expands spherically outwards from the Sun, must decrease as $1/r^2$ where r is the distance from the Sun. At a heliocentric distance r_h the pressure in the solar wind must balance the pressure of the interstellar medium, and the solar wind must again be subsonic. The change from a supersonic plasma to a subsonic plasma is expected to be related to a shock wave, and at this shock the interplanetary magnetic field terminates, forming the boundary of *the heliosphere*, the heliopause. It is at the moment impossible to determine the distance to this shock, but it is believed to be situated between 60 and 100 AU. As the Sun is moving through the interstellar medium, the heliosphere will probably take an oval shape (Fig. 2.3).

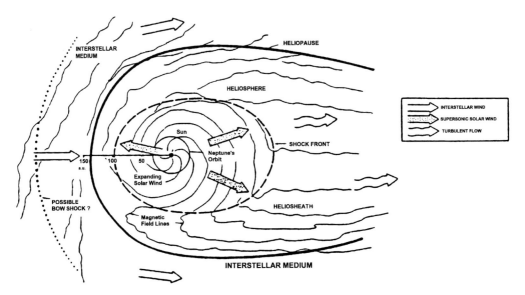

Fig. 2.3. A possible model of the heliosphere as due to interaction between the interplanetary hydrogen gas and the hot solar plasma as the Sun moves through space. The plane of the figure is in the Sun's equatorial plane. (From Dryer, 1987.)

2.2 LAVAL'S NOZZLE

To get an idea of the mechanism behind the acceleration of the solar wind, we will first discuss Laval's nozzle which is actually the principle for the jet engine.

Let an incompressible gas with density ρ stream through a tube with a varying cross-section A (Fig. 2.4). Since the mass flux through any cross-section of the tube must be constant, we have:

$$\phi_m = A \cdot \rho \cdot v = \text{const.} \tag{2.4}$$

when v is the velocity of the gas through the cross-section.

According to the equation of Bernoulli the pressure gradient will be balanced by the inertia force when other forces like gravity and collisions are neglected.

$$\frac{dp}{dr} = -\rho \frac{dv}{dt} = -\rho \frac{dv}{dr} \cdot v \tag{2.5}$$

$$dp = -\rho v \, dv \tag{2.6}$$

Laval's nozzle

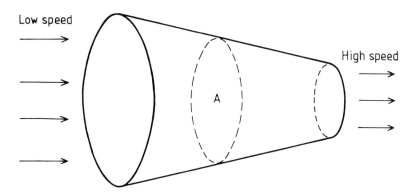

Fig. 2.4. Mass flow through a nozzle used to explain the acceleration of the flow speed.

or rewritten by another partial derivative:

$$\frac{dp}{\rho} = \frac{dp}{d\rho} \cdot \frac{d\rho}{\rho} = -v \cdot dv \tag{2.7}$$

If we now assume that this process of the gas flow is adiabatic, we have:

$$p \cdot \rho^{-\gamma} = \text{const.} \tag{2.8}$$

where γ is the adiabatic constant. Then

$$\frac{dp}{d\rho} = \gamma \frac{p}{\rho} = c_s^2 \tag{2.9}$$

where c_s is the speed of sound in the gas. It then follows by inserting (2.9) into (2.7):

$$\frac{d\rho}{\rho} = -\frac{v}{c_s^2} dv \tag{2.10}$$

According to the conservation of mass flux we have:

$$\rho v \, dA + v A \, d\rho + \rho A \, dv = 0 \tag{2.11}$$

or

$$\frac{dA}{A} + \frac{d\rho}{\rho} + \frac{dv}{v} = 0 \tag{2.12}$$

By inserting for $d\rho/\rho$ from (2.10) we get:

$$\frac{dA}{A} - \frac{v}{c_s^2} dv + \frac{dv}{v} = 0 \tag{2.13}$$

and

$$\frac{dA}{A} = \left(\frac{v}{c_s^2} - \frac{1}{v}\right) dv = \left(\frac{v^2}{c_s^2} - 1\right) \frac{dv}{v} \qquad (2.14)$$

As long as the cross-section A decreases and the velocity increases (dA/A; negative and dv/v; positive), the velocity v must be smaller than c_s, i.e., the flow is subsonic. When v finally reaches c_s, dA/A must be zero and the cross-section will not decrease anymore.

If, on the other hand, v is going to be larger than c_s, i.e. supersonic speed, dA/A must be positive and the cross-section must increase again. This is illustrated in Fig. 2.5. This tube is called Laval's nozzle. If, however, the gas is streaming so slowly that it does not reach the speed of sound at the narrowest part of the tube, it is called a Venturi tube. The speed of the gas will then decrease as it passes the throttle. Whether the tube will act as a Laval or Venturi tube depends on the pressure ratio between the two terminals of the tube.

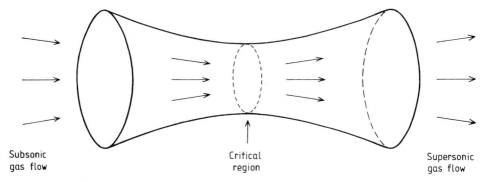

Fig. 2.5. Mass flow through a nozzle with a minimum cross-section to explain the presence of a critical region in the mass flow in order for the flow speed to become supersonic.

2.3 THE SOLAR WIND EQUATION

We will now assume that the coronal gas is an ideal gas and that we can treat it as a hydrodynamic fluid. This means that the collisions are frequent enough that local thermodynamic equilibrium is maintained. We will in the following treatment neglect viscosity and magnetism and limit ourselves only to internal forces and gravity. Again the mass conservation law must apply.

$$\rho v A = \phi_m = \text{const.} \qquad (2.15)$$

The Bernoulli equation will have an extra term due to gravity:

$$dp = -\rho v \, dv - \rho \frac{GM_\odot}{r^2} dr \qquad (2.16)$$

The solar wind equation

where r is the radial distance from the Sun, G the constant of gravity ($= 6.67 \times 10^{-11}$ N m^2 kg^{-2}) and $M_\odot$ is the solar mass. Since the mass flux is conserved

$$\frac{d\phi_m}{\phi_m} = \frac{dA}{A} + \frac{d\rho}{\rho} + \frac{dv}{v} = 0 \tag{2.17}$$

and $dp/d\rho = c_s^2$ or $dp/\rho = c_s^2 (d\rho/\rho)$ where c_s again is the speed of sound. From (2.16) we find:

$$\frac{dp}{\rho} = -v\, dv - \frac{GM_\odot}{r^2} dr \tag{2.18}$$

and substituting dp/ρ from (2.10) we have:

$$c_s^2 \frac{d\rho}{\rho} = -v\, dv - \frac{GM_\odot}{r^2} dr \tag{2.19}$$

Rearranging (2.17) and inserting (2.18):

$$\frac{d\rho}{\rho} = -\left(\frac{dA}{A} + \frac{dv}{v}\right) = -\frac{v}{c_s^2} dv - \frac{GM_\odot}{c_s^2} \frac{dr}{r^2} \tag{2.20}$$

we finally have:

$$\frac{dA}{A} = \left(\frac{v^2}{c_s^2} - 1\right) \frac{dv}{v} + \frac{GM_\odot}{c_s^2} \frac{dr}{r^2} \tag{2.21}$$

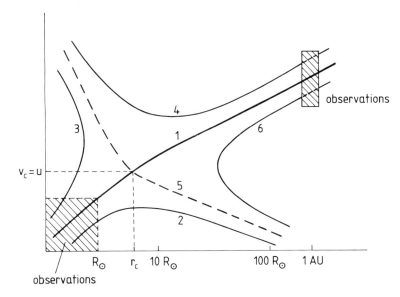

Fig. 2.6. Alternative solutions of the solar wind equation explained in the text.

The solar wind is expanding spherically and symmetrically so that:

$$\frac{dA}{A} = \frac{2}{r} dr \qquad (2.22)$$

This leads us to the following relationship between v and r:

$$\left(2 - \frac{GM_\odot}{c_s^2}\frac{1}{r}\right)\frac{dr}{r} = \left(\frac{v^2}{c_s^2} - 1\right)\frac{dv}{v} \qquad (2.23)$$

There are several solutions to this equation (Fig. 2.6).

1. If the wind starts out from the Sun with an increasing velocity which initially is smaller than c_s, then both sides are negative. When r equals the critical distance $r_c = GM_\odot/2c_s^2$, the left-hand side is zero and therefore the right-hand side has to be zero also, i.e., $v = c_s$. If $v = c_s$ at the critical point, then v can continue to accelerate into a supersonic speed. For a temperature of $T_\odot = 3.25 \times 10^4$ K, $r_c = 1\ R_\odot$. In order for the critical point to be well outside the Sun, the temperature has to be lower or the adiabatic constant must be lower.

2. The velocity is lower than u at the surface of the Sun, but is less than u at the critical point (it is not accelerated fast enough). The left-hand side will then be positive past the critical point, and the right-hand side can only become positive here if the velocity decreases. The wind ends up as a breeze.

3. The velocity is subsonic at the Sun's surface, but is accelerated so fast that it is supersonic before it reaches the critical point. The right-hand side then becomes positive, and the left-hand side can only be positive if r decreases. The wind blows back to the Sun with a supersonic flow.

4. The wind blows out from the Sun with a supersonic speed. The left-hand side of the equation is negative, and the only way the right-hand side can be negative is for the velocity to decrease. At the critical point the left-hand side of the equation changes sign. The only way the right-hand side can change sign, as long as v is supersonic, is when the velocity increases again.

5. If the wind starts supersonically at the Sun, the left-hand side is negative, and in order for the right-hand side to become negative, the velocity must decrease. At the critical point the right-hand side changes sign and becomes positive. If at the same instance the velocity has decreased to c_s, it must continue to decrease in order for the right-hand side to become positive.

6. Finally there is a solution for inward flow in the corona. If the flow starts at infinity blowing subsonically toward the Sun, the left-hand side is negative, and the right-hand side can only become negative if the velocity increases. When the velocity reaches c_s at some distance from the Sun, larger than r_c, the right-hand side changes sign, and for the left-hand side to become positive, r must increase again.

The solar wind equation

The one solution that connects the observations at the Sun and at the Earth's distance from the Sun is solution 1 which is the supersonic solar wind.

It is worth noticing that the solar temperature is just within the range which makes the supersonic solar wind possible. If the solar atmosphere is too cool, it will remain static. If it is too hot, the corona will expand subsonically as a solar breeze.

If the atmosphere is extremely cool, we notice from (2.19) that as long as the velocity is small compared to the speed of sound, the first term on the right-hand side can be neglected and then:

$$c_s^2 \frac{d\rho}{\rho} = - \frac{GM_\odot}{r^2} dr \qquad (2.24)$$

The atmosphere is in static equilibrium.

For a very hot atmosphere, however, the gravity term can be neglected compared to the left-hand side, and (2.23) then becomes:

$$2\frac{dr}{r} = \left(\frac{v^2}{c_s^2} - 1\right) \frac{dv}{v} \qquad (2.25)$$

For outward motion dr and the left-hand side are positive. If the wind starts out subsonically from the Sun, it will decrease outward in order to maintain the same signs in the equation, and the solar wind will be a "breeze".

We should, however, realize that in the expression for the sound speed in the solar atmosphere we have made an oversimplification as the solar wind gas is a composition of both protons and electrons with equal numbers of density, but different temperatures. Let m_e and m_p be the masses of the electron and the proton respectively, and n_e, n_p, T_e and T_p the respective densities and temperatures for the two species. Now we have a more proper expression for c_s^2 when we remember that $n_e = n_p$ and $m_p \gg m_e$:

$$c_s^2 = \gamma \cdot \frac{n_e \kappa T_e + n_p \kappa T_p}{n_e m_e + n_p m_p} \approx \gamma \cdot \frac{\kappa(T_e + T_p)}{m_p} \qquad (2.26)$$

If we assume that $T_e = T_p = T_\odot$, the sound speed will be

$$c_s^2 = 2\gamma \frac{\kappa T_\odot}{m_p} \qquad (2.27)$$

This would decrease the temperature further for which the solar wind can be supersonic outside the atmosphere. There are indications that the adiabatic constant is closer to $\frac{4}{3}$ than $\frac{5}{3}$ in the solar wind plasma.

For a temperature of $T_\odot = 10^4$ K and $\gamma = \frac{4}{3}$ the critical distance r_c would be about $2\,R_\odot$ when corrected for the composition of the gas.

We notice that the escape velocity from the Sun is given by:

$$v_{\odot e} = \left(\frac{2GM_\odot}{R_\odot}\right)^{1/2} \tag{2.28}$$

Therefore we find that

$$\frac{r_c}{R_\odot} = \frac{GM_\odot}{2c_s^2} \cdot \frac{1}{R_\odot} = \left(\frac{v_{\odot e}}{c_s}\right)^2 \tag{2.29}$$

If r_c is situated at $1.4\ R_\odot$, we then have that the speed of sound is $\frac{1}{2}\ v_{\odot e}$.

2.4 INDICATIONS OF MAGNETIC FIELDS ON THE SUN

As seen from photographs of the Sun during solar eclipses, the structure of the corona shows features that indicate the presence of a magnetic field on the Sun (Fig. 2.7). Other observed features that also support the existence of the presence of such solar magnetic fields, are the loop prominences observed in the H_α line (Fig. 2.8). Tentative measurements give magnetic fields of the order of 50 gauss (1 gauss = 1 Γ = 10^{-4} tesla). It is believed that the solar magnetic field is rather complicated and that local features can change dramatically within minutes. Loop prominences, for example, can be seen to last for only 20 minutes.

Fig. 2.7. The solar corona photographed during a total solar eclipse two years before sunspot minimum (Feb. 25, 1952). (From van Biesbroeck, 1953.)

Sec. 2.4] Indications of magnetic fields on the Sun 43

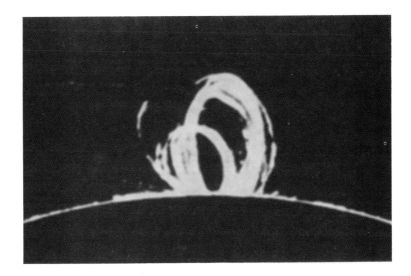

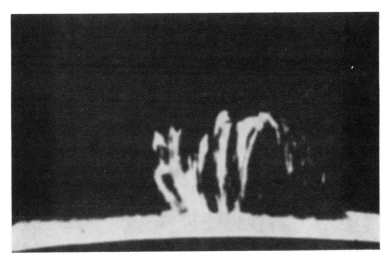

Fig. 2.8. Loop prominences on the Sun photographed in the H_α emission. (From Zirin, 1966.)

An attempt to draw the gross field configuration of the Sun is shown in Fig. 2.9. What is important to notice here is the separated regions where the magnetic field points toward or away from the solar surface. In the region between inward and outward pointing magnetic field lines there must be concentrated layers of electrical currents. At other regions a slow solar wind may occur or coronal streamers may be formed.

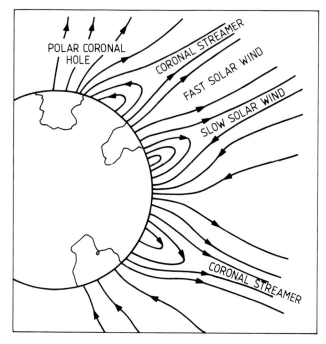

Fig. 2.9. A schematic illustration of the different magnetic field structures believed to be present at the Sun. Some of the field lines are pulled into space by the solar wind, while the boundary between neighbouring inward and outward pointing field lines must be penetrated by concentrated layers of electric currents. (After Dryer, 1987.)

2.5 THE FROZEN-IN FIELD CONCEPT

As already mentioned the plasma or gas of the solar wind has a high electrical conductivity, and therefore it carries the magnetic field along as a "glue" in the plasma. We will now explain this phenomenon of "frozen-in" magnetic fields more in details. We will do this because the "frozen-in" magnetic field concept turns out to be a useful tool in understanding plasma flow in the solar wind, magnetosphere and upper ionosphere. For a true conducting fluid the current density is given by $\mathbf{j}' = \sigma \mathbf{E}'$ where $\mathbf{j}'$ and $\mathbf{E}'$ are the currents and the electric fields in the frame moving with the plasma. Since the plasma is highly conductive, the conductivity takes on this very simple scalar form, σ, and not the usual tensor form. From another frame of reference where the plasma is moving with the velocity $\mathbf{v}$, the electric field is $\mathbf{E} = \mathbf{E}' - \mathbf{v} \times \mathbf{B}$, the current, however, is conserved, $\mathbf{j} = \mathbf{j}'$, and so is the magnetic field, $\mathbf{B} = \mathbf{B}'$, if the velocity is small enough ($\mathbf{v} \ll \mathbf{c}$). Thus

$$\mathbf{j} = \sigma \mathbf{E}' = \sigma (\mathbf{E} + \mathbf{v} \times \mathbf{B}) \tag{2.30}$$

which is the generalized Ohm's law. From this equation we find:

$$\mathbf{E} + \mathbf{v} \times \mathbf{B} = \frac{\mathbf{j}}{\sigma} \tag{2.31}$$

and notice that when $\sigma \to \infty$

$$\mathbf{E} + \mathbf{v} \times \mathbf{B} = 0 \tag{2.32}$$

which is the so-called "frozen-in" condition and which implies that no current is flowing in the medium.

We now want to investigate further what other physical consequences the "frozen-in" conditions implies. From Maxwell's equations we have

$$\frac{\partial \mathbf{B}}{\partial t} = -\nabla \times \mathbf{E} \tag{2.33}$$

$$\nabla \cdot \mathbf{B} = 0 \tag{2.34}$$

$$\nabla \times \mathbf{B} = \mu_0 \mathbf{j} + \epsilon_0 \mu_0 \frac{\partial \mathbf{E}}{\partial t} \tag{2.35}$$

where μ_0 $(= 1.257 \times 10^{-6}$ H/m) is the permeability in vacuum. When introducing $\mathbf{E}$ from (2.31) we find:

$$\frac{\partial \mathbf{B}}{\partial t} = -\nabla \times \left(\frac{1}{\sigma} \mathbf{j} - \mathbf{v} \times \mathbf{B} \right) \tag{2.36}$$

If we now neglect the vacuum displacement current, i.e.

$$\left| \epsilon_0 \mu_0 \frac{\partial \mathbf{E}}{\partial t} \right| \ll |\nabla \times \mathbf{B}| \tag{2.37}$$

we have

$$\mathbf{j} = \frac{1}{\mu_0} \nabla \times \mathbf{B} \tag{2.38}$$

and find since σ is isotropic

$$\frac{\partial \mathbf{B}}{\partial t} = -\frac{1}{\mu_0 \sigma} \nabla \times (\nabla \times \mathbf{B}) + \nabla \times (\mathbf{v} \times \mathbf{B}) \tag{2.39}$$

Using now the vector relationship

$$\nabla \times (\nabla \times \mathbf{B}) = \nabla(\nabla \cdot \mathbf{B}) - \nabla^2 \mathbf{B} \tag{2.40}$$

and applying (2.34) we find:

$$\frac{\partial \mathbf{B}}{\partial t} = \frac{1}{\mu_0 \sigma} \nabla^2 \mathbf{B} + \nabla \times (\mathbf{v} \times \mathbf{B}) \tag{2.41}$$

For an infinite value of σ we have from (2.41):

$$\frac{\partial \mathbf{B}}{\partial t} = \nabla \times (\mathbf{v} \times \mathbf{B}) \tag{2.42}$$

If we now consider the magnetic flux through a surface moving with velocity v, this is given by:

$$\frac{d\phi}{dt} = \iint_S \frac{\partial \mathbf{B}}{\partial t} \cdot d\mathbf{s} + \oint_L \mathbf{B} \cdot (\mathbf{v} \times d\boldsymbol{\ell}) \tag{2.43}$$

where S is the surface and L is a closed loop encircling this surface. The first integral yields the change in ϕ due to the time variation of $\mathbf{B}$, while the second integral yields the change in ϕ due to the motion of the surface across $\mathbf{B}$.

The second term can be rewritten:

$$\oint_L \mathbf{B} \cdot (\mathbf{v} \times d\boldsymbol{\ell}) = -\oint_L (\mathbf{v} \times \mathbf{B}) d\boldsymbol{\ell} = -\iint_S \nabla \times (\mathbf{v} \times \mathbf{B}) d\mathbf{s} \tag{2.44}$$

and therefore $d\phi/dt$ becomes

$$\frac{d\phi}{dt} = \iint_S \left[\frac{\partial \mathbf{B}}{\partial t} - \nabla \times (\mathbf{v} \times \mathbf{B}) \right] d\mathbf{s} \tag{2.45}$$

We have, however, just shown that when σ is very large and the vacuum displacement current is negligible, the expression in the integral vanishes so that $d\phi/dt = 0$ and ϕ is a constant. The flux through the surface is remained, and the magnetic field is therefore "frozen-in".

Let us look back on the condition we forced on the vacuum displacement current $\epsilon_0 \mu_0 (\partial \mathbf{E}/\partial t)$ in (2.37). From a simple dimensional analysis we see that

$$\mu_0 \epsilon_0 \frac{E}{T} \ll \frac{B}{L} \tag{2.46}$$

where T and L are the characteristic time and distance in the plasma respectively.

From Maxwell's equation (2.33) we obtain a similar dimensional relationship

$$\frac{B}{T} = \frac{E}{L} \tag{2.47}$$

and therefore

$$\mu_0 \epsilon_0 \frac{E}{T} \ll \frac{ET}{L^2} \tag{2.48}$$

or finally

$$V^2 = \frac{L^2}{T^2} \ll \epsilon_0 \mu_0 = c^2 \tag{2.49}$$

where V is a characteristic velocity. This means that the relationship we have found for the "frozen-in" magnetic field, is valid as long as the highly conducting plasma is moving with a speed which is much less than the speed of light.

The concept of the "frozen-in" magnetic field is a very useful one. If a plasma sticks to a magnetic field line, then it will continue to do so as long as the "frozen-in" condition is valid. This allows visualization of complex plasma flow if we know the magnetic field configuration, and conversely if the motion of the plasma is known, the field geometry can be deduced. On the other side, if a highly conducting plasma is approaching a volume with a magnetic field, the field cannot penetrate into the plasma, and the plasma pushes the field ahead. This is illustrated in Fig. 2.10.

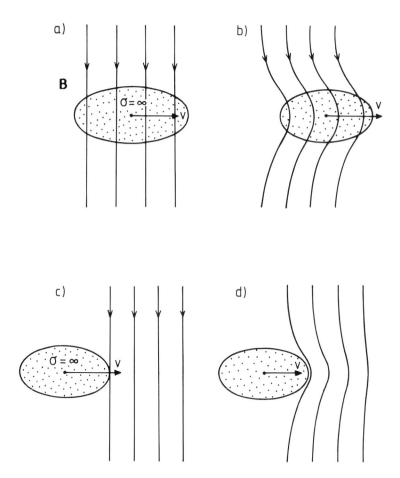

Fig. 2.10. An illustration of the "frozen-in" field concept. (a) A magnetic field **B** is assumed to be penetrating a region of highly conducting plasma. (b) When the plasma starts to move, the magnetic field lines will be "frozen-in" and follow the motion of the plasma. (c) A highly conducting plasma is approaching an area of magnetic field. (d) Due to the high conductivity the field cannot penetrate into the plasma and is pushed ahead of the plasma blob.

If $\sigma \neq \infty$ the field can slip through the surface of the fluid, and we have

$$\frac{\partial \mathbf{B}}{\partial t} - \nabla \times (\mathbf{v} \times \mathbf{B}) = \frac{1}{\mu_0 \sigma} \nabla^2 \mathbf{B} \qquad (2.50)$$

For a stationary case ($\mathbf{v} = 0$) this reduces to

$$\frac{\partial \mathbf{B}}{\partial t} = \frac{1}{\mu_0 \sigma} \nabla^2 \mathbf{B} \qquad (2.51)$$

This is a diffusion equation of the magnetic field with a time constant given by

$$\tau = \mu_0 \sigma L^2 \qquad (2.52)$$

where L is a typical distance in the plasma. Because of the high values of σ and the large distances in the solar wind plasma τ is very high.

Another concept which should be noticed in this context is the magnetic Reynolds number R_m given by:

$$R_m = LV\mu_0\sigma \qquad (2.53)$$

where L and V are characteristic length and velocity, respectively, and the other parameters are as defined earlier. Within the characteristic time $T = L/V$ the magnetic flux through a surface S has to change very little if the "frozen-in" concept is going to hold. Then

$$\left|\frac{d\phi}{dt}\right| \ll \frac{\phi}{T} = \frac{BS}{L/V} = \frac{BSV}{L} \qquad (2.54)$$

By dividing (2.38) by σ we can form:

$$\left|\frac{\mathbf{j}}{\sigma}\right| = \left|\frac{1}{\mu_0\sigma} \nabla \times \mathbf{B}\right| \approx \frac{B}{\mu_0 \sigma L} \qquad (2.55)$$

Since we can show that

$$\frac{d\phi}{dt} = -\iint_S \left(\nabla \times \frac{\mathbf{j}}{\sigma}\right) ds \qquad (2.56)$$

we find that

$$\left|\frac{d\phi}{dt}\right| = \iint_S \left(\nabla \times \frac{\mathbf{j}}{\sigma}\right) ds = \frac{j}{\sigma} \cdot \frac{S}{L} = \frac{B}{\mu_0 \sigma} \cdot \frac{S}{L^2} \qquad (2.57)$$

Finally, by equating (2.54) and (2.57) we have

$$\frac{B}{\mu_0 \sigma} \frac{S}{L^2} \ll \frac{BSV}{L} \qquad (2.58)$$

and

$$LV\mu_0\sigma = R_m \gg 1 \qquad (2.59)$$

2.6 THE ELECTRIC FIELD IN THE SOLAR WIND

The solar wind plasma is electrically neutral with the same amount of positive and negative charges, therefore $n_e = n_i = n$. Furthermore, the velocities have to be the same for the two charges unless they will accumulate and form space charges, therefore:

$$\mathbf{v}_e = \mathbf{v}_i = \mathbf{v} \tag{2.60}$$

We will now only consider radial motions and assume that all quantities depend only on the radial distance from the Sun. Thus the equation of motion for the electrons can be written as:

$$m_e n_e \frac{dv_e}{dt} = m_e n_e v_e \frac{dv_e}{dr} = -\frac{dp_e}{dr} - n_e m_e \frac{GM_\odot}{r^2} - n_e eE \tag{2.61}$$

where

$$\frac{dv_e}{dt} = \frac{dv_e}{dr}\frac{dr}{dt} = v_e \frac{dv_e}{dr} \tag{2.62}$$

m_e is the electron mass, p_e is the pressure of the electron gas, and E is an electric field in the plasma due to the different charges there. For the ions the equation of motion will be:

$$m_i n_i v_i \frac{dv_i}{dr} = -\frac{dp_i}{dr} - n_i m_i \frac{GM_\odot}{r^2} + n_i eE \tag{2.63}$$

By adding these two equations we find

$$(m_e n_e + m_i n_i) v \frac{dv}{dr} = -\frac{d}{dr}(p_e + p_i) - (n_e m_e + n_i m_i)\frac{GM_\odot}{r^2} \tag{2.64}$$

Since $m_i \gg m_e$ we can simplify the equation and find:

$$m_i n_i v \frac{dv}{dr} = -\frac{d}{dr}(p_e + p_i) - n_i m_i \frac{GM_\odot}{r^2} \tag{2.65}$$

We have then neglected the terms $m_e n_e (dv_e/dr)$ and $n_e m_e (GM_\odot/r^2)$. From the equation of motion for the electrons (equation (2.61)) we can therefore find the electric field:

$$E = -\frac{1}{n_e \cdot e}\frac{dp_e}{dr} \tag{2.66}$$

The electric field in the solar wind can be interpreted as a polarization field set up to balance the pressure gradients of the electron gas.

By assuming that the electron density decreases exponentially by distance with the scale height of the solar atmosphere, and that the atmosphere is isothermal, the electron density can be expressed as function of radial distance from the Sun by:

$$n_e = n_{e0} \exp\left(-\frac{r}{H_\odot}\right) \tag{2.67}$$

Here $H_\odot$ ($= \kappa T_\odot/m_p g_\odot$) is the scale height of the solar atmosphere, and $g_\odot = GM_\odot/R_\odot^2$. The electric field is then found from (2.66):

$$E = -\frac{1}{n_e \cdot e} \frac{d}{dr}(n_e \kappa T_\odot) = +\frac{\kappa T_\odot}{e \cdot H_\odot} = \frac{m_p g_\odot}{e} \tag{2.68}$$

The electric force is therefore balanced by the gravity force on the proton at the solar surface. By inserting numerical values we find:

$$E = 2.74 \times 10^{-6} \frac{V}{m} = 2.74 \; \mu V/m \tag{2.69}$$

A very small electric field indeed, but extending over large distances it becomes a significant voltage drop. The energy density of the electric field is

$$\varepsilon_E = \frac{1}{2}\epsilon_0 E^2 = 3.32 \times 10^{-24} \; J/m^3$$

Compared to the energy density of the magnetic field in the solar wind when $B = 5 \; nT$:

$$\varepsilon_B = \frac{B^2}{2\mu_0} = 10^{-11} \; J/m^3$$

The electric field due to the pressure gradient is therefore negligible as far as energy is concerned, but it may still have important effects on the drift of the charged particles.

2.7 THE GARDEN HOSE EFFECT

We have now seen that the magnetic field can be carried along in the solar wind from the Sun and through interplanetary space in a "frozen-in" manner. As the plasma is streaming radially out from the rotating Sun, it will move in spirals as the water from a spinning nozzle of a garden hose.

Let us assume that the plasma leaves the Sun in the equatorial plane at a distance r_0 from the solar centre. Then after a time the position of the plasma in the same plane can be described as

$$r = v \cdot t + r_0 \tag{2.70}$$

$$\phi = \Omega \cdot t + \phi_0 \tag{2.71}$$

where ϕ_0 is the longitude on the Sun from which the plasma emerges, and Ω is the angular velocity of the Sun, and v is the solar wind velocity.

Sec. 2.7] The garden hose effect

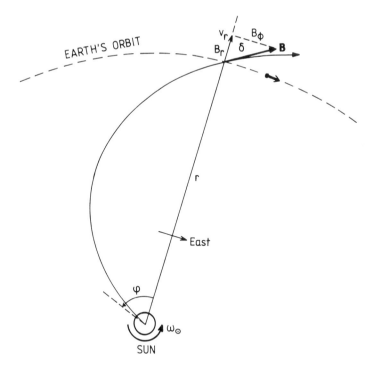

Fig. 2.11. The solar wind plasma streams radially out from a rotating Sun, and its motion can be described as an Archimedean spiral (garden hose). At the position of the Earth the angle (δ) between the plasma velocity and the Sun–Earth line is close to 45°. The Earth's orbit and the eastward direction are indicated.

Eliminating the time gives:

$$r = v \cdot \frac{\phi - \phi_0}{\Omega} + r_0 \tag{2.72}$$

which is the equation for an Archimedean spiral and illustrated in Fig. 2.11.

As the plasma now will drag the magnetic field along, it will follow the same trajectory and form similar spirals. Introducing polar coordinates and still confining ourselves to the equatorial plane, we have:

$$\mathbf{v} = (v_r, v_\phi) \tag{2.73}$$

$$\mathbf{B} = (B_r, B_\phi) \tag{2.74}$$

We will assume that

$$B = B(r) \tag{2.75}$$

and

$$v = v(r) \tag{2.76}$$

i.e. the magnitude of the solar wind velocity and the magnetic field depend only on the radial distance from the Sun. We then have in spherical coordinates:

$$\nabla \cdot \mathbf{B} = \frac{1}{r^2} \frac{\partial}{\partial r}(r^2 B_r) = 0 \tag{2.77}$$

or

$$r^2 B_r = r_0^2 B_0 = \text{const.} \tag{2.78}$$

The magnetic flux through spherical shells is conserved. The magnetic field therefore decreases as

$$B_r = B_0 \left(\frac{r_0}{r}\right)^2 \tag{2.79}$$

where B_0 is the magnetic field in the reference point at a distance r_0 from the Sun. Since $\partial B/\partial t = 0$ and the "frozen-in" concept applies we have from (2.42):

$$\nabla \times (v \times \mathbf{B}) = 0 \tag{2.80}$$

and in spherical coordinates

$$\frac{1}{r}\frac{\partial}{\partial r}(r(v_\phi B_r - v_r B_\phi)) = 0 \tag{2.81}$$

$$r(v_\phi B_r - v_r B_\phi) = \text{const.} \tag{2.82}$$

If we assume that $\mathbf{B}$ is radial at the reference point, then $B_{\phi 0} = 0$ and $B_{r_0} = B_0$ there.

$$r_0 v_{\phi 0} B_0 = r v_\phi B_r - r v_r B_\phi \tag{2.83}$$

Since the gas is rotating with the rotation speed of the Sun, we find at the surface of the Sun:

$$v_{\phi 0} = r_0 \Omega \tag{2.84}$$

By inserting into (2.83)

$$r_0^2 \Omega B_0 = r v_\phi B_r - r v_r B_\phi \tag{2.85}$$

and solving for B_ϕ:

$$B_\phi = \frac{r v_\phi B_r - r_0^2 \Omega B_0}{r v_r} = \frac{v_\phi B_r - r \Omega \left(\frac{r_0}{r}\right)^2 B_0}{v_r} = \frac{v_\phi - r\Omega}{v_r} B_r \tag{2.86}$$

The garden hose effect

For very large r; $r\Omega > v_\phi$, then:

$$B_\phi = -\frac{r\Omega}{v_r} B_r = -\frac{r_0^2 \Omega}{r v_r} B_0 \qquad (2.87)$$

The azimuthal component therefore decreases with the distance from the Sun as $1/r$, that is more slowly than the radial component as expressed by (2.79).

The angle the magnetic field will form with the radius vector to the Sun is given by (see Fig. 2.11):

$$\tan \delta = \frac{B_\phi}{B_r} \qquad (2.88)$$

For large distances $(r\Omega > v_\phi)$:

$$\tan \delta = \frac{r_0^2 \Omega}{r v_r} B_0 \frac{1}{B_0 \left(\frac{r_0}{r}\right)^2} = \frac{\Omega r}{v_r} \qquad (2.89)$$

At the Earth's orbit ($r = 1$ AU) and $\Omega = 2\pi/T$ and $T = 24.7$ days for the rotation time at solar equator, we find when assuming a solar wind speed equal to 400 km/s:

$$\tan \delta \approx 1 \qquad (2.90)$$

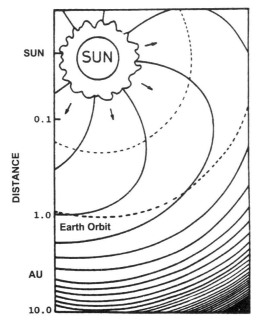

Fig. 2.12. The structure of the interplanetary magnetic field as viewed above the north pole of the Sun. Arrows indicate the flow pattern of the solar wind in interplanetary space. The Earth's orbit is indicated. (From Sakurai, 1987.)

At the Earth's orbit the "garden hose" angle is therefore close to 45°, in good agreement with the observations. From observations of the motion of charged particles it has been inferred that those leaving the Sun at the western heliospheric hemisphere are more often observed reaching the Earth than particles leaving the eastern hemisphere as if the particles are guided toward the Earth along Archimedean spirals.

As the solar wind plasma emerges from the Sun whether the solar magnetic field points toward or away from the Sun, it will carry magnetic field with different polarity into interplanetary space. Therefore we find that the interplanetary field is divided up into sectors where the field points predominantly away from or toward the Sun. This is illustrated in Fig. 2.12.

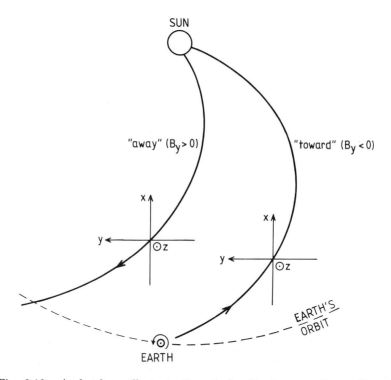

Fig. 2.13. A sketch to illustrate the relationship between "toward" and "away" sectors of the interplanetary field and the positive and negative values of the y-component. The z-axis is perpendicular to the ecliptic plane. The x-axis is in the noon–midnight meridional plane pointing positive toward the Sun. The y-axis is pointing positive toward the east.

Fig. 2.13 shows a Cartesian coordinate system often used to describe the interplanetary magnetic field (IMF). The z-axis is vertical to the ecliptic plane in the northward direction, i.e. parallel to the Earth's rotation axis. x is directed towards the Sun parallel to the midday–midnight meridian plane, and y is directed eastward in order to complete the right-handed orthogonal system.

Because of the spiral structure of the interplanetary field we notice that when the field is directed away from the Sun, the y-component is positive, and for a "toward" directed IMF the B_y component is negative.

We have in our discussion neglected any component of the magnetic field perpendicular to the equatorial plane of the Sun. This is of course an over-simplification. The component vertical to the equatorial plane, or rather to the ecliptic plane, will be later shown to have a profound influence on the shape of the magnetosphere of the Earth.

Furthermore, the magnetic field lines do not only emanate from the ecliptic plane but at any latitude of the Sun. Therefore the sector structure has a rather complicated shape in 3 dimensions, and the demarcation sheets between the different structure regions will wobble around the Sun like the skirt of a ballerina. The Earth will move in and out of the pleats in this skirt (Fig. 2.14). During one solar revolution the Earth will usually encounter 2 sector reversals, but quite often 4 and sometimes more. During the crossing of such sector boundaries disturbances in the Earth's atmosphere is likely to change character. Remember, however, that the solar wind effects on Earth will usually be delayed by about 4 days from a sector crossing at the central meridian of the Sun.

While the magnetic field component perpendicular to the heliospheric equatorial plane (or the ecliptical plane) appears to have a profound influence on the

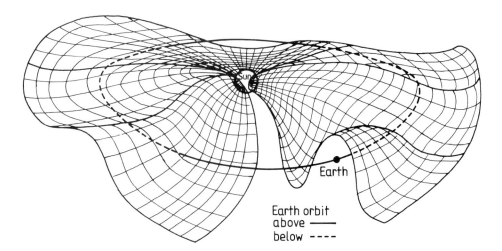

Fig. 2.14. A three-dimensional sketch of the solar equatorial current sheet and associated with the magnetic field line configuration in interplanetary space. The current sheet is here portrayed as lying near the solar equator and containing folds or flutes. When the Sun rotates, an observer near the ecliptic such as on the Earth, will alternately lie above and below this current sheet and will see a changing sector pattern. (After Kelley, 1989.)

shape of the magnetosphere, the azimuthal components of this field might have important influence on the Sun–weather relationship.

2.8 HYDROMAGNETIC WAVES IN INTERPLANETARY SPACE

We will now study the possibility of waves in the interplanetary plasma. In order to do this we will treat the background plasma as an infinite homogeneous gas of uniform density ρ_0 and pressure p_0, in equilibrium ($\mathbf{v}_0 = 0$) and permeated by a uniform magnetic field $\mathbf{B}_0$.

The following general equations can be applied to describe this medium:

$$\frac{\partial \rho}{\partial t} + \nabla \cdot (\rho \mathbf{v}) = 0 \tag{2.91}$$

$$\rho \left[\frac{\partial \mathbf{v}}{\partial t} + (\mathbf{v} \cdot \nabla) \mathbf{v} \right] = -\nabla p - \mathbf{B} \times \mathbf{j} \tag{2.92}$$

$$\frac{\partial \mathbf{B}}{\partial t} = -\nabla \times \mathbf{E} \tag{2.93}$$

$$\nabla \times \mathbf{B} = \mu_0 \mathbf{j} \tag{2.94}$$

Due to the very high conductivity in the interplanetary plasma we also apply the "frozen-in" condition from (2.32)

$$\mathbf{E} + \mathbf{v} \times \mathbf{B} = 0 \tag{2.95}$$

We can then rewrite the mobility equation (2.92) by introducing $\mathbf{j}$ from Maxwell's equation (2.94):

$$\rho \left[\frac{\partial \mathbf{v}}{\partial t} + (\mathbf{v} \cdot \nabla) \mathbf{v} \right] = -\nabla p - \frac{1}{\mu_0} \mathbf{B} \times (\nabla \times \mathbf{B}) \tag{2.96}$$

and the electric field can be left out from Maxwell's equation (2.93) by using the "frozen-in" condition (equation (2.96)) and we get equation (2.42):

$$\frac{\partial \mathbf{B}}{\partial t} = +\nabla \times (\mathbf{v} \times \mathbf{B}) \tag{2.97}$$

We will now let a wave propagate through the medium that will create perturbations in the velocity, pressure, density and magnetic as well as electric field. The perturbed parameters can then be written as:

Hydromagnetic waves in interplanetary space

$$\mathbf{v} = \mathbf{v}_1 \quad \text{(a)}$$

$$\rho = \rho_0 + \rho_1 \quad \text{(b)}$$

$$p = p_0 + p_1 \quad \text{(c)} \quad (2.98)$$

$$\mathbf{B} = \mathbf{B}_0 + \mathbf{B}_1 \quad \text{(d)}$$

$$\mathbf{E} = \mathbf{E}_0 + \mathbf{E}_1 \quad \text{(e)}$$

where $\mathbf{v}_1$, ρ_1, p_1, $\mathbf{B}_1$ and $\mathbf{E}_1$ are the perturbed quantities of the velocity, density, pressure, magnetic and electric field respectively. Introducing these quantities to equations (2.91), (2.96), and (2.97), respectively, we find when neglecting higher order terms than the linear term in the perturbation (also remember that the background plasma is uniform and in equilibrium):

$$\frac{\partial \rho_1}{\partial t} + \nabla(\rho_0 \mathbf{v}_1) = 0 \qquad (2.99)$$

$$\rho_0 \frac{\partial \mathbf{v}_1}{\partial t} = -\nabla p_1 - \frac{1}{\mu_0} \mathbf{B}_0 \times (\nabla \times \mathbf{B}_1) \qquad (2.100)$$

$$\frac{\partial \mathbf{B}_1}{\partial t} = \nabla \times (\mathbf{v}_1 \times \mathbf{B}_0) \qquad (2.101)$$

To complete our set of equations for describing the medium we need an equation of state and choose the adiabatic gas law expressed as:

$$p \cdot \rho^{-\gamma} = S \qquad (2.102)$$

where S is the entropy and γ is the adiabatic constant given by:

$$\gamma = C_p / C_v \qquad (2.103)$$

Here C_p and C_v are the specific heat capabilities for constant pressure and volume, respectively. We then get:

$$(p_0 + p_1) = S \cdot (\rho_0 + \rho_1)^\gamma \qquad (2.104)$$

and

$$\nabla p_1 = S \cdot \gamma (\rho_0 + \rho_1)^{\gamma - 1} \cdot \nabla \rho_1 \qquad (2.105)$$

when applying that $\nabla p_0 = \nabla \rho_0 = 0$. Solving for ∇p_1 gives:

$$\nabla p_1 = \gamma \cdot \frac{p_0 + p_1}{\rho_0 + \rho_1} \nabla \rho_1 \qquad (2.106)$$

Since $p_1 \ll p_0$ and $\rho_1 \ll \rho_0$ we find:

$$\nabla p_1 = \gamma \frac{p_0}{\rho_0} \nabla \rho_1 = c_s^2 \nabla \rho_1 \qquad (2.107)$$

where $c_s^2 = \gamma(p_0/\rho_0)$ is the speed of sound in the background plasma. Introducing this to the mobility equation (2.100), we find:

$$\rho_0 \frac{\partial \mathbf{v}_1}{\partial t} = -c_s^2 \nabla \rho_1 - \frac{1}{\mu_0} \mathbf{B}_0 \times (\nabla \times \mathbf{B}_1) \qquad (2.108)$$

We will now let the parameters related to the wave, $\mathbf{v}_1$, ρ_1, p_1, $\mathbf{B}_1$ and $\mathbf{E}_1$, all have the form $\propto \exp[i(\mathbf{k} \cdot \mathbf{r} - \omega t)]$, and since we then can substitute $\partial/\partial t \to -i\omega$ and $\nabla \to i\mathbf{k}$, we get:

$$-i\omega \rho_1 + i\mathbf{k} \cdot \mathbf{v}_1 \rho_0 = 0 \qquad (2.109)$$

$$-i\omega \rho_0 \mathbf{v}_1 = -c_s^2 i\mathbf{k}\rho_1 - \frac{1}{\mu_0} \mathbf{B}_0 \times (i\mathbf{k} \times \mathbf{B}_1) \qquad (2.110)$$

$$-i\omega \mathbf{B}_1 = i\mathbf{k} \times (\mathbf{v}_1 \times \mathbf{B}_0) \qquad (2.111)$$

By introducing (2.109) and (2.110) into (2.111), we reduce our system of equations to one:

$$-\rho_0 \omega \mathbf{v}_1 + c_s^2 \frac{\rho_0}{\omega} (\mathbf{k} \cdot \mathbf{v}_1)\mathbf{k} - \frac{1}{\mu_0 \omega} \mathbf{B}_0 \times (\mathbf{k} \times (\mathbf{k} \times (\mathbf{v}_1 \times \mathbf{B}_0))) = 0 \qquad (2.112)$$

Dividing by ρ_0 and multiplying by ω when introducing the Alfvén velocity

$$\mathbf{c}_A = (\mu_0 \rho_0)^{-1/2} \mathbf{B}_0 \qquad (2.113)$$

to (2.112) we find

$$\omega^2 \mathbf{v}_1 - c_s^2 (\mathbf{k} \cdot \mathbf{v}_1)\mathbf{k} - \mathbf{c}_A \times (\mathbf{k} \times (\mathbf{k} \times (\mathbf{v}_1 \times \mathbf{c}_A))) = 0 \qquad (2.114)$$

When completing the last vector product and reorganizing, we get:

$$\omega^2 \mathbf{v}_1 - (c_s^2 + c_A^2)(\mathbf{k} \cdot \mathbf{v}_1)\mathbf{k} - (\mathbf{c}_A \cdot \mathbf{k}) \times \\ [(\mathbf{c}_A \cdot \mathbf{k})\mathbf{v}_1 - (\mathbf{c}_A \cdot \mathbf{v}_1)\mathbf{k} - (\mathbf{k} \cdot \mathbf{v}_1)\mathbf{c}_A] = 0 \qquad (2.115)$$

Here different solutions may occur, and we will split our discussion into the transversal and longitudinal modes:

1. *The transversal mode.*

   If $\mathbf{k} \cdot \mathbf{v}_1 = 0$, i.e. $\mathbf{v}_1$ is perpendicular to the wave vector $\mathbf{k}$, then from (2.111):

   $$\omega \mathbf{B}_1 = -\mathbf{k} \times (\mathbf{v}_1 \times \mathbf{B}_0) = -(\mathbf{k} \cdot \mathbf{B}_0)\mathbf{v}_1 \qquad (2.116)$$

   and $\mathbf{B}_1$ is parallel to $\mathbf{v}_1$ and therefore perturbations $\mathbf{B}_1$ related to the wave are perpendicular to $\mathbf{k}$. If also $\mathbf{c}_A \cdot \mathbf{v}_1 = 0$ or $\mathbf{B}_0 \cdot \mathbf{v}_1 = 0$ which means that the perturbations $\mathbf{v}_1$ and $\mathbf{B}_1$ caused by the wave are perpendicular to the background magnetic field, then from (2.115):

   $$\omega^2 \mathbf{v}_1 - (\mathbf{c}_A \cdot \mathbf{k})^2 \mathbf{v}_1 = 0 \qquad (2.117)$$

Hydromagnetic waves in interplanetary space

and

$$\omega^2 = (\mathbf{c}_A \cdot \mathbf{k})^2 = c_A^2 k^2 \cos^2 \theta \tag{2.118}$$

where θ is the angle between the wave vector $\mathbf{k}$ and the background magnetic field $\mathbf{B}_0$.

Since the group velocity of the wave is given by:

$$\mathbf{v}_g = \frac{\partial \omega}{\partial \mathbf{k}} = \mathbf{c}_A = (\mu_0 \rho_0)^{-1/2} \mathbf{B}_0 \tag{2.119}$$

the wave energy propagate along $\mathbf{B}_0$ with the Alfvén speed.

From Maxwell's equation (2.93) we find:

$$\mathbf{B}_1 = \frac{1}{\omega} \mathbf{k} \times \mathbf{E}_1 \tag{2.120}$$

and the electric field in the wave is also perpendicular to $\mathbf{B}_1$ and therefore also perpendicular to $\mathbf{v}_1$ and $\mathbf{c}_A$. Simultaneous observations of $\mathbf{E}_1$ and $\mathbf{B}_1$ in space therefore give information about the direction of the propagation of energy. Transversal modes are illustrated in Figs. 2.15a and 2.15b.

2. *The longitudinal case.*

By multiplying (2.115) by $\mathbf{k}$ and $\mathbf{c}_A$, respectively, we find:

$$[\omega^2 - (c_s^2 + c_A^2)k^2](\mathbf{k} \cdot \mathbf{v}_1) + (\mathbf{c}_A \cdot \mathbf{k})(\mathbf{c}_A \cdot \mathbf{v}_1)k^2 = 0 \tag{2.121}$$

$$\omega^2 (\mathbf{c}_A \cdot \mathbf{v}_1) - c_s^2 (\mathbf{k} \cdot \mathbf{v}_1)(\mathbf{c}_A \cdot \mathbf{k}) = 0 \tag{2.122}$$

Inserting (2.122) into (2.121), we find:

$$\left[\omega^2 - (c_s^2 + c_A^2)k^2 + \frac{k^2 c_s^2}{\omega^2}(\mathbf{c}_A \cdot \mathbf{k})^2\right](\mathbf{k} \cdot \mathbf{v}_1) = 0 \tag{2.123}$$

For longitudinal propagation ($\mathbf{k} \cdot \mathbf{v}_1 \neq 0$), the dispersion relation is now given by:

$$\frac{\omega^4}{k^4} - \frac{\omega^2}{k^2}(c_s^2 + c_A^2) + \frac{c_s^2}{k^2}(\mathbf{c}_A \cdot \mathbf{k})^2 = 0 \tag{2.124}$$

and the solution is

$$\left(\frac{\omega}{k}\right)^2 = \frac{1}{2}\left\{c_s^2 + c_A^2 \pm \left[(c_s^2 + c_A^2)^2 - 4\frac{c_s^2}{k^2}(\mathbf{c}_A \cdot \mathbf{k})^2\right]^{1/2}\right\} \tag{2.125}$$

We notice that for $\mathbf{c}_A \parallel \mathbf{k}$ or $\mathbf{k}$ parallel to $\mathbf{B}_0$ we have two different modes:

$$A: \quad \omega_A^2 = k^2 c_A^2 \quad (a)$$
$$B: \quad \omega_B^2 = k^2 c_s^2 \quad (b)$$
(2.126)

Mode A has the same dispersion relation ($\theta = 0$) as the transversal wave already discussed, while the B mode has the same dispersion relation as sound waves.

For propagation vertically to $\mathbf{B}_0$ ($\mathbf{k} \cdot \mathbf{c}_A = 0$) we find only one mode, namely

$$C: \quad \omega_C^2 = k^2 \cdot (c_s^2 + c_A^2) \tag{2.127}$$

This is an acoustic mode modified by the presence of the magnetic field $\mathbf{B}_0$. It has a larger group velocity than the transversal mode. Longitudinal modes are illustrated in Figs. 2.15c and 2.15d.

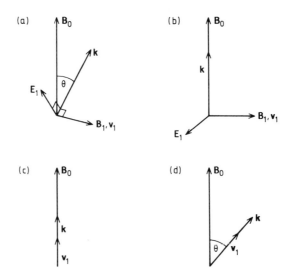

Fig. 2.15. (a) A transverse wave $\mathbf{B}_1$, $\mathbf{E}_1$ and $\mathbf{v}_1$ are all perpendicular to $\mathbf{k}$. The wave is propagating at an angle θ to the background magnetic field, $\mathbf{B}_0$. The dispersion relation corresponds to Alfvén waves. (b) A transverse wave is propagating along the background magnetic field $\mathbf{B}_0$ ($\theta = 0$). (c) A longitudinal wave, $\mathbf{v}_1$ is parallel to $\mathbf{k}$ is propagating along the background magnetic field. The dispersion relation corresponds either to an Alfvén wave and is equivalent to b or to a sound wave. (d) A longitudinal wave is propagating perpendicular to the background magnetic field, $\mathbf{B}_0$. The dispersion relation corresponds to an acoustic wave.

Sec. 2.8] Hydromagnetic waves in interplanetary space

We notice that in the case of the transversal mode $\mathbf{k} \cdot \mathbf{v}_1 = 0$, the density fluctuations ρ_1 is equal to zero. This indicates that the wave which is not associated with density perturbations, can propagate long distances without being damped. Such waves, called Alfvén waves, probably play an important role in transporting energy out from the photosphere to the solar corona and interplanetary space. Variations observed in interplanetary space in the pressure, density, velocity and field parameters do indicate that such waves are of fundamental importance in the solar wind–magnetosphere interaction.

Fig. 2.16 illustrates the spectral distribution in the low-frequency hydrodynamic waves observed in interplanetary space. The amplitude is seen to decay rather rapidly with increasing frequency.

These waves which by correlation analysis between $\mathbf{B}_1$ and $\mathbf{v}_1$ are identified as Alfvén waves, can for the transversal case when $\mathbf{v}_1 \parallel \mathbf{B}_0$, propagate along the magnetic field lines from interplanetary space through the magnetosphere and into the high-latitude ionosphere, where a variety of pulsations in the geomagnetic field often are observed.

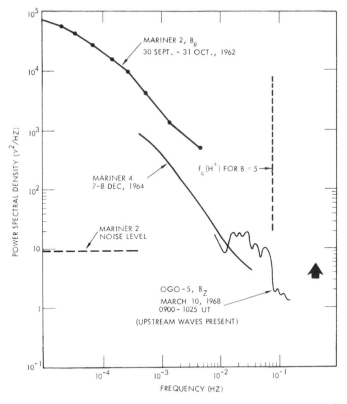

Fig. 2.16. The spectral density of low-frequency hydromagnetic and electromagnetic waves observed in interplanetary space in the frequency range 10^{-5}–1 Hz. (From Scarf, 1970.)

The longitudinal wave can also propagate perpendicularly to $\mathbf{B}_0$ and thereby reach the ionosphere at lower latitudes than the transversal waves. Since the longitudinal waves imply that ρ_1 and therefore p_1 are different from zero, one can imagine these waves as pressure waves generating variations in the magnetic field pressure which then can propagate perpendicularly to B_0. The magnetic field oscillations observed on ground is therefore a mixture of different wave modes contributing with different strength depending on the position of observation.

Fig. 2.17 shows the field strength of such pulsations as observed on ground in the frequency range 10^{-4}–10^4 Hz. The different peaks in the spectrum are associated with characteristic wave features mostly determined by their period, but also by their visual appearance.

Fig. 2.18 demonstrates an example of well developed so-called giant micropulsations as observed on ground at the high geomagnetic latitude station of Tromsø ($\sim 67°$N geomagnetic).

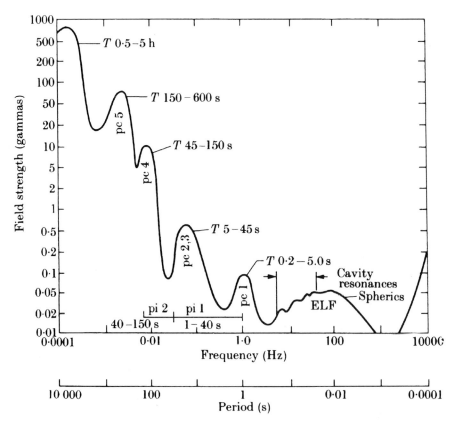

Fig. 2.17. The field strength of geomagnetic micropulsations in the frequency range 10^{-4}–10^4 Hz. (From Campbell, 1967.)

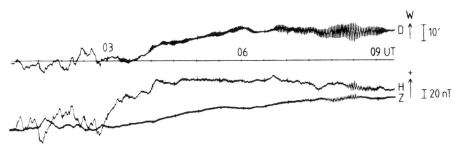

Fig. 2.18. Examples of giant micropulsations as observed at the high geomagnetic latitude station of Tromsø ($\sim 67°$N geom.). (From Brekke et al., 1987.)

2.9 REFERENCES

Brekke, A., Feder, T. and Berger, S. (1987) *J. Atmos. Terr. Phys.*, **49**, 1027–1032.

Campbell, W. H. (1967) in *Physics of Geomagnetic Phenomena*, Vol. 2, Matsushita, S. and Campbell, W. H. (Eds.), pp. 821–909, Academic Press, New York.

Dryer, M. (1987) in *The Solar Wind and the Earth*, Akasofu, S.-I. and Kamide, Y. (Eds.), pp. 19–35, Terra Sci. Publ. Comp., Tokyo.

Fälthammar, C.-G. (1973) in *Cosmical Dynamics,* Egeland, A., Holter, Ø. and Omholt, A. (Eds.), p. 102, Universitetsforlaget, Oslo.

Hundhausen, A. J., Bame, S. J., Ashbridge, J. R. and Sydoriak, S. J. (1970) *J. Geophys. Res.*, **75**, 4643–4657.

Kelley, M. C. (1989) *The Earth's Ionosphere. Plasma Physics and Electrodynamics*, Academic Press, Inc., New York.

Ness, N. F. (1969) Preprint X-616-69-334, Goddard Space Flight Centre.

Sakurai, K. (1987) in *The Solar Wind and the Earth*, Akasofu, S.-I. and Kamide, Y. (Eds.), pp. 39–53, Terra Sci. Publ. Comp., Tokyo.

Scarf, F. L. (1970) *Space Sci. Rev.*, **11**, 234–270.

van Biesbroeck, G. (1953) in *The Sun*, Kuiper, G. P. (Ed.), p. 601, Chicago University Press, Chicago.

Zirin, H. (1966) *The Solar Atmosphere*, Ginn Blaisdell, Waltham, Mass.

2.10 EXERCISES

1. Show that for a plasma with a very high conductivity, σ, and moving with a velocity $v \ll c$ the magnetic flux through an area following the plasma can be expressed as:

$$\frac{d\phi}{dt} = -\iint_S \left(\nabla \times \frac{\mathbf{j}}{\sigma}\right) \cdot d\mathbf{s}$$

where S is the surface and $\mathbf{j}$ is the current density.

2. Assume that the magnetic flux at the Sun is $B_\odot = 10^{-4}$ tesla at the solar surface ($r = 1\ R_\odot$). What will be the strength of this field at the trajectory of the Earth?

(Assume that the solar wind has a radial velocity of 400 km/s and that the rotation speed of the Sun is 24.7 days.)

3. Assume that in addition to (2.15) and (2.16) the energy conservation equation for the solar wind plasma can be expressed as

$$A\rho v \left[\frac{v^2}{2} + \frac{5}{2}\frac{p}{\rho} - \frac{GM_\odot}{r}\right] + A \cdot \phi = \mathcal{E}$$

(a) Find a solution for the equation

$$-A\kappa_0 T^{5/2} \frac{dT}{dr} = \text{const.}$$

that satisfies the conditions.
$T = T_1$ when $r = r_1$ and $T = 0$ when $r \to \infty$.

(b) Assume that $T = 10^6$ K when $r = 1\ R_\odot$. Determine the temperature in the solar wind at 1 AU.

Far out in the solar wind the heat conduction flux ϕ can be expressed by

$$\phi = \alpha n k T v$$

where α is a constant and n is the particle number density.

(c) Show that the temperature now can be expressed as:

$$T = T_2 \left(\frac{n}{n_2}\right)^{2/(\alpha+3)}$$

when $T = T_2$ when $n = n_2$ at $r = r_2$.

3

The atmosphere of the Earth

3.1 NOMENCLATURE

The atmosphere of the Earth is an ocean of gas encircling the globe. It stretches out into far distances from the surface; how far out is a question of definition.

In Fig. 3.1 we have illustrated schematically the variation of some characteristic parameters up to an altitude of 450 km. There are rather strong variations in the upper atmosphere above 100 km during a solar cycle, and this is illustrated in the figure by curves representing average solar maximum and minimum conditions. The cross-hatched areas between these curves illustrate the variability of the parameters. The number density (n) of the atmosphere decreases monotonically by height from 10^{25} m^{-3} at ground level to 10^{14} m^{-3} at 400 km. The atomic mass number is constant and close to 30 from the ground and up to about 100 km, above that region it decreases gradually toward 15 at 400 km. According to the variations in the composition the atmosphere is divided into two main regions, *the homosphere* and *the heterosphere*, indicating that below 100 km the gas constituents are fully mixed into a homogeneous gas. Above this height, however, the different constituents are behaving independently, and the atmosphere is heterogeneous.

The temperature (T) of the atmosphere has a more complicated behaviour by height. It starts out by decreasing in *the troposphere* from about 290 K at the ground and reaching a minimum close at 215 K at 15–20 km, called *the tropopause*.

Above the tropopause is *the stratosphere*, and here the temperature increases again up to a maximum of close to 280 K, called *the stratopause*, usually situated close to 50 km. Above the stratopause the temperature decreases again in *the mesosphere* and reaches the lowest temperature in the atmosphere in *the mesopause*, usually situated at about 70–90 km. The temperature in the mesopause may be as low as 160 K or even lower at occasions. As a curiosity, however, the mesopause temperature in the polar regions is higher in winter than in summer. The production of heat in winter time is well understood and due to chemical reactions; recombination of atomic oxygen. The seasonal change in the downward

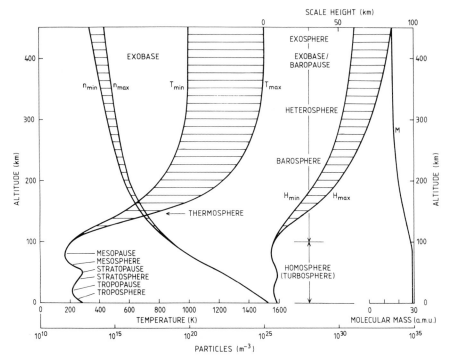

Fig. 3.1. Model height profiles of the temperature T, density n, molecular mass M, and scale height H, distributions in the Earth's atmosphere below 450 km. The different regions are also indicated by their characteristic names according to temperature or composition. The variability in the different parameters with respect to solar activity is indicated by the hatched areas.

transport of O atoms from the thermosphere to the mesopause is still a challenge for atmospheric modelling.

Another seasonal phenomenon which also is difficult to account for in modelling of the stratosphere–mesosphere region is the so-called sudden polar warming which is observed once or twice in the winter season. It is revealed from global temperature measurements that it is a wave phenomenon redistributing the temperature in the polar region. It is also established a correlation between the occurrence of stratospheric warming and radio wave absorption in the ionospheric D-region (see Chapter 5). Thus there is a clear relationship between stratospheric parameters such as temperature and wind and the chemistry of the D-region. For energetic reasons, however, it is believed that it is the wave phenomena in the stratosphere that influences the chemistry of the D-region and not the reverse (Hargreaves, 1979).

Above the mesopause the temperature increases dramatically in the thermosphere where temperatures of more than 1000 K can be found above the *exobase* indicated in Fig. 3.1 at 400 km. Above this region the temperature is fairly constant by height, but may vary considerably by time.

Sec. 3.1]   Nomenclature   67

The nomenclature used for characterizing different areas in the atmosphere is referring to either temperature, composition or dynamics. At the ground we find that the atmosphere is composed of close to 80% N_2 and 20% O_2, while the contribution from other gases is less than 1%. This mixture holds all the way up to about 100 km. This region below 100 km is therefore called the *homosphere* or the *turbosphere*. The latter is reflecting the fact that turbulence is causing the mixture. Above 100 km the molecules are starting to dissociate and become more independent of each other; they are more heterogeneous. This region is therefore called the *heterosphere* or the *barosphere*; the latter reflecting to the fact that the different species have different scale heights or barometric heights.

In Fig. 3.2 we show a more detailed presentation of the altitude profiles of the different atmospheric species between 1000 km and the ground for solar minimum and maximum conditions. While atomic oxygen dominates the upper atmosphere above 250 km during solar maximum conditions, hydrogen is more dominant above 400 km during solar minimum conditions. Below 200 km, however, molecular oxygen and nitrogen together with argon are the dominant species independent of solar activity.

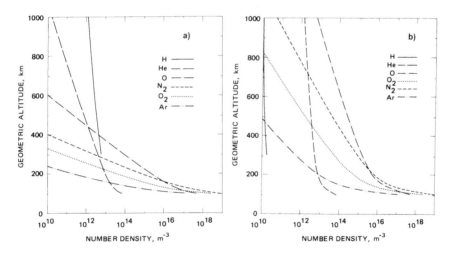

Fig. 3.2. Composition changes in the atmosphere with respect to (a) solar minimum, and (b) solar maximum condition showing the dominance of heavier constituents at larger altitudes during solar maximum phase. (From U.S. Standard Atmosphere, 1976.)

Due to the composition at ground (80% N_2 and 20% O_2) the average molecular mass will be 28.8 a.m.u. Since the molecules start to dissociate above about 100 km and the nitrogen molecules dissociate faster than the oxygen molecules, the molecular weight will decrease and oxygen atoms will be the dominant species from 400 km to above 1000 km depending on the solar cycle. See Fig. 3.2 for references.

At 600 km where we might have 84% O and 16% He, the molecular mass is 14 a.m.u. While the number density at the ground is about 2.5×10^{25} m^{-3}, it is about 10^{19} m^{-3} at 100 km, and at 200 km it has been reduced to about 10^{16} m^{-3}. The density therefore has decreased by more than 10^9 from the ground and up to typical rocket trajectories. Even so, the atmosphere at these heights can have a devastating effect on rockets, satellites and instruments carried by them.

3.2 THE GREENHOUSE EFFECT

It is the heat radiating from the Sun that is the main heat source for the Earth and her atmosphere. To hold a stable temperature in the atmosphere and to stop the Earth from being heated above all limits there has to be a comparable heat radiation away from the Earth. We have seen that the Sun radiates approximately as a black body at a temperature close to 6000 K. The Earth, however, has an average global temperature of 288 K. The radiated heat from the Earth is therefore in a different wavelength regime compared to the solar radiation, namely in the infrared (Fig. 3.3).

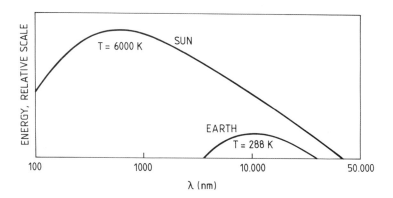

Fig. 3.3. A schematic diagram showing the difference between the solar blackbody radiation spectrum corresponding to 6000 K between 100 and 50 000 nm and the Earth blackbody radiation spectrum corresponding to 288 K. The spectral radiances are in arbitrary scales.

According to the Stephan–Boltzmann law (equation (1.13)) the Earth as a black body will radiate a total energy per unit time given by:

$$\dot{Q}_e = 4\pi \cdot R_e^2 \cdot \sigma \cdot T_e^4 \qquad (3.1)$$

where R_e is the Earth radius ($= 6.371 \times 10^6$ m) and T_e is the mean temperature of the Earth. σ is the Stephan–Boltzmann constant given by (1.14). The Earth will absorb the radiation from the Sun:

$$\dot{Q}_{Se} = (1-\alpha) \cdot E_e \cdot \pi \cdot R_e^2 \qquad (3.2)$$

where α is the albedo or reflection coefficient (≈ 0.3) and E_e is the solar constant which we, according to (1.17), set equal to 1380 W/m² at 1 AU.

For thermal equilibrium (heat in = heat out) we have:

$$\dot{Q}_{Se} = \dot{Q}_e \tag{3.3}$$

and inserted from (3.1) and (3.2):

$$(1-\alpha) \cdot E_e \cdot \pi \cdot R_e^2 = \sigma \cdot 4\pi R_e^2 T_e^4 \tag{3.4}$$

this gives us

$$T_e = \left(\frac{(1-\alpha) \cdot E_e}{4 \cdot \sigma}\right)^{1/4} \tag{3.5}$$

By inserting the appropriate numbers

$$T_e = 255 \text{ K}$$

This is a temperature that on the average is about 40 K lower than the present day average global temperature. We notice that even if we assume that $\alpha = 0$, i.e., all solar energy is absorbed by the Earth, the temperature of the Earth would still be only 278 K as compared to 288 K as it is on the average.

The albedo of the Earth is a difficult parameter in itself as it depends on the ice coverage, the clouds and the amount of ozone at any moment. Table 3.1 gives a list of the albedo for different surfaces. In Fig. 3.4 the variation with latitude of the planetary albedo averaged around the Earth along parallels of latitude is shown. The high values in polar latitudes is due to snow cover there. In middle latitudes intermediate values correspond to the relatively large amounts of cloudiness in the region of storminess associated with the polar front to be mentioned in Chapter 6. The low values in low latitudes are due to the infrequency of clouds in the belt of subtropical high pressure. The slightly increased values just north of equator correspond to the greater cloudiness in the intertropical convergence zone where the trade winds from north and south merge.

Table 3.1. Albedo of various surfaces

Surface	%
Fresh snow, high sun	80–85
Fresh snow, low sun	90–95
Old snow	50–60
Sand	20–30
Grass	20–25
Dry earth	15–25
Wet earth	10
Forest	5–10
Water (sun near horizon)	50–80
Water (sun near zenith)	3–5
Thick cloud	70–80
Thin cloud	25–50
Planetary albedo	30

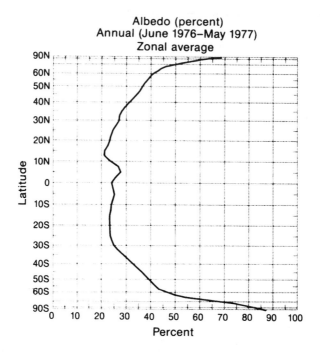

Fig. 3.4. Variation with latitude of the planetary albedo averaged around the Earth along parallels of latitude for the period June 1976 – May 1977 according to measurements made by NOAA polar orbiting satellites. (From Earth–Atmosphere Radiation Budget Analyses Derived from NOAA Satellite Data June 1974 – February 1978, Washington, D.C., NOAA–NESS, 1979.)

Independent of what albedo we choose we need a "blanket" around the Earth to keep us from freezing to death.

This comes about because the solar radiation which is mainly distributed in the visible region of the spectrum, passes through the Earth's atmosphere with little absorption. The Earth, on the other hand, radiates as a black body at an average temperature of 288 K. According to the Wien displacement rule (equation (1.11)) this gives the wavelength for maximum radiation:

$$\lambda_m = \frac{2.89 \times 10^{-3} \text{ m K}}{T} = 10 \ \mu\text{m} \qquad (3.6)$$

which is in the infrared. It is then very fortunate for life on Earth that H_2O, CO_2 and O_3 among other gases in the atmosphere absorb the infrared emission from ground and reradiate this heat in all directions and part of it back to Earth.

Let us therefore assume that the atmosphere is forming a thin layer encircling the Earth at a height h which is much smaller than the Earth's radius. Let this layer be transparent to the solar radiation, but opaque to the infrared radiation from the ground (Fig. 3.5). If this atmosphere has a temperature T_a, it will radiate heat both outward from the atmosphere and downward towards the Earth. The total heat radiated per unit time from the atmosphere will be:

$$\dot{Q}_a = 2 \cdot \sigma \cdot 4\pi \cdot R_e^2 \cdot T_a^4 \tag{3.7}$$

when we have neglected the height as well as the thickness of the atmosphere compared to the Earth radius. If the atmosphere is only heated by the radiation from the Earth, the balance in heat radiation gives

$$\sigma \cdot 4\pi \cdot R_e^2 \cdot T_e^4 = 2 \cdot \sigma \cdot 4\pi \cdot R_e^2 \cdot T_a^4 \tag{3.8}$$

and

$$T_a^4 = \frac{1}{2} T_e^4 \tag{3.9}$$

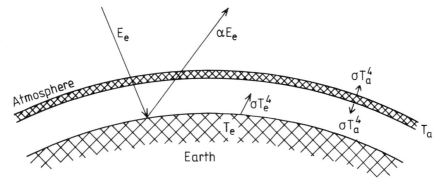

Fig. 3.5. A simple illustration of the greenhouse effect in the Earth's atmosphere showing the incoming solar radiation E_e and the reflected part αE_e. The blackbody temperature of the Earth is represented by T_e. An atmospheric layer is indicated having a blackbody temperature T_a. The system is supposed to be in thermal equilibrium.

For the Earth receiving heat radiation from the Sun as well as from the atmosphere the energy balance is now expressed by:

$$(1-\alpha)E_e \cdot \pi \cdot R_e^2 + \sigma \cdot 4\pi R_e^2 \cdot T_a^4 = \sigma \cdot 4 \cdot \pi \cdot R_e^2 T_e^4 \tag{3.10}$$

and

$$T_e^4 = \frac{(1-\alpha)E_e}{2\sigma} \tag{3.11}$$

Inserting the proper values we find

$$T_e = \left[\frac{(1-\alpha)E_e}{2\sigma}\right]^{1/4} = 304 \text{ K} \tag{3.12}$$

and the temperature increases by almost 50 °C compared to a situation without any atmosphere acting as a blanket around the globe. It would, however, in this simplified model be about 16 °C too high, but we can definitely conclude that without a greenhouse effect in the atmosphere the globe would not be a place for the life we know.

To treat the atmosphere in a more realistic way is of course far more difficult than this simple illustration indicates. But to prove the point the arguments used above are quite valid.

To give a full treatment of the problem the radiation through the atmosphere has to be computed by the equation of radiation transfer.

3.3 THE TEMPERATURE STRUCTURE OF THE ATMOSPHERE

As we have seen does the temperature in the atmosphere decrease quite monotonically up to the tropopause. This is due to the fact that the infrared radiation from the ground which is absorbed in the atmosphere, is rather constant because the surface temperature of the globe is constant, and therefore the heat expands out in the atmosphere in radial directions. The heat will then be distributed into larger and larger volumes and therefore the temperature must decrease.

Due to the ozone layer situated between 15 and 40 km above ground (Fig. 3.6) which absorbs a large portion of the solar radiation between 200 and 300 nm, the atmosphere becomes heated in the stratosphere and the temperature increases again. Above the stratopause, however, the heat balance results in an excess radiation outwards again, and the temperature decreases rapidly in the mesosphere until the sharp minimum occurs at the mesopause. In Fig. 3.6 a typical temperature profile in the atmosphere below 100 km is illustrated. Also shown is the expected profile if the ozone was not present. The temperature at the stratopause would then drop to about 150 K. The presence of the ozone layer must have had a profound influence on the development on life on Earth, since many species that do exist, are very vulnerable to UV radiation at wavelengths below 300 nm. Above the mesopause solar radiation in the UV band is strongly absorbed due to dissociation of molecules like O_2, N_2 and NO and ionization of atomic oxygen and other molecules and atoms. This leads to a new temperature increase which is particularly strong up to about 400 km.

The very strong absorption of the UV radiation in the thermosphere, however, is accompanied by a strong variability in the temperature of this region as illustrated in Fig. 3.1. This is because the solar UV radiation itself is so variable. As the temperature above 400–600 km appears fairly constant as function of altitude, it is often referred to as the exospheric temperature (T_∞). Fig. 3.7 which is an alternative presentation of Fig. 3.1 above 100 km, shows the range of variation the thermospheric temperature can display at different altitudes in the course of a solar cycle due to variations in the exospheric temperature. The intensity index of the solar radio emission at 10.7 cm (equation (1.26)) MHz is used as a reference parameter to the solar cycle. The value 50 ($\times 10^{-22}$ W m^{-2} Hz^{-1})

Sec. 3.3] The temperature structure of the atmosphere

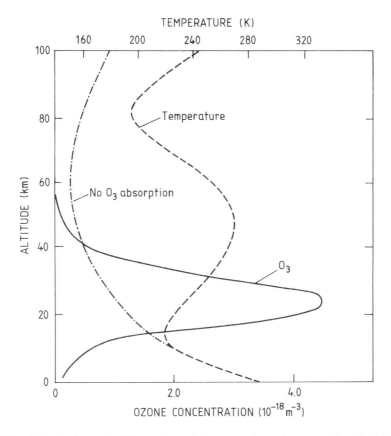

Fig. 3.6. A schematic illlustration of the ozone layer between 15 and 40 km showing a peak at about 25 km. The concentration is given in molecules per cubic metre. An average temperature model below 100 km is compared to a temperature profile expected if the ozone layer was absent.

represents solar minimum conditions while the value 200 represents the solar maximum conditions.

The exospheric temperature can change by 600 K or more during a solar cycle according to similar model calculations as presented in Fig. 3.7.

The thermospheric temperature, however, is not constant all over the globe but exhibits a seasonal variation. Presented in Fig. 3.8 is the thermospheric temperature at 300 km as function of latitude for solstice and equinox conditions. Also shown in Fig. 3.8 is the mean molecular mass (a.m.u.) for the corresponding conditions. Especially at solstice are the latitudinal variations large as the temperature changes from close to 1400 K at the summer pole to slightly above 900 K at the winter pole, and the molecular mass changes from 21 a.m.u. to 17 a.m.u. in the same region.

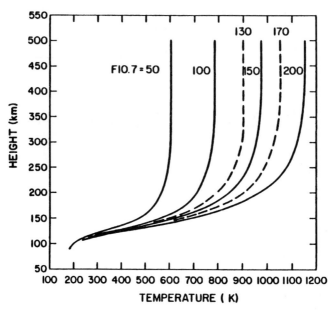

Fig. 3.7. The variability in the thermospheric temperature for different values of the solar 10.7 cm radio flux index ($F_{10.7}$) in units of 10^{-22} W m^{-2} Hz^{-1} reduced to 1 AU. For average solar minimum and maximum conditions the $F_{10.7}$ index is 50 and 200 respectively. (From Roble, 1987.)

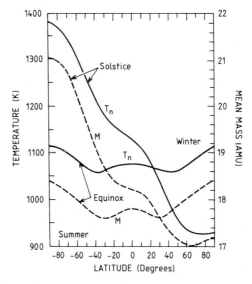

Fig. 3.8. The latitudinal distribution of the neutral temperature T_n and mean molecular mass M at 300 km for equinox and solstice conditions. (After Roble, 1987.)

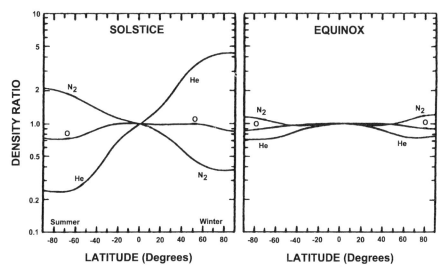

Fig. 3.9. The latitudinal distribution of molecular nitrogen (N_2), atomic oxygen (O_2) and helium (He) for solstice and equinox conditions. (From Roble, 1987.)

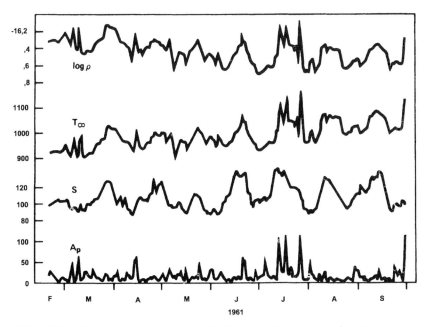

Fig. 3.10. A comparison between the thermospheric mass density ρ, exospheric temperature T_∞, the solar 10.7 cm radio flux index S and the Ap index as function of time for the year of 1961. (From Giraud and Petit, 1978.)

The consequence of this is that heavier molecules are brought up to higher altitudes from below in the summer hemisphere and downward in the winter hemisphere (Fig. 3.9). The summer thermosphere at 300 km and above is therefore dominated by N_2 molecules, while the winter thermosphere in the same height region has a large contribution of helium atoms.

The thermospheric temperature is also responding to solar variations on a shorter time scale than solar cycles or seasonal periods. Fig. 3.10 presents a comparison of variations in the exospheric temperature T_∞, the atmospheric mass density ρ, the solar radio emission flux, S, at 10.7 cm and a geomagnetic index Ap. The latter is representing variations in the Earth's magnetic field presumably due to ionospheric currents.

The 27-day period in the solar radio emission, S, is clearly reflected in T_∞ as well as ρ, as if the atmosphere is expanding and contracting as the solar flux increases and decreases.

3.4 ATMOSPHERIC DRAG ON SATELLITES

Fig. 3.11 gives a schematic presentation of the variability of the atmospheric density between 100 and 1000 km for solar maximum and minimum conditions as well as for extreme solar maximum daytime conditions. Density variations of 2 orders of magnitude can in fact take place during a solar cycle above 400 km altitude.

Satellites at these altitudes will experience great differences in the friction force due to atmospheric drag in response to these density variations, and during severe

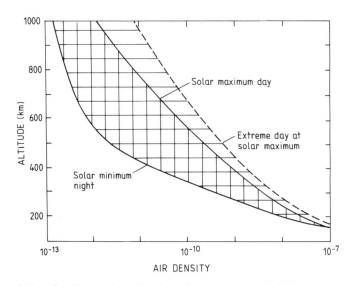

Fig. 3.11. An illustration showing the extreme variability in the neutral density of the thermosphere between solar minimum and solar maximum conditions.

disturbances the satellite trajectories can be altered significantly. It is therefore of great importance to know the behaviour of the upper atmosphere when planning space missions, because an unexpected atmospheric disturbance can reduce the lifetime of a satellite by several months, if not destroy the mission completely.

We also notice from Fig. 3.10 that the geomagnetic activity due to enhanced ionospheric currents can give rise to abrupt changes in the exospheric temperature as well as density, again is this one of the main reasons for the interest in geomagnetic disturbances as it has practical importance for satellite trajectories. These disturbances are also the most difficult to predict because geomagnetic disturbances occur in a very erratic manner especially at high latitudes.

Let us therefore consider the effect of the variations in the atmospheric density on a satellite orbit. For a sphere with mass m moving with velocity v with respect to the atmospheric gas, there will be a drag force F_D acting on the sphere due to collisions with the atmospheric particles which can be expressed in the following way:

$$F_D = \frac{1}{2} \rho v^2 C_D \tag{3.13}$$

Here ρ is the atmospheric gas density and C_D is what is called the ballistic coefficient. It is proportional to the cross-section of the sphere, and depends on the surface conditions of the sphere's material.

For a satellite moving in a circular orbit with radius r around the Earth the total energy is:

$$E = \frac{1}{2} mv^2 - \frac{GM_e m}{r} \tag{3.14}$$

where G is the constant of gravity and M_e the mass of the Earth.

If the atmospheric drag is small, the circular orbit will be maintained and the total energy of the satellite moving in a circular orbit in a central force field therefore is given by:

$$E = -\frac{GM_e m}{2r} \tag{3.15}$$

The rate of change of energy for the satellite due to atmospheric drag can be expressed as:

$$\frac{dE}{dt} = -F_D v = -\frac{1}{2} \rho v^3 C_D \tag{3.16}$$

By deriving dE/dt from (3.15) and equating it to (3.16) we have:

$$\frac{dE}{dt} = \frac{GM_e m}{2r^2} \frac{dr}{dt} = -\frac{1}{2} \rho v^3 C_D \tag{3.17}$$

and the rate of change of the radius of the orbit is

$$\frac{dr}{dt} = -\frac{\rho v C_D \cdot r}{m} \tag{3.18}$$

By observing the rate of change of the orbit's radius one could now derive the atmospheric density, or vice versa, when the atmospheric density is known, the expected rate of change of the orbital radius could be obtained. Since the variations in r from orbit to orbit is very small, it is not so practical to use this last relationship to study the effects on satellite orbits from atmospheric drag. It turns out that the orbital period ($T = 2\pi r/v$) is a better parameter for this.

According to Kepler's third law we have:

$$T^2 = \frac{4\pi^2 r^3}{GM_e} \tag{3.19}$$

The rate of change of T can then be found by taking the derivative (3.19):

$$2T\frac{dT}{dt} = \frac{12\pi^2 r^2}{GM_e}\frac{dr}{dt} = \frac{12\pi^2 r}{v^2}\frac{dr}{dt} \tag{3.20}$$

and by inserting (3.18) and solving for dT/dt we find:

$$\dot{T} = \frac{dT}{dt} = -\frac{3\pi C_D r}{m}\rho \tag{3.21}$$

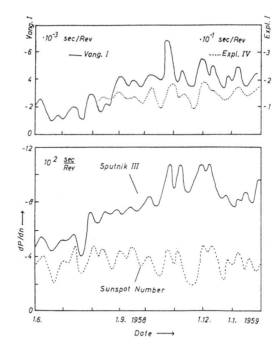

Fig. 3.12. Upper panel: The rate of change in orbital period per period for Explorer IV and Vanguard I satellites between June 1958 and Feb. 1959. Lower panel: A correlation between the variations in orbital periods for Sputnik III and sunspot numbers for the same period as above. (From Paetzold and Zschörner, 1960.)

The period therefore decreases as the density increases. This can appear as an acceleration of the satellite, but since the energy is constant, the gain in kinetic energy must be lost by a reduction in potential energy due to the reduction in orbital radius by increasing neutral density. The fractional change in the orbital period per period is given by:

$$\frac{\dot{T}}{T} = -\frac{3C_D G M_e}{4\pi m r^2} \rho \tag{3.22}$$

Measuring the amount of change of the rotation period of the satellite per period is much easier than measuring the rate of change of the radius, and it can actually be used to derive the atmospheric density at the satellite altitude.

In Fig. 3.12 the variation in the rotation periods of the satellites Explorer IV and Vanguard I is compared for a few months in 1958–59. A close correlation in the variations are shown indicating that the effect is not local but global. Furthermore, it is demonstrated in the lower panel of Fig. 3.12 that these variations in the case of Sputnik III is correlated with variations in the solar sunspot number. These observations are interpreted as due to variations in atmospheric density above altitudes of 300 km due to expansion and contraction caused by variations in the solar heat input.

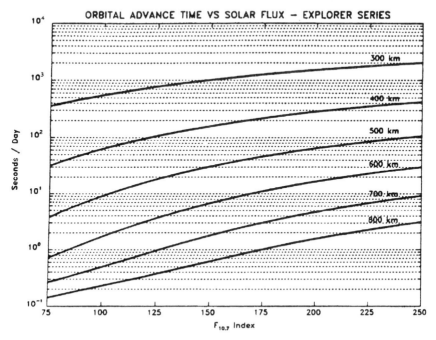

Fig. 3.13. The rate of change in orbital period per day as function of solar flux at 10.7 cm radio emission for satellites at different heights between 300 and 800 km. (From Walterscheid, 1989.)

Fig. 3.13 illustrates the rate of changes in seconds per day of the orbital period as function of solar heat input for satellites at different altitudes. The 10.7 cm radio emission flux is used as a parameter for the solar heat input. At lower altitudes the rate of change becomes more severe for higher solar fluxes.

3.5 THE ATMOSPHERE AS AN IDEAL GAS

From the ideal gas law we know that

$$pV = N'R_0T \tag{3.23}$$

where p is the pressure (N/m^2), V is the volume (m^3), T is the temperature (K), N' is the number of moles, and R_0 is the universal gas constant ($= 8.314$ J/mole K). If m is the mass of the molecule, N is the number of molecules and M' is the molecular mass (mass per mole), then

$$N' = \frac{N \cdot m}{M'} \tag{3.24}$$

and the ideal gas law can be expressed as

$$pV = \frac{Nm}{M'} R_0 T \tag{3.25}$$

The pressure is then given by

$$p = \frac{Nm}{M'V} R_0 T = \frac{\rho R_0}{M'} T = \rho R T \tag{3.26}$$

where ρ is the mass density of the gas (kg/m^3) and $R = R_0/M'$ is the gas constant per unit mass for the specific gas.

Another form of the ideal gas law is:

$$pV = N'R_0T = \frac{N}{N_A} R_0 T = N\kappa T \tag{3.27}$$

where N_A ($= N/N'$) is Avogadro's number ($= 6.02 \times 10^{23}$ molecules/mole) and κ ($= R_0/N_A$) is the Boltzmann constant. The pressure is now given by

$$p = \frac{N}{V} \kappa T = n\kappa T \tag{3.28}$$

where n is the number density of the particles (m^{-3}).

The normalized velocity distribution in a gas in thermal equilibrium is according to the Maxwell–Boltzmann distribution law

$$f(v) = \frac{N(v)}{N_T} = 4\pi \left(\frac{m}{2\pi\kappa T}\right)^{3/2} v^2 \exp\left(-\frac{mv^2}{2\kappa T}\right) \tag{3.29}$$

where $N(v)$ is the number of particles with speed between v and $v + dv$ and N_T is the total number of particles.

Sec. 3.5] The atmosphere as an ideal gas 81

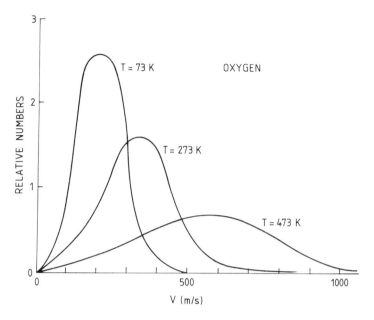

Fig. 3.14. Velocity distribution of O_2 molecules in relative numbers for 3 different temperatures.

In Fig. 3.14 distribution functions for O_2 molecules for 3 different temperatures are illustrated.

We can also show that from (3.29) the average speed $\bar{v}$ is given by:

$$\bar{v} = \int_0^\infty f(v) v \, dv = \left(\frac{8\kappa T}{\pi m}\right)^{1/2} \tag{3.30}$$

The root mean square velocity $v_{\text{r.m.s.}}$ is given by:

$$v_{\text{r.m.s.}} = \sqrt{\bar{v^2}} = \left[\int_0^\infty f(v) v^2 \, dv\right]^{1/2} = \left(\frac{3\kappa T}{m}\right)^{1/2} \tag{3.31}$$

and the most probable speed $v_{\text{m.p.}}$ is given by:

$$v_{\text{m.p.}} = \left(\frac{2\kappa T}{m}\right)^{1/2} \tag{3.32}$$

For a mixture of gases in thermal equilibrium their kinetic energies must be

$$\frac{1}{2} m\bar{v}^2 = \frac{3}{2} \kappa T \tag{3.33}$$

for all particles, and therefore

$$m_1 \bar{v}_1^2 = m_2 \bar{v}_2^2 \tag{3.34}$$

where m_1 and m_2 are masses for the two different molecules and v_1 and v_2 are the different speeds respectively. The lighter gas then has a higher root mean square velocity for the same temperature.

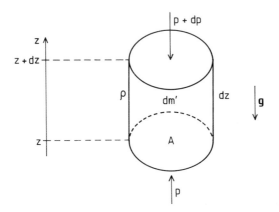

Fig. 3.15. A volume element of air used to illustrate the balance between the pressure and gravity forces.

Consider the forces on a small mass element dm' of air (Fig. 3.15) at a height z above the ground. Let the mass element take the form of a small cylinder with horizontal cross-section A and height dz. The air mass in this cylinder can be expressed as

$$dm' = n \cdot m \cdot A \cdot dz = n \cdot M \cdot m_0 \cdot A \, dz \tag{3.35}$$

where n is the number density of air molecules and m is the mass of each molecule. m_0 is one atomic unit ($= 1$ a.m.u. $= 1.660 \times 10^{-27}$ kg) and

$$m = M \cdot m_0 \tag{3.36}$$

where M is the molecular mass number. This air mass will be acted upon by gravity, and this force can be expressed as

$$df = -dm'g = -nmgA dz \tag{3.37}$$

where the minus sign indicates that the force is directed downward against positive z. In static equilibrium the gravity force must be balanced by the net pressure force which can be expressed by:

$$[p - (p + dp)] A - nmgA dz = 0 \tag{3.38}$$

which gives:

$$\frac{dp}{dz} = -nmg = -\rho g \tag{3.39}$$

where $\rho = nm$ is the mass density. This is called the barometric equation. Assuming that the atmosphere is an ideal gas which is a very good assumption at least for the lower parts of the atmosphere, then we can apply (3.28) and derive

$$\frac{1}{p}\frac{dp}{dz} = -\frac{nmg}{n\kappa T} = -\frac{mg}{\kappa T} = -\frac{1}{H} \tag{3.40}$$

The parameter $H = \kappa T/mg$ is called the scale height.

We recall from kinetic gas theory that according to the equipartition principle each degree of freedom gives rise to a kinetic energy equal to $\frac{1}{2}\kappa T$. We then notice that if all this kinetic energy in the vertical direction is converted to potential energy, the molecule must be lifted a height equal to the half of the scale height

$$\frac{1}{2}mv_z^2 = mgh = \frac{1}{2}\kappa T \tag{3.41}$$

$$h = \frac{1}{2}\frac{\kappa T}{mg} = \frac{1}{2}H \tag{3.42}$$

Whether a particle will reach this height or not, depends on its velocity and rate of collision. All in all, in thermal equilibrium there will always be a balance between the kinetic and potential energies of the particles so that the velocity distribution is kept constant at all heights. Only the density will decrease in an isothermal atmosphere as will be shown below.

The pressure at any height can now be found by integrating (3.40):

$$\ln p = -\frac{z}{H} + \text{const.} \tag{3.43}$$

If $p = p_0$ at a reference height z_0, we have:

$$p = p_0 \exp\left(-\frac{z - z_0}{H}\right) \tag{3.44}$$

For a constant temperature and a constant molecular mass the scale height is constant as long as the acceleration of gravity is constant. Close to Earth therefore, where $T = 288$ K and $M' = 28.8$, we find

$$H = \frac{\kappa T}{mg} = 8.43 \text{ km} \tag{3.45}$$

Since the temperature between the ground and 100 km altitude vary quite markedly, the scale height will vary between 5.0 and 8.3 km (see Fig. 3.1) in this region. Above 100 km, however, the temperature increases drastically, and the molecules dissociate so that the molecular mass decreases. The scale height increases and the pressure therefore does not decrease as rapidly above 100 km as below. At 300 km, for example, at a temperature of 980 K and with molecular mass number of 16 we find $H \approx 50$ km. Since $p = n\kappa T$, we find an expression for the mass density by inserting n:

$$\rho = n \cdot m = \frac{m}{\kappa T} \cdot p \tag{3.46}$$

We see that for a constant temperature and molecular mass, the mass density will decrease exponentially with the same rate as the pressure.

Let us consider for a while the number density when the atmosphere is isothermal. From the expression of the pressure we have

$$p = n\kappa T = p_0 \exp\left[-\frac{z-z_0}{H}\right] = n_0 \kappa T_0 \exp\left[-\frac{z-z_0}{H}\right] \quad (3.47)$$

where n_0 and T_0 are the number density and temperature at the reference height z_0 respectively. Since the atmosphere is assumed isothermal, we see that

$$n = n_0 \exp\left[-\frac{z-z_0}{H}\right] \quad (3.48)$$

When the reference height is set at the ground level $z_0 = 0$, the total sum of all particles from the ground to infinity above a unit area on Earth is given by:

$$\mathcal{N} = \int_0^\infty n\, dz = n_0 \int_0^\infty \exp\left(-\frac{z}{H}\right) dz = n_0 H \quad (3.49)$$

We therefore see that the scale height is equivalent to the height we would get if the atmosphere encircling the Earth had a constant density by height. It would then reach only 8.4 km above our heads, and outside there would be vacuum. This is the reason why people in the old days believed that the atmosphere was 8–10 km thick. They knew that the air pressure balanced a water column of about 10 m, and by knowing the ratio between the density of water (10^3 kg/m^3) and the density of air at ground level (1.2 kg/m^3), the height of the atmospheric "lid" would be

$$h_a = \frac{10^3 \text{ kg/m}^3}{1.2 \text{ kg/m}^3} \cdot 10 \text{ m} = 8.3 \text{ km} \quad (3.50)$$

Realizing that the atmosphere was not finite, really opened the universe for the human being.

Let us derive an estimate of the total mass of the atmosphere and compare it with the mass of the Earth.

For an isothermal atmosphere the mass density is given by:

$$\rho = \rho_0 \exp(-z/H) \quad (3.51)$$

The total mass above a unit area of the Earth is then given by:

$$m_A = \int_0^\infty \rho\, dh = H \cdot \rho_0 = 1.03 \times 10^4 \text{ kg/m}^2 \quad (3.52)$$

and the total mass of the atmosphere is

$$M_A = 4 \cdot \pi R_e^2 \cdot \rho_0 \cdot H = 5.25 \times 10^{18} \text{ kg} \quad (3.53)$$

where R_e is the radius of the Earth. Compared with the mass of the Earth which is $M_e = 6 \times 10^{24}$ kg, the mass of the atmosphere is less than one fraction in 10^6.

The atmosphere as an ideal gas

We will now examine how the vertical density distribution changes in a column of air if it is heated. We will neglect any horizontal variation and assume that any additional vertical acceleration is much less than g, the acceleration of gravity. Therefore the hydrostatic equilibrium can be assumed to hold.

Let the base of the column be at ground level. Then for an isothermal atmosphere with fixed molecular mass the density distribution as function of height can be given by:

$$n = n_0 \exp(-z/H) \tag{3.54}$$

Here n_0 is the number density at ground level and H is the scale height as before. If the temperature is increased by a factor k so that

$$T' = kT \tag{3.55}$$

then the new scale height will be

$$H' = \frac{\kappa \cdot T'}{m \cdot g} = k \cdot H \tag{3.56}$$

The density distribution n' can then be expressed by

$$n' = n'_0 \exp(-z/H') = n'_0 \exp(-z/kH) \tag{3.57}$$

where n'_0 now is the number density at ground level.

Since no air can disappear from the column, the number of particles above a unit area at the ground level must be conserved and:

$$\int_0^\infty n \, dz = \int_0^\infty n' \, dz \tag{3.58}$$

It follows that

$$n_0 H = n'_0 H' \tag{3.59}$$

or that

$$\frac{n_0}{n'_0} = \frac{H'}{H} = k \tag{3.60}$$

We then notice that the density at the base decreases in the same proportion as the scale height increases. There will be an altitude z', however, where n does not change and this altitude can be found by solving

$$n = n_0 \exp(-z'/H) = n' = n'_0 \exp(-z'/H') \tag{3.61}$$

The solution for z' is given by

$$z' = -\ln \frac{n_0}{n'_0} \cdot \frac{1}{\frac{1}{H'} - \frac{1}{H}} \tag{3.62}$$

or by introducing the factor k:

$$z' = \frac{k}{k-1} \ln k \cdot H \tag{3.63}$$

We notice that when H is reduced by $1/k$, the density is unchanged at the height z'' given by

$$z'' = \frac{\ln k}{k-1} H \tag{3.64}$$

and therefore

$$\frac{z'}{z''} = k \tag{3.65}$$

Fig. 3.16 shows the height variation of n and n' when the altitude scale is normalized to H. Also presented in the figure is the height variation of n'' when the scale height is reduced to $(1/k)H$, i.e. the temperature is reduced to $1/k$.

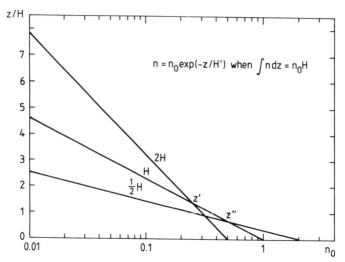

Fig. 3.16. The distribution of the number density in a column of air in an atmosphere with scale height H. Also illustrated is this distribution if the total number of particles in the column is kept constant while the scale height is doubled or halved. The height scale is normalized to H.

3.6 THE EXOSPHERE

The ordinary gas law can only be applied to the atmospheric gas as long as the molecules make enough collisions to establish statistical equilibrium with their surroundings. Let us assume that a molecule travelling with the most probable velocity:

$$v_{\text{m.p.}} = \sqrt{\frac{2\kappa T}{m}} \tag{3.66}$$

Moving upward from a level in the atmosphere without experiencing collisions, it would reach a height h_e given by

$$\frac{1}{2} m v_{\text{m.p.}}^2 = \kappa T = m g h_e \tag{3.67}$$

or

$$h_e = \frac{\kappa T}{mg} = H \tag{3.68}$$

where H is the scale height. If this distance h_e was less than the mean free path, l, which is the distance between two collisions, it could on the average be considered to move freely. The part of the atmosphere where the scale height is of the same order as, or larger than, the mean free path is called the exosphere.

The mean free path is given approximately as

$$l = \frac{1}{\sigma \cdot n} \tag{3.69}$$

where σ is the cross-section for collisions and n is the number density of the atmosphere. We then notice at the exobase, which is the bottom of the exosphere and where $H = l$, that:

$$H \cdot n = \frac{1}{\sigma} \tag{3.70}$$

From (3.49) we find that:

$$H \cdot n = N \tag{3.71}$$

where N is the total number of particles per unit area above the height where the number density is n. We therefore have at the exobase that:

$$N \cdot \sigma = 1 \tag{3.72}$$

One such particle will therefore experience exactly one collision on its way up to the exobase. If $H < l$, the particle will not experience such a collision at all.

Most particles entering the exosphere from below travel in gravity controlled orbits (ballistic motion) without making collisions until they either escape or return back to the atmosphere below. If a particle is escaping from the Earth's gravitation field, its kinetic energy must be larger than its potential energy at the height of escape. Therefore

$$\frac{1}{2} m v^2 = \frac{3}{2} \kappa T > m g_r r \tag{3.73}$$

where g_r is the acceleration of gravity at the distance of escape r measured from the centre of the Earth.

$$v > \sqrt{2 g_r \cdot r} = \sqrt{2 g_0 R_e^2 \frac{1}{r}} = v_{\text{esc}} \tag{3.74}$$

where g_0 is the acceleration of gravity at the Earth surface ($= 9.80$ m/s^2). At the Earth's surface $v_{esc} = 11.2$ km/s, while for larger distances v_{esc} becomes smaller. At about 2000 km $v_{esc} \approx 9.7$ km/s.

3.7 HEIGHT-DEPENDENT TEMPERATURE

We have demonstrated in Figs. 3.1 and 3.6 that the temperature in the lower atmosphere (below say 100 km) is varying by height, and therefore it is strictly not legitimate to assume that T is constant. Let us therefore express T as a linearly varying function with height

$$T = T_0 + \alpha \cdot z \tag{3.75}$$

α is often called the linear "lapse rate". α can be positive as in the stratosphere and thermosphere and negative as in the troposphere and the mesosphere.

Introducing T in the barometric equation, but still assuming an ideal gas, we find:

$$\frac{dp}{p} = -\frac{mg}{\kappa T} dz = -\frac{T_0}{H_0} \frac{dz}{T_0 + \alpha z} \tag{3.76}$$

where $H_0 = \kappa T/mg$ is the scale height referring to $z = 0$. Solving the equation for p we find

$$\ln \frac{p}{p_0} = -\frac{T_0}{H_0} \int_0^z \frac{dz}{T_0 + \alpha z} = -\frac{T_0}{H_0 \alpha} \ln \left(\frac{T}{T_0}\right) = -\frac{mg}{\kappa \alpha} \ln \frac{T}{T_0} \tag{3.77}$$

Therefore

$$\frac{p}{p_0} = \left(\frac{T}{T_0}\right)^{-\frac{mg}{\kappa \alpha}} \tag{3.78}$$

From (3.46) we have

$$\rho = \frac{m}{\kappa T} p \tag{3.79}$$

and by inserting (3.78):

$$\rho = \frac{m}{\kappa T} \cdot p_0 \left(\frac{T}{T_0}\right)^{-\frac{mg}{\kappa \alpha}} = \rho_0 \left(\frac{T}{T_0}\right)^{-1-\frac{mg}{\kappa \alpha}} \tag{3.80}$$

Therefore

$$\frac{\rho}{\rho_0} \neq \frac{p}{p_0} \tag{3.81}$$

The pressure and the density will not have the same variation by altitude.

Another complication to be mentioned here is the variation in the acceleration of gravity

$$g(z) = g_0 \left(\frac{R_e}{R_e + z}\right)^2 \tag{3.82}$$

3.8 THE ADIABATIC LAPSE RATE

Assume that when a volume element in the atmosphere moves in altitude, the motion will occur without any exchange of heat with the surrounding atmosphere. This can happen if the motion is rapid enough. We have then the following adiabatic relationship for p and T:

$$Tp^{\frac{1-\gamma}{\gamma}} = \text{const.} \tag{3.83}$$

where γ ($= c_p/c_v$) is the adiabatic constant. c_v and c_p are the specific heat for constant volume and constant pressure, respectively. By differentiating (3.83) with respect to z we find

$$\frac{\partial T}{\partial z} = \frac{\gamma - 1}{\gamma} \frac{T}{p} \frac{\partial p}{\partial z} \tag{3.84}$$

By introducing (3.40)

$$\frac{\partial T}{\partial z} = -\frac{\gamma - 1}{\gamma} \frac{mg}{\kappa} = \alpha^* \tag{3.85}$$

where α^* is called the adiabatic "lapse rate".

At the Earth's surface $\gamma = 1.4$ and therefore

$$\alpha^* = -9.8 \text{ K/km} \tag{3.86}$$

The temperature decreases by almost 1 K per 100 m elevation at ground level.

Let us now assume that we have an atmosphere where the temperature within a certain height region decreases more rapidly than the adiabatic lapse rate (Fig. 3.17).

If we now imagine a small bubble of air ascending from the height z_0 to the height z_1 without heat exchange with the environment, then the temperature of the bubble will follow the adiabatic temperature illustrated by T_a to the temperature $T_{a,1}$ which is above the temperature T_1 in the atmosphere itself at height z_1. Therefore the air bubble will be lighter than the surrounding air and the bubble will continue to ascend.

If, on the other hand, the bubble at z_0 starts to descend to z_2 without heat exchange with the surroundings, the temperature in the bubble will be $T_{a,2}$ according to the adiabatic temperature. The temperature in the bubble will therefore be less than in the surrounding air, and the bubble becomes heavier and continues to sink. In a situation where the temperature of the air decreases more rapidly than the adiabatic lapse rate, the air is unstable.

For the opposite sense, when the temperature in the atmosphere decreases more slowly than the adiabatic lapse rate as illustrated in Fig. 3.18, the atmosphere becomes stable.

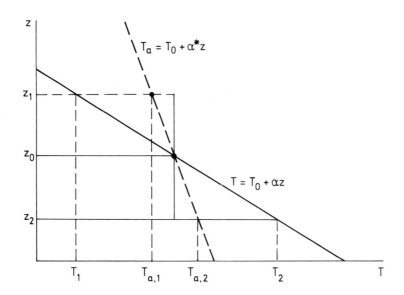

Fig. 3.17. The atmospheric temperature T as function of height with a lapse rate α as compared to the temperature representing the adiabatic lapse rate α^* for an unstable atmosphere.

STABLE ATMOSPHERE

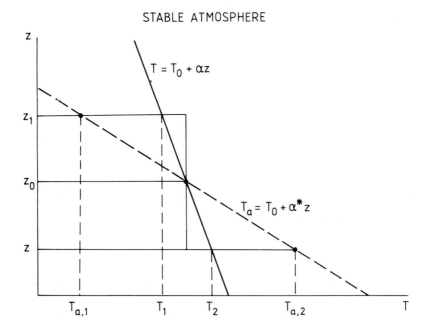

Fig. 3.18. The same as in Fig. 3.17 except the atmosphere is stable.

Assume again that an air bubble rises from z_0 to z_1. The temperature in the bubble $T_{a,1}$ will now become less than T_1, the temperature in the surroundings, and the bubble is heavier than the atmosphere around and it will sink downwards. When it reaches the height z_2, the temperature in the bubble will be $T_{2,a}$ which is higher than in the surrounding atmosphere. The bubble becomes lighter and rises up again.

Small disturbances in the atmosphere are usually adiabatic because the inertia of the atmosphere is too great for any heat exchange can take place. The most likely areas for an unstable atmosphere, however, are the regions of the atmosphere where the temperature decreases, i.e., in the troposphere and the mesosphere that are also the regions where turbulence is most prominent.

3.9 THERMAL STRUCTURE OF THE ATMOSPHERE

The thermal structure of the atmosphere is generally controlled by the following factors:

1. External energy inputs from heating sources such as solar radiation, precipitating of energetic particles from the magnetosphere and interplanetary space, plasma waves, travelling waves etc.

2. Internal sources of energy such as heating or cooling associated with chemical reactions, Joule heating due to ionospheric currents especially at high latitudes etc.

3. Heat transfer by atmospheric radiation such as thermal radiation in the infrared, airglow etc.

4. Transport of thermal and potential energy by thermal conduction and advection.

5. Transformation of thermal energy to large scale motions such as general circulation and atmospheric tides etc.

If we now neglect the transport effects due to air motion as particularly mentioned in point 5, we can write the rate of change of thermal energy (u) of the air per unit volume in the following manner:

$$\frac{du}{dt} = q_T - L_T - \nabla \phi \tag{3.87}$$

where q_T is the primary production rate of thermal energy due to external and internal sources. L_T represents the thermal loss due to radiation and ϕ denotes the flux associated with heat conduction and advection. The internal energy per unit volume is given by

$$u = nc_v T \tag{3.88}$$

where n is the number density. Then the following equation for the temperature must hold:

$$nc_v \frac{dT}{dt} = q_T - L_T - \nabla\phi \qquad (3.89)$$

Generally the terms on the right-hand side in the above equation are rather complex and nonlinear and will therefore not be treated in any detail. It is, however, customary to discuss the energy balance equation in terms of the heating rate

$$\left(\frac{dT}{dt}\right)_{\text{source}} = \frac{q_T}{nc_v} \qquad (3.90)$$

and the cooling rates

$$\left(\frac{dT}{dt}\right)_{\text{radiation}} = -\frac{L_T}{nc_v} \qquad (3.91)$$

$$\left(\frac{dT}{dt}\right)_{\text{conduction}} = -\frac{\nabla\phi}{nc_v} \qquad (3.92)$$

The heat source q_T is essentially the photoabsorption of the solar radiation that will be dealt with when discussing the formation of the ionosphere. One can write $q_T(z)$ in the following manner:

$$q_T(z) = \sum_i \int F_\lambda(\lambda, z)\eta_i \sigma_i(\lambda) n_i(z) d\lambda \qquad (3.93)$$

where $q_T(z)$ is the source function at a height z, $F_\lambda(\lambda, z)$ is the radiation flux per unit wavelength (or spectral irradiance) of photons in units of m^{-2} sec^{-1} Å^{-1} at height z, $\sigma_i(\lambda)$ is the absorption cross-section and η_i is the amount of heat energy liberated per unit radiation energy absorbed (i.e. thermal efficiency). i represents each species of the gas constituents. In any absorption process a great amount of the radiation energy is not transformed into thermal energy but is lost from the atmosphere through reemission. The transformation efficiency of radiation energy to thermal energy can be calculated by knowing the detailed quantum processes associated with photoabsorption, such as photoionization, photodissociation and recombination. Here is defined the heating efficiency η_i which is known from empirical methods to vary between 0.1 and 0.6 in the thermosphere below 500 km (Fig. 3.19).

High-energy particles may also contribute significantly to the heat source, especially at auroral latitudes. If a particle enters the top of the atmosphere with an initial kinetic energy K, it will be decelerated to a smaller kinetic energy E at a height z by losing part of its energy through collisional ionization and excitation with atmospheric molecules. Its momentum, however, may be so large that the deflection of its trajectory due to collisions may be neglected. We express the loss rate of kinetic energy per unit length for such a particle by introducing the stopping cross-section $\sigma_s(E)$ in the following way:

$$\frac{dE}{dz} = \sigma_s(E) n(z) \qquad (3.94)$$

Sec. 3.9]   Thermal structure of the atmosphere   93

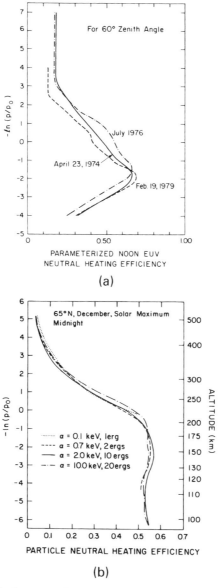

Fig. 3.19. (a) The heating efficiency for solar UV radiation in the neutral thermosphere. The reference pressure level p_0 is 5×10^{-4} μb corresponding to an altitude between 200 and 275 km depending on the solar conditions and time of day. (From Roble and Emery, 1983.) (b) The heating efficiency profiles for auroral electron fluxes with a Maxwellian spectrum and for several characteristic energies penetrating the neutral atmosphere during solar maximum conditions at 65° N and midnight in December. (From Rees et al., 1983.)

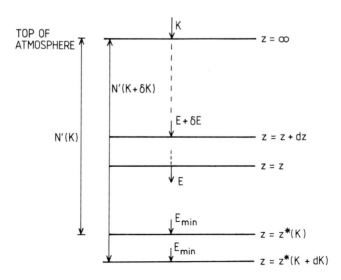

Fig. 3.20. Illustrating the deceleration of an energetic particle reaching the upper atmosphere with a kinetic energy K. At an altitude z the energy is reduced to E and at an altitude z^* the energy is reduced to a minimum E_{min}. No energetic particle exists with lower energy than this, and they are therefore stopped at this altitude. $N(K)$ represents the atmospheric range in particles per unit area through which the energetic particles penetrate. Particles with a higher energy $K + dK$ at the top of the atmosphere will reach down to a lower altitude z before the energy is reduced to E_{min}. Then the atmospheric range is $N(K + dK)$. (After Tohmatsu, 1990.)

$n(z)$ is the number density at altitude z in the atmosphere. Obviously $\sigma_s(E)$ depends on the type of molecules forming the collisional targets as well as the energy of the incoming charged particle. We notice that by forming

$$n(z)dz = \frac{dE}{\sigma_s(E)} \tag{3.95}$$

and integrating:

$$N(z) = \int_z^\infty n(z)dz = \int_E^K \frac{dE}{\sigma_s(E)} \tag{3.96}$$

$N(z)$, given in particles per unit area, represents the atmospheric depth in terms of atmospheric particles within a unit column above the height z where the kinetic energy of the precipitating particle has been reduced to E (Fig. 3.20). The deceleration of the charged particle is most effective below a certain energy $E \approx E_{min}$ where the deflection of the particle trajectory becomes extremely important. E_{min} is therefore a threshold below which the energetic particles disappear. Practically E_{min} is set to be equal to 15 eV which is close to the ionization potential of most atmospheric molecules (see Table 5.1).

The range $N(z^*)$ and the penetration height $z^*(K)$ for a particle with initial kinetic energy K is then defined as:

$$N(z^*) = \int_{E_{\min}}^{K} \frac{dE}{\sigma_s(E)} \tag{3.97}$$

If the energy spectrum of the flux of the incoming particles at the top of the atmosphere is given by $F(\infty, K)$ (in particles m^{-2} sec^{-1} eV^{-1}), then the corresponding spectrum at height z can be expressed as $F(z, E)$. If we further assume that any charged particle in the energy range K to $K + dK$ at the top of the atmosphere ends up in the energy range E to $E + dE$ at height z, then due to the continuation of particle flux as long as $E \geq E_{\min}$ (Tohmatsu, 1990):

$$F(\infty, K)dK = F(z, E)dE \tag{3.98}$$

An incoming particle with an initial energy $K + \delta K$ slightly larger than K will reach the height z with an energy $E + \delta E$ slightly larger than E, and therefore

$$N(z) = \int_{E+\delta E}^{K+\delta K} \frac{dE}{\sigma_s(E)} = \int_{E}^{K} \frac{dE}{\sigma_s(E)} \tag{3.99}$$

We now have (Tohmatsu, 1990)

$$\frac{\delta E}{\sigma_s(E)} = \frac{\delta K}{\sigma_s(K)} \tag{3.100}$$

and

$$F(z, E) = \begin{cases} \frac{\sigma_s(K) \cdot F(\infty, K)}{\sigma(E)} & z \geq z^*(K) \\ 0 & z < z^*(K) \end{cases} \tag{3.101}$$

The heating rate at height z in the atmosphere due to the charged particles is now given by:

$$q_T(z) = n(z) \int_{z_{\min}}^{\infty} \eta(E) \cdot \sigma_s(E) \cdot F(z, E) dE \tag{3.102}$$

where $\eta(E)$ is the heating efficiency of the corpuscular absorption varying between 0.1 and 0.6 in the thermosphere (Fig. 3.19b). Allowing for the flux of incident particles to depend on direction we can modify the initial flux spectrum as $F(\infty, K, \Omega)$ (in particles m^{-2} sec^{-1} ster^{-1} eV^{-1}) and the spectrum at height z as $F(z, E, \Omega)$ and finally

$$q_T(z) = n(z) \int_{4\pi} d\Omega \int_{E_{\min}}^{\infty} \eta(E) \sigma_s(E) F(z, E, \Omega) dE \tag{3.103}$$

z^* will now depend on the incident angle for the particles.

3.10 DIFFUSION

We have seen from Fig. 3.2 that above 100 km the densities of different species decay according to individual scale heights. This comes about because above the mesopause the temperature is increasing and the atmosphere is stabilized. Turbulence does not take place, and the different species are no longer homogeneously mixed, they will therefore distribute themselves according to the barometric law with individual scale heights as if they were the only gas.

Sometimes, however, it is of interest to study the situation where the main gas, the major constituent, is distributed according to its specific scale height while another gas, a minor constituent, represented by a smaller density is not. The minor constituent will then move through the major one with a velocity which is determined by diffusion.

Assume first that there is no gravitational field and the majority gas is at rest and uniformly distributed. Let the minority gas with molecular mass m be distributed by its density n, where there is a gradient dn/dx along the x-axis. Because of this gradient in the gas density the molecules will move down the gradient by a speed v, and the particle flux $\phi = nv$ will be proportional to this gradient (Fick's law):

$$\phi = nv = -D \cdot \frac{dn}{dx} \tag{3.104}$$

where D is what we will call the diffusion coefficient with dimension m^2/s.

In 3 dimensions the expression would be:

$$n\mathbf{v} = -D \cdot \nabla n \tag{3.105}$$

From the equation of continuity for a gas we have:

$$\frac{\partial n}{\partial t} + \nabla \cdot (n\mathbf{v}) = 0 \tag{3.106}$$

or in one dimension when inserting (3.104)

$$\frac{\partial n}{\partial t} = -\frac{\partial}{\partial x}(nv) = -\frac{\partial}{\partial x}\left(-D\frac{\partial n}{\partial x}\right) \tag{3.107}$$

For a constant diffusion coefficient with respect to x we get from (3.107):

$$\frac{\partial n}{\partial t} = D\frac{\partial^2 n}{\partial x^2} \tag{3.108}$$

which is the rate of change of the gas density at a given point in space. Again in 3 dimensions

$$\frac{\partial n}{\partial t} = D\nabla^2 n \tag{3.109}$$

This is the diffusion equation for the gas density n.

Diffusion

Because of the gradient in the minority gas density, there will be a pressure force acting on the gas although the temperature is constant, and this is given by:

$$F_p = -\frac{\partial p}{\partial x} = -\kappa T \frac{\partial n}{\partial x} \tag{3.110}$$

If now each of the minority particles experiences ν collisions per unit time with the majority gas which is at rest, there will be a restoring force:

$$F_\nu = nm\nu v \tag{3.111}$$

which, when no other force is acting, must balance the pressure force. Thus

$$-\kappa T \frac{\partial n}{\partial x} = nm\nu v \tag{3.112}$$

and

$$nv = -\frac{\kappa T}{m\nu}\frac{\partial n}{\partial x} = -D\frac{\partial n}{\partial x} \tag{3.113}$$

A simple expression for the diffusion coefficient now emerges:

$$D = \frac{\kappa T}{m\nu} \tag{3.114}$$

Since the collision frequency between the minority and the majority gas is proportional to the density of the majority gas, n_M, and the square root of the temperature

$$\nu \propto n_M T^{1/2} \tag{3.115}$$

the diffusion coefficient obeys the following proportionality

$$D \propto T^{1/2} n_M^{-1} \tag{3.116}$$

The diffusion in a neutral gas is therefore higher at a higher temperature and a lower density.

Let now the space coordinate be vertical and assume that the gravity force mg is acting downward on each volume of the minority constituent, then the collisions must balance the sum of the pressure and gravity forces as follows:

$$-\frac{\partial p}{\partial z} - nmg = nm\nu w \tag{3.117}$$

where w is the vertical velocity. Then

$$-\kappa T \frac{\partial n}{\partial z} - nmg = nm\nu w \tag{3.118}$$

and solving for the vertical flux we get

$$nw = -\frac{kT}{m\nu}\left(\frac{\partial n}{\partial z} - \frac{gm}{kT}n\right) = -D\left(\frac{\partial n}{\partial z} + \frac{n}{H_m}\right) \tag{3.119}$$

D is given by (3.114) and $H_m = \kappa T/mg$ is the scale height of the minority constituent. We notice that H_m enters the equation only as a constant and does not need to be equal to the distribution height $[-(1/n)(\partial n/\partial z)]^{-1}$ of the minority gas.

Now when applying (3.107) for vertical motion and inserting (3.119):

$$\frac{\partial n}{\partial t} = -\frac{\partial}{\partial z}(nw) = \frac{\partial}{\partial z}\left\{D\left(\frac{\partial n}{\partial z} + \frac{n}{H_m}\right)\right\} \tag{3.120}$$

From (3.114) we have that $D = \kappa T/m\nu$, and since ν must be proportional to the density n_M of the majority constituent which is distributed according to the scale height H_M, we get:

$$\nu \propto n_M = n_{M_0}\exp(-z/H_M) \tag{3.121}$$

The diffusion coefficient can then be expressed as:

$$D = D_0\exp(z/H_M) \tag{3.122}$$

It increases exponentially with height in an isothermal atmosphere. We now find from (3.120):

$$\begin{aligned}\frac{\partial n}{\partial t} &= \left(\frac{\partial n}{\partial z} + \frac{n}{H_m}\right)\frac{\partial D}{\partial z} + D\left(\frac{\partial^2 n}{\partial z^2} + \frac{1}{H_m}\frac{\partial n}{\partial z}\right) \\ &= D\left\{\frac{\partial^2 n}{\partial z^2} + \left(\frac{1}{H_M} + \frac{1}{H_m}\right)\frac{\partial n}{\partial z} + \frac{n}{H_M H_m}\right\}\end{aligned} \tag{3.123}$$

If it is assumed that at a particular height z the density of the minority constituent is distributed according to the exponential function

$$n = n_0\exp(-z/\delta) \tag{3.124}$$

then at this height the density will change by time at a rate:

$$\begin{aligned}\frac{\partial n}{\partial t} &= D\left\{+\frac{1}{\delta^2} - \left(\frac{1}{H_m} + \frac{1}{H_M}\right)\frac{1}{\delta} + \frac{1}{H_M H_m}\right\}n \\ &= D\left\{\left(\frac{1}{\delta} - \frac{1}{H_m}\right)\left(\frac{1}{\delta} - \frac{1}{H_M}\right)\right\}n \\ &= \Gamma \cdot n\end{aligned} \tag{3.125}$$

As long as the density of the minority constituent remains approximately exponentially distributed by distribution height δ, its concentration will vary in time at altitude z by the rate $\Gamma \cdot n$ where Γ is given by:

$$\Gamma = D\left\{\left(\frac{1}{\delta} - \frac{1}{H_m}\right)\left(\frac{1}{\delta} - \frac{1}{H_M}\right)\right\} \tag{3.126}$$

This result, however, deserves some comments. In Table 3.2 we show the different values that δ can take in relation to H_M and H_m and the following values of Γ/D.

Chemistry of an oxygen atmosphere

Table 3.2. (From Ratcliffe, 1972)

Value of δ	Approximate value of Γ/D
$\delta > H_m$ and $\delta > H_M$	$1/H_M \cdot H_m$
$\delta = H_m$	0
$\delta > H_M$ and $\delta < H_m$	$-1/\delta \cdot H_M$
$\delta < H_M$ and $\delta > H_m$	$-1/\delta \cdot H_m$
$\delta = H_M$	0
$H_m > \delta$ and $H_M > \delta$	$1/\delta^2$
$\delta < 0$	$> 1/H_M \cdot H_m$

In most cases δ, H_m and H_M are roughly of the same order of magnitude so that (Ratcliffe, 1972)

$$\Gamma = D/\delta^2, \quad H_M^2 \quad \text{or} \quad H_m^2$$

There are two situations, however, where $\partial n/\partial t = 0$, i.e. steady state. First, when $\delta = H_m$ or the minority constituent is distributed according to its natural scale height. There will then according to (3.119) be no vertical motion because

$$w = -D\left(-\frac{1}{H_m} + \frac{1}{H_m}\right) = 0 \tag{3.127}$$

everywhere. The second case appears when $\delta = H_M$; the minority gas is distributed according to the scale height of the majority gas. Then, however,

$$w = -D\left(-\frac{1}{H_M} + \frac{1}{H_m}\right) \neq 0 \tag{3.128}$$

and w increases by height if the $H_m > H_M$, otherwise it decreases. The number of particles crossing a unit area per unit time, however, is:

$$nw = -nD\left(-\frac{1}{H_M} + \frac{1}{H_m}\right) = n_0 D_0 \left(\frac{1}{H_M} - \frac{1}{H_m}\right) \tag{3.129}$$

The upward decrease in n ($\approx \exp(-z/H_M)$) is just equal to the upward increase in D ($\approx \exp(z/H_M)$) so that there is a steady flow of gas upwards or downwards depending on whether H_m is greater or smaller than H_M. In spite of the fact that $\partial n/\partial t = 0$ the situation is therefore not one of equilibrium.

3.11 CHEMISTRY OF AN OXYGEN ATMOSPHERE

Since the atmosphere contains such a variety of different neutral as well as ionized atomic and molecular species, it appears as an insurmountable task to treat the system of chemical reactions in a self-consistent way. We will not aim at accomplishing such an enterprise but rather illustrate some fundamental problems of neutral atmospheric chemistry by presenting the classical oxygen atmosphere theory due to Chapman. In this model O, O_2 and O_3 are the principal elements in the chemical reaction system, and the purpose is to attempt to describe the time and spatial variations of these elements in a qualitative sense.

The photochemical processes assumed to be controlling the densities of the oxygen species are:

$$O_2 + h\nu \rightarrow O + O, \quad J_2 \quad \text{(a)}$$

$$O_3 + h\nu \rightarrow O_2 + O, \quad J_3 \quad \text{(b)}$$

$$O + O + M \rightarrow O_2 + M, \quad k_{11} \quad \text{(c)} \quad (3.130)$$

$$O + O_2 + M \rightarrow O_3 + M, \quad k_{12} \quad \text{(d)}$$

$$O + O_3 \rightarrow O_2 + O_2, \quad k_{13} \quad \text{(e)}$$

where the dissociation rates (J_2 and J_3) and the reaction rates (k_{11}, k_{12} and k_{13}) are indicated at the corresponding chemical equations, respectively. Here M represents an arbitrary neutral species, mainly N_2.

The equation of continuity for the different oxygen species can now be expressed as:

$$\frac{Dn_1}{Dt} + n_1 \nabla \cdot \mathbf{v}_1 = 2J_2 n_2 + J_3 n_3 - 2k_{11} n_M n_1^2 - \quad (3.131)$$

$$k_{12} n_M n_2 n_1 - k_{13} n_3 n_1 \quad \text{(a)}$$

$$\frac{Dn_2}{Dt} + n_2 \nabla \cdot \mathbf{v}_2 = -J_2 n_2 + J_3 n_3 - k_{11} n_M n_1^2 -$$

$$k_{12} n_1 n_2 + 2k_{13} n_1 n_3 \quad \text{(b)}$$

$$\frac{Dn_3}{Dt} + n_3 \nabla \cdot \mathbf{v}_3 = -J_3 n_3 + k_{12} n_M n_1 n_2 - k_{13} n_1 n_3 \quad \text{(c)}$$

where n_1, n_2, n_3 and n_M are the number densities of O, O_2, O_3 and M, respectively, and $\mathbf{v}_1$, $\mathbf{v}_2$ and $\mathbf{v}_3$ are the velocities of the respective oxygen species.

By assuming that the only other species in the atmosphere at the region of interest (< 110 km) is molecular nitrogen N_2 except for the oxygen species, then

$$n_1 + n_2 + n_3 + n_{N_2} = n_M \quad (3.132)$$

where n_{N_2} and n_M are the number density of molecular nitrogen and of all constituents, respectively. Since O_2 is the most abundant oxygen species below 110 km at any time $n_2 \gg n_1, n_3$ and by introducing the relative number α of O_2 molecules, then:

$$\alpha = \frac{n_2}{n_M} \quad (3.133)$$

When neglecting n_1 and n_3 and inserting into (3.132):

$$n_{N_2} = (1 - \alpha) n_M \quad (3.134)$$

Chemistry of an oxygen atmosphere

We will now assume, however, that the relative number of the total oxygen and nitrogen atoms are constant. Then

$$\frac{n_1 + 2n_2 + 3n_3}{2n_{N_2}} = \frac{\alpha}{1-\alpha} = \text{const.} \tag{3.135}$$

This condition in fact implies that the density of the molecular oxygen does not change due to the reactions listed. Below 100 km $\alpha = 0.209$ in the atmosphere.

When now neglecting time and spatial variations in the oxygen species, we obtain the steady-state condition for the oxygen species from (3.131):

$$2J_2 n_2 + J_3 n_3 - 2k_{11} n_M n_1^2 - k_{12} n_M n_2 n_1 - k_{13} n_3 n_1 = 0 \quad \text{(a)}$$

$$-J_2 n_2 + J_3 n_3 + k_{11} n_M n_1^2 - k_{12} n_M n_2 n_1 + 2k_{13} n_1 n_3 = 0 \quad \text{(b)} \tag{3.136}$$

$$-J_3 n_3 + k_{12} n_M n_1 n_2 - k_{13} n_1 n_3 = 0 \quad \text{(c)}$$

By adding (3.136a) and (3.136c) the equation for n_1 follows:

$$k_{11} n_M n_1^2 + k_{13} n_3 n_1 - J_2 n_2 = 0 \tag{3.137}$$

Since the amount of ozone will always be less than either molecular or atomic oxygen, we neglect n_3 ($n_3 \ll n_2$ and $n_3 \ll n_1$). Then we find from (3.132) and (3.135)

$$n_1 + n_2 + n_{N_2} = n_M \tag{3.138}$$

$$n_1 + 2n_2 = \frac{2\alpha}{1-\alpha} n_{N_2} = \frac{2\alpha}{1-\alpha}(n_M - n_1 - n_2) \tag{3.139}$$

By solving for n_2 in (3.138) and (3.139) we find:

$$n_2 = \alpha n_M - \frac{1+\alpha}{2} n_1 \tag{3.140}$$

Introducing this in (3.137) gives:

$$k_{11} n_M n_1^2 + k_{13} n_3 n_1 - J_2 \alpha n_M + \frac{1+\alpha}{2} J_2 n_1 = 0 \tag{3.141}$$

Since the loss of atomic oxygen will be dominated by the reaction rate k_{11}, we can neglect the second term in this equation ($k_{11} n_M n_1^2 \gg k_{13} n_3 n_1$) and finally we have:

$$k_{11} n_M n_1^2 + \frac{1+\alpha}{2} J_2 n_1 - J_2 \alpha n_M = 0 \tag{3.142}$$

The solution for n_1 ($n_1 > 0$) will be

$$n_1 = \frac{J_2(1+\alpha)}{4k_{11} n_M} \left[\left(1 + \frac{16\alpha k_{11} n_M^2}{J_2(1+\alpha)^2}\right)^{1/2} - 1 \right] \tag{3.143}$$

J_2 depends on the altitude profile of n_2, therefore n_1 can only be derived by a stepwise numerical process when an altitude profile of n_2 is given. Starting out at the uppermost height where $J_2 = J_2(\infty) = \text{const.}$, n_1 and n_2 can be calculated for this altitude. Then J_2 will change accordingly for the next height step and so on.

From the continuity equation of ozone (equation (3.136c)) we derive

$$n_3 = \frac{k_{12} n_1 n_2 n_M}{k_{13} n_1 + J_3} \tag{3.144}$$

At night when the dissociation by solar irradiation ceases and $J_3 = 0$, the ozone loss is completely determined in the steady state by dissociative recombination with O ($k_{13} n_1 \gg J_3$)

$$n_3 = \frac{k_{12} n_2 n_M}{k_{13}} \quad \text{(night)} \tag{3.145}$$

In the daytime, however, (for $z \geq 60$ km) when the radiative dissociation J_3, dominates over the ozone loss by recombination with O ($J_3 \gg k_{13} n_1$), we have:

$$n_3 = \frac{k_{12} n_1 n_2 n_M}{J_3} \quad \text{(day above 60 km)} \tag{3.146}$$

Lower down in the atmosphere in the ozone layer between 30 and 80 km the number density of O_2 is so much larger than n_1 and n_3 that we can assume that

$$n_2 + n_{N_2} = n_M \tag{3.147}$$

and since $n_2 = \alpha n_M$ we find from (3.137):

$$k_{11} n_M n_1^2 + k_{13} n_3 n_1 - J_2 \alpha n_M = 0 \tag{3.148}$$

In the daytime below 80 km the radiative dissociation of O_3 will dominate above radiative recombination with O as a loss mechanism for O_3 and therefore from (3.144):

$$n_3 = \frac{k_{12} n_M n_1 n_2}{J_3} = \frac{k_{12} \alpha n_M^2 n_1}{J_3} \quad \text{(day below 80 km)} \tag{3.149}$$

as above 60 km. By inserting this in (3.148) above we get:

$$k_{11} n_M n_1^2 + \frac{k_{13} k_{12} \alpha n_1^2 n_M^2}{J_3} - J_2 \alpha n_M = 0 \tag{3.150}$$

$$n_1 = \left(\frac{J_2 J_3}{k_{12} k_{13} n_M} \right)^{1/2} \left(1 + \frac{J_3 k_{11}}{k_{12} k_{13} \alpha n_M} \right)^{-1/2} \tag{3.151}$$

and for the ozone number density we have

$$n_3 = \alpha n_M \left(\frac{J_2 k_{12} n_M}{J_3 k_{13}} \right)^{1/2} \left(1 + \frac{J_3 k_{11}}{k_{12} k_{13} \alpha n_M} \right)^{-1/2} \tag{3.152}$$

J_2 as well as J_3 depend upon the distribution of n_2 and n_3, therefore a stepwise solution has to be performed by starting at the upper height where some reasonable conditions can be set such that $n_2 = \alpha n_M$. Then J_2 can be estimated at the upper height and assuming $J_3 = J_3(\infty)$ above 60 km, J_3 can be derived for lower altitudes as well.

Serious disagreements are observed between the measured ozone densities at different heights and those derived from these classical pure oxygen models. The theory gives a too large density compared to the measurements at almost all heights (Tohmatsu, 1990).

Furthermore, since $J_2 = J_3 = 0$ at night, the ozone density should increase at night as also observed except that the densities again are much lower than the one derived from this simple theory.

There are at least two reasons for these disagreements. Firstly must the chemical reactions with HO_x and NO_x be included because the chemistry of atomic oxygen and ozone does not constitute a closed reaction system among O, O_2 and O_3, but a photochemical system for the H–N–O atmosphere. Secondly the dynamical transport effects which effectively redistribute the minor species such as ozone must be taken into account. The latter includes the advection of air mass due to atmospheric circulation and oscillation.

3.12 THE OZONE LAYER

As already shown in Fig. 3.6 and mentioned in the last section, the ozone layer is situated in the height region between 15 and 40 km. By observing the ozone density profile at different latitudes, however, the peak altitude is found to decrease and the peak magnitude to increase by increasing latitude. More ozone is therefore present at lower altitudes (< 25 km) in the polar region (Fig. 3.21).

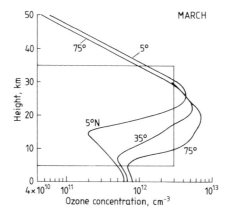

Fig. 3.21. Ozone density profiles in the atmosphere at different latitudes in the northern hemisphere. The concentration is given in molecules per cm^3. Also indicated is a slab about 20 km wide in altitude with an average concentration of 3×10^{12} cm^{-3} or 3×10^{18} m^{-3} as used in the text. (After Shimazaki, 1987.)

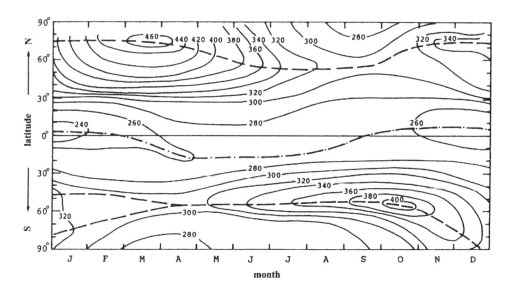

Fig. 3.22. The latitudinal averaged ozone distribution for February shown as function of height and latitude. The labels added to the isolines indicate the partial pressure of ozone in mb. A strong maximum is observed below 20 km in the Arctic region. (From Dütsch, 1978.)

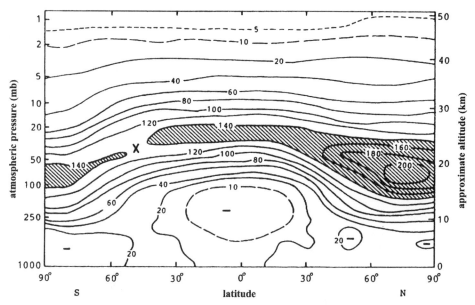

Fig. 3.23. Latitudinal and seasonal variation of the total ozone content. The labels added to the isolines are given in Dobson units. (From London, 1985.)

This is even more clearly brought out in Fig. 3.22 where the latitudinal average of the ozone content at different altitudes below 50 km are presented versus latitude for the average February month. The winter pole has a rather high ozone content at a low altitude (< 20 km) while the ozone density above equator is much smaller and maximized above 20 km. This latter effect cannot be explained by the simple photochemical equilibrium models of the atmosphere since we from these would expect the ozone content to be high at lower heights in the equatorial region where the solar radiation and the radiative dissociation of O_2 is highest. Transport processes must therefore be of fundamental importance for the global ozone distribution.

In Fig. 3.23 are the latitudinal distribution of the total ozone content for different seasons of the year presented. Again there are clear maxima above the polar regions, but it is seen that these maxima occur in spring. The maximum in the northern hemisphere in March is about 15% higher than the corresponding maximum in the southern hemisphere in October.

In the northern hemisphere the amplitudes of the seasonal variation are also larger, varying from a maximum of close to 460 Dobson units in April at times to about 280 Dobson units in November. That is a variation of more than 20% around the mean value.

The total ozone content in a vertical column above a ground-based observer has been measured for more than 50 years at different places on Earth. By choosing two wavelengths at which the intensity of incoming solar radiation is about the same, but where one is subject to greater absorption by the ozone than the other (for example 306 nm and 325 nm), the total amount of ozone between the observer and the Sun can be derived from the two intensities measured (Dobson photospectrometer).

The term "total ozone", sometimes referred to as the total layer thickness, thus represents the height integral of the column density and is often abbreviated atm-cm or cm (STP) (Table 3.3). This corresponds to the thickness of the layer if the pressure and density were reduced to standard atmospheric values throughout the layer. The Dobson unit is derived in a similar manner and abbreviated m-atm-cm which equals 10^{-3} atm-cm. 300 Dobson units therefore represent an ozone content which, when reduced to standard temperature and pressure throughout the layer, would correspond to a column of 3 mm.

One of the early findings in the research of ozone was the smaller amount of ozone in Antarctic spring than in the Arctic spring. Another early result of such measurements was the occasional decrease observed in the total ozone during the early springtime.

By assuming the ozone layer is forming a slab as indicated in Fig. 3.21, where the number density is constant and equal to $n = 3.0 \times 10^{18}$ molecules/m^3 and the slab thickness is $\Delta h = 30$ km (corresponding to $55°$ (latitude)). We find the total mass of the ozone in the atmosphere to be

$$M_{O_3} = 4\pi R_e^2 \cdot n \cdot m_{O_3} \cdot \Delta h = 3.6 \times 10^{12} \text{ kg} \tag{3.153}$$

Table 3.3. Different units used in presenting the ozone content in the atmosphere and the conversion factors between these units. (From Herman and Goldberg, 1978)

Derived quantity	Basic quantity	
	Mass density ρ_3	Column density ϵ_3
Number density n_3 (molecules) m^{-3}	$(N_A/M_3) \cdot \rho_3$ $1.25467(10^{25}) \cdot \rho_3$	$10^{-5} \cdot (N_A/V_0) \cdot \epsilon_3$ $2.68684(10^{20}) \cdot \epsilon_3$
Column density ϵ_3 atm-cm km^{-1}	$10^5 \cdot (V_0/M_3) \cdot \rho_3$ $4.66968(10^4) \cdot \rho_3$	ϵ_3
Mass density ρ_3 kg m^{-3}	ρ_3	$10^{-5} \cdot (M_3/V_0) \cdot \epsilon_3$ $2.14148(10^{-5}) \cdot \epsilon_3$
Partial pressure p_3 N m^{-2}	$(R^*/M_3) \cdot T_S \cdot \rho_3$ $1.73222(10^2) \cdot T_S \cdot \rho_3$	$10^{-5} \cdot (R^*/V_0) \cdot T_S \cdot \epsilon_3$ $3.70951(10^{-3}) \cdot T_S \cdot \epsilon_3$
mb	$1.73222 \cdot T_S \cdot \rho_3$	$3.70951(10^{-5}) \cdot T_S \cdot \epsilon_3$
Mass mixing ratio r_3 dimensionless	ρ_3/ρ_S	$10^{-5} \cdot (M_e \cdot \epsilon_3/V_0 \cdot \rho_S)$ $2.14148(10^{-5}) \cdot \epsilon_3/\rho_S$
Volume mixing ratio r_3' dimensionless	$\rho_3 \cdot M/\rho_S \cdot M_3$ $6.03448(10^{-1}) \cdot \rho_3/\rho_S$	$10^{-5} \cdot (M \cdot \epsilon_3/V_0 \cdot \rho_S)$ $1.29227(10^{-5}) \cdot \epsilon_3/\rho_S$

Avogadro's number $\qquad N_A = 6.022169(10^{26})$ (molecules) kmol^{-1}
Universal gas constant $\qquad R^* = 8.31434(10^3)$ N m K^{-1} kmol^{-1}
Volume of ideal gas at STP $\qquad V_0 = 22.4136$ m^3 kmol^{-1}
Molecular mass of O$_3$ $\qquad M_3 = 47.9982$ kg kmol^{-1}
Molecular mass of air $\qquad M = 28.9644$ kg kmol^{-1}
Temperature of the U.S. Standard Atmosphere T_S (K at height Z)
Density of the U.S. Standard Atmosphere ρ_S (kg m^{-3} at height Z)
1.0 N m^{-2} = 0.01 mb

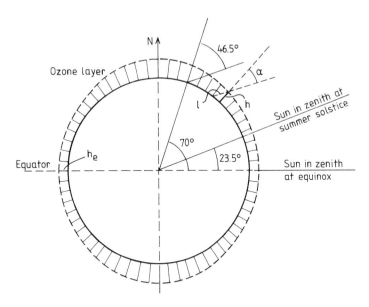

Fig. 3.24. A schematic illustration of the global ozone distribution showing a thinning of the layer at lower latitudes where also the solar irradiation can have vertical impact on the atmosphere. The height of the maximum in the layer, however, is larger at equatorial regions than at the poles.

where $m_{O_3} = 48 \times 1.673 \times 10^{-27}$ kg and R_e is the Earth's radius. We have found that the total mass of the atmosphere is $M_A = 5.25 \times 10^{18}$ kg. Therefore the total ozone of the atmosphere represents only one portion in 10^6 with respect to mass of the total atmosphere – a small portion indeed, but essential for life on Earth.

As is well known, the ozone layer is shielding the Earth from the UV radiation which can be a health hazard to some people. The shielding, however, is most effective in the polar region where the ozone layer is at its thickest and the Sun has a large angle to the zenith, while the layer is more shallow at lower latitudes where the Sun is close to overhead. This is illustrated in Fig. 3.24.

For a station at 70° latitude, for example, the Sun can never make an angle α with the zenith less than 46.5°. Therefore the distance that the solar UV ray must pass through the ozone layer will be (Fig. 3.25)

$$l = \frac{h}{\cos \alpha} \qquad (3.154)$$

if h is the thickness of the layer. At the tropics where the Sun can be in the zenith and $\alpha = 0$, the ray path through the layer is equal to the layer thickness. In order for the UV intensity observed at 70° latitude to be equal to the intensity observed at equator the layer thickness at 70° is given by:

$$h' = h_e \cos \alpha \qquad (3.155)$$

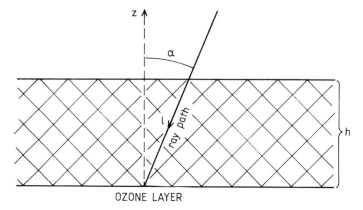

Fig. 3.25. An illustration of the variation in the length of the ray path through the ozone layer with a varying solar zenith angle.

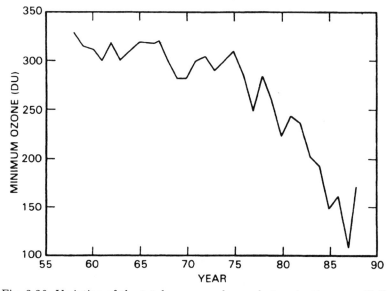

Fig. 3.26. Variation of the total ozone as observed at springtime over Halley Bay from 1957 to 1989. The content is given in Dobson units. (After Farman et al., 1985 and modified by Aikin, 1992.)

Since the present thickness of the ozone layer at high latitudes is about $1.5\ h_e$ we notice that the layer thickness must be reduced to about half its present value:

$$\frac{h'}{h} = \frac{h_e}{1.5\ h_e} \cos 46.5 \approx 0.5 \qquad (3.156)$$

At 50° N the corresponding reduction of the height of the ozone layer will be about 0.7 its present value. What has upset most of the population lately, is an apparent

Sec. 3.12] The ozone layer 109

steady decrease of the total ozone content above the Antarctic continent as first discovered at Halley Bay. It is in fact the Antarctic spring minimum ozone content observed in October–November every year which has been deepening (Fig. 3.26). It is reduced from slightly above 300 Dobson units in the early 1950s to less than 200 Dobson units in the late 1980s.

The dispute now goes on worldwide why this happens. Some argue that this is due to anthropogenic effects such as the heavy use of carbon chloride and freon gases. Others would tend to claim that it is all due to natural variations.

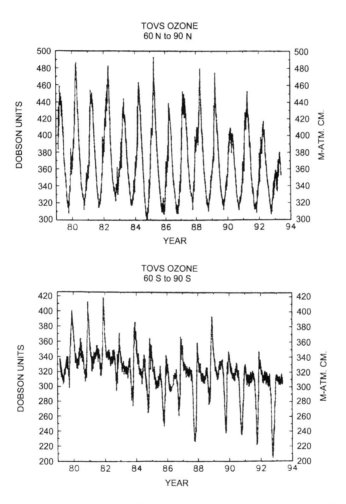

Fig. 3.27. Daily mean values of the total ozone content for the period 1979 to 1993 as derived from TIROS Operational Vertical Sounder (TOVS) for the Arctic region (upper panel) and the Antarctic region (lower panel). (From Neuendorffer, 1994.)

Several campaigns have been carried through in the northern hemisphere in order to try to disclose a similar ozone depletion in the Arctic as observed in the Antarctic spring. So far we can note that if an ozone depletion is developing in the Arctic, it is much less pronounced than in the Antarctic. A comparison between the annual variations of the daily averaged total ozone content above 60° latitude in the Arctic and the Antarctic regions are presented in Fig. 3.27. The data are derived from the NOAA TIROS Operational Vertical Sounder (TOVS) and presented in Dobson units (DU). The measurements which are made between 1980 and 1993 show a slow decline in the Antarctic region, and the minima are in particular getting deeper and deeper by the years and reaching an overall minimum of close to 210 DU in October 1992.

The strong maxima occurring in midsummer in the early 1980s have almost completely disappeared about 10 years later.

In the Arctic region, however, the late winter minima are rather constant throughout the observation period and close to 310 DU. The spring maxima appear to vary from year to year and may have had a tendency to decrease during the latest years.

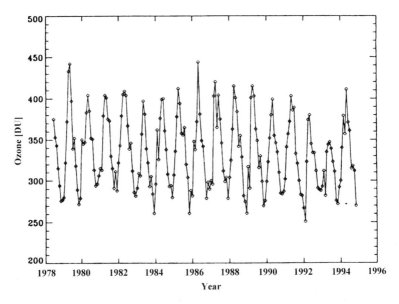

Fig. 3.28. Monthly mean values as observed from the ground in Oslo (60°N) between 1978 and 1995. (From Henriksen, 1994.)

Observations from the ground of the total ozone content have been regularly carried out at Oslo about 60° N since 1978. Monthly mean values have been derived and they are shown in Dobson units in Fig. 3.28 for the time period between 1978 and 1995. In Oslo the winter minima are also fairly constant but about 70 DU higher than for the average Arctic region above 60° N, as shown in Fig. 3.27. The maxima are also fairly constant and close to 400 DU except for the years 1992 and 1993 when they are only 380 and 350 DU, respectively. In 1994 the summer maximum is above 400 DU, and the ozone situation appears quite normal in Oslo.

The results presented in Fig. 3.27 for the Arctic region and in Fig. 3.28 for Oslo demonstrate that the behaviour of the ozone layer in the northern hemisphere is much less dramatic than for the ozone layer in the southern hemisphere. These results also indicate that there is a good agreement between the observations made by the modern satellite technique, TOVS, and the more conventional ground-based technique introduced by Dobson in the 1920s.

3.13 THE DESTRUCTION OF OZONE

As we have noticed in section 3.11 ozone is formed in the atmosphere by dissociation of O_2 molecules due to the solar UV emission at shorter wavelengths than 242 nm followed by a three-body process which rapidly attach the released oxygen atoms to other oxygen molecules.

$$O_2 + h\nu \ (< 242 \text{ nm}) \longrightarrow O + O \tag{3.157}$$

$$O + O_2 + M \longrightarrow O_3 + M \tag{3.158}$$

The natural loss of ozone is supposed to take place by a number of processes, the most important probably being

$$O_3 + h\nu \longrightarrow O_2 + O \tag{3.159}$$

A large fraction of the O atoms thus produced can react with O_2 very quickly and form new O_3 molecules. Excited O atoms, however, react very effectively with O_3 as follows:

$$O(^3P) + O_3 \longrightarrow O_2 + O_2 \tag{3.160}$$

This has the end result of destroying, odd numbers of oxygen compounds

$$2 O_3 + h\nu \longrightarrow 3 O_2 \tag{3.161}$$

The wavelength regime here at play is in the range of 450–650 nm and above 310 nm. It is, however, the absorption of wavelengths below 310 nm by ozone which represents the major heat source to the stratosphere.

Table 3.4. Some reaction rates related to the chemistry of oxygen in the neutral atmosphere

Reaction	Index	Value
$O + O + M \rightarrow O_2 + M$	k_{11}	2.7×10^{-45} m^6 s^{-1}
$O + O_2 + M \rightarrow O_3 + M$	k_{12}	$5.8 \times 10^{-47} \exp(0.89/R_M T)$ m^6 s^{-1}
$O + O_3 \rightarrow O_2 + O_2$	k_{13}	$3.3 \times 10^{-17} \exp(-4.2/R_M T)$ m^3 s^{-1}

$R_M = 1.9861 \times 10^{-3}$ (kcal/mol deg.)

Another possible natural ozone sink mechanism involves nitric oxide (NO) as follows:

$$NO + O_3 \longrightarrow NO_2 + O_2 \tag{3.162}$$

$$NO_2 + O \longrightarrow NO + O_2 \tag{3.163}$$

These processes both destroy ozone and atomic oxygen which otherwise could react with molecular oxygen and create ozone. The NO content is, however, balanced. Part of the nitric dioxide can again be photolysed by radiation of wavelengths shorter than 400 nm

$$NO_2 + h\nu \; (< 400 \text{ nm}) \longrightarrow NO + O \tag{3.164}$$

which liberates an oxygen atom that can take part in the formation of ozone again. It is therefore clear that NO plays an important role in the natural chemical scheme balancing the ozone content.

In the literature the notation NO_x often refers to NO and NO_2, and if the concentration of NO_x is specified, it indicates the sum of the concentrations of these two oxides. NO_y is also often found in the literature and it means "odd nitrogen" and is defined as the sum of NO_x and all oxidized nitrogen species that represents sources or sinks of NO_x.

The point of controversy, however, is the so-called "ozone hole" observed in the Antarctic and the role chlorine and bromine can play in this region of the atmosphere. Notice, however, that the term "ozone hole" which is introduced as a bogey of a natural phenomenon known to have been occurring for years, was originally coined to describe a large ozone decrease which was observed over Boulder, Colorado, in conjunction with a volcanic event in 1964 (Pittoc, 1965).

Chlorine, Cl, is a fallout of the CFC gases which is heavily in use in refrigerators, spray cans, air conditioning equipment etc. and is widely used in factories making insulation material. If Cl reaches the ozone layer, the following reactions may take place:

$$Cl + O_3 \longrightarrow ClO + O_2 \tag{3.165}$$

$$ClO + O \longrightarrow Cl + O_2 \tag{3.166}$$

The net result is that two oxygen molecules are produced out of one oxygen atom and an ozone molecule while the Cl atom can start over again destroying a new ozone molecule (Fig. 3.29). Since Cl is a gas which will have a long lifetime at ozone heights, it may represent a threat to the ozone layer.

The chemical reactions related to bromine, Br, being an ozone destructor also depend on chlorine and are as follows:

$$Br + O_3 \longrightarrow BrO + O_2 \tag{3.167}$$

$$Cl + O_3 \longrightarrow ClO + O_2 \tag{3.168}$$

$$\text{ClO} + \text{BrO} \longrightarrow \text{Cl} + \text{Br} + \text{O}_2 \tag{3.169}$$

The net result is that three oxygen molecules are produced out of two ozone molecules, and the Cl and Br atoms are unchanged and free to start over on a new process as catalysts.

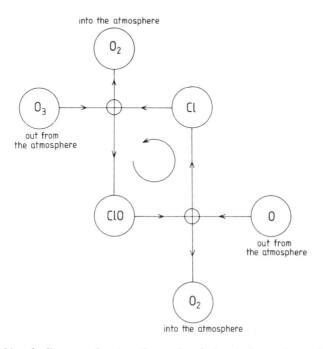

Fig. 3.29. A diagram showing the cycle of chemical reactions which can destroy odd oxygen species, leaving Cl as catalyst. An ozone molecule and a Cl atom react to form ClO and O_2. The ClO molecule reacts with an oxygen atom to form O_2 and Cl. The Cl atom can repeat the cycle which destroys one ozone molecule and an oxygen atom, leaving two oxygen molecules per cycle.

In order to find out whether the natural loss mechanism as controlled by naturally occurring NO or the anthropogenic mechanism as controlled by Cl is the more dominant in the Antarctic, airborne experiments have been carried through to measure the concentrations of different gas constituents. Results from such flights are shown in Fig. 3.30. The observations which are made at 20 km altitude between 64° S and 70° S, show the variation in ozone and ClO content for two different flights before and after the minimum ozone content has been developed. In the flight before the minimum is reached the ClO content is low and close to 0.4 particles per thousand in volume (pptv) and behaves fairly independently of the ozone. After the minimum is developed, however, the ClO content is high, and close to 1.2 pptv at lower latitudes and varies in antiphase with variations in the ozone content. The latter is reduced at all latitudes with respect to the situation before the minimum developed. This very strong anticorrelated behaviour between the ClO and O_3 content is suggestive of a causal linkage. In Fig. 3.31 similar observations made later during the minimum phase are presented. Here the ozone content is compared with the content of ClO, NO_y and H_2O at the same altitude between 58° S and 72° S. NO_y represents the total of odd nitrogen or the total of reactive nitrogen. Again there is an anticorrelation between the variations in ozone and ClO, while the content of NO_y and H_2O decreases together with the ozone decrease. If, for instance, the NO_y compounds were causing the ozone depletion in a natural sense, then we could expect the content of NO_y to be high when the content of O_3 is low. As this is not the case, these results as a whole suggest that ClO plays an important role in the depletion of ozone in the Antarctic at spring. These are the observed facts which have led the world into such a hectic dispute.

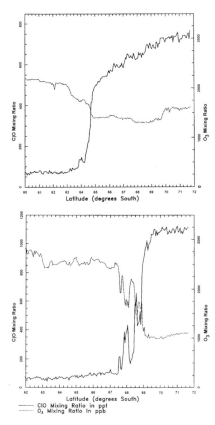

Fig. 3.30. Simultaneous observations of the content of ozone and ClO as a function of latitude in the southern hemisphere, before (upper panel) and during (lower panel) the development of the ozone spring depletion in the Antarctic. (From Anderson et al., 1989.)

Sec. 3.14] The height profile of ozone 115

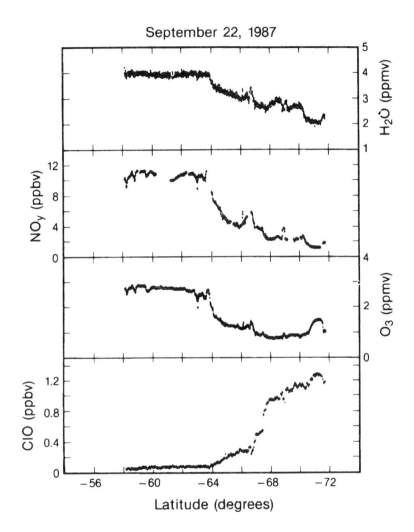

Fig. 3.31. Simultaneous observations of H_2O, NO_y, O_3 and ClO during the formation of a spring ozone depletion in the Antarctic. (From Fahey et al., 1989.)

3.14 THE HEIGHT PROFILE OF OZONE

A comparison is made in Fig. 3.32 between an ozone density profile observed in the height region below 35 km at Aug. 29, 1987 just before the onset of a springtime ozone depletion above the Antarctic and a similar profile observed about 2 months later during the main phase of the depletion. A marked reduction takes place between 12 and 20 km – this is the signature of an ozone hole. Also shown in the figure are the temperature profiles on the corresponding days. The temperature is enhanced above 10 km during the main phase of the depletion due to increased solar radiation later in the spring. Of special importance are the 5 bean-shaped figures to the right in the picture, these indicate typical heights of polar stratospheric clouds (PCSs).

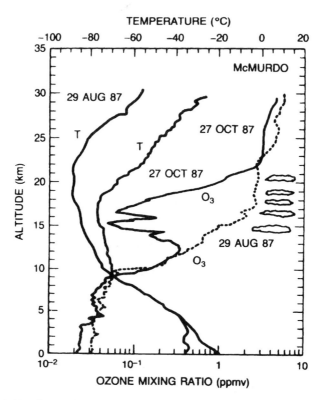

Fig. 3.32. Ozone density profiles below 35 km, given in terms of the ozone mixing ratio (ppmv), for two days in the Antarctic, one (Aug. 29, 1987) before the onset of a springtime ozone depletion, and one (Oct. 27, 1987) in the main phase of the depletion. Also shown are the simultaneously observed temperature profiles. (From Hofmann et al., 1989.) To the right in the figure are 5 different polar stratospheric clouds (PCSs) indicated for reference. (After Aikin, 1992).

The formation of these clouds is believed to have an intimate connection with the strong ozone depletion observed in spring over the Antarctic. There are two possible sources for these clouds: one is that they are formed by nitric acid trihydrate and the other that they are formed by pure water. The former starts to grow at 188 K at a pressure of 30 mb and at 192 K at 50 mb, temperatures observed between 10 and 25 km before the depletion starts in Fig. 3.32. The point is then that when the stratosphere is cold enough, NO_3 and N_2O_5 are formed which reduces the amount of NO responsible for the natural ozone reduction and therefore the ozone can grow. When the Sun returns in the spring, however, the stratosphere is rich in N_2O_5 which is radiatively dissociated by sunlight as follows:

$$N_2O_5 + h\nu \longrightarrow NO_3 + NO_2 \tag{3.170}$$

$$NO_3 + h\nu \longrightarrow NO + O_2 \tag{3.171}$$

and NO and NO_2 are reintroduced to the scene and can start attacking the ozone.

With respect to the height distribution of ozone in the Arctic two ozone profiles observed from Scandinavia in Jan.–Feb., 1990 are demonstrated in Fig. 3.33; one profile is observed inside and the other one outside the polar vortex, respectively. A clear depletion in the ozone content is observed inside the vortex compared to outside between altitudes of 18 and 26 km. Whether this is an indication of a growing ozone depletion in the Arctic during early springtime is too early to decide on.

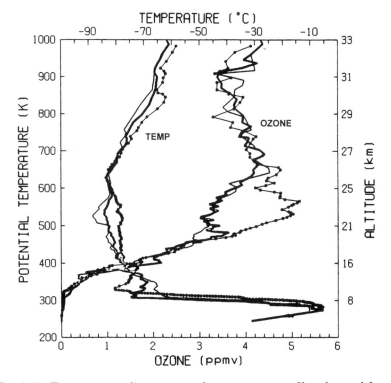

Fig. 3.33. Two corresponding ozone and temperature profiles observed from Scandinavia in Jan.–Feb., 1990 inside (heavy line) and outside (thin line) the polar vortex indicating an ozone depletion between 18 and 26 km. (From Koike et al., 1991.)

3.15 THE ROLE OF THE CFC GASES

The question is now why the ozone depletion effect is observed most strongly in the Antarctic, where the consumption of freon gases is nil and not in the atmosphere above urban areas, if the effect is caused by human activity. The transport processes in the atmosphere are of course important, so wherever the CFC gases are deployed in the atmosphere, they can be transported up to the stratosphere and far away if their lifetime is long enough, and it is. Eventually they will also reach the Antarctic and other extreme areas. In the Antarctic atmosphere the situation is quite special because the temperature in the stratosphere can be as low as -90 °C or 183 K. This is about 10 °C lower than is usually observed in the Arctic. In such extremely low temperatures ice crystals can be formed which can create so-called polar stratospheric clouds (PSCs), and within these clouds the following reactions can occur:

$$HCl + ClONO_2 \longrightarrow HNO_3 + Cl_2 \tag{3.172}$$

$$H_2 + ClONO_2 \longrightarrow HNO_3 + HOCl_2 \tag{3.173}$$

These processes will lead to an accumulation of Cl bindings during the dark period. When the Sun comes back in the spring, the Cl_2 bounds and molecules are broken up by solar radiation, for example as follows:

$$Cl_2 + h\nu \longrightarrow Cl + Cl \tag{3.174}$$

Then the chlorine can start to react with and destroy ozone as already explained. The reason why the Arctic stratosphere is not cold enough to build up a chlorine reservoir during the winter as in the Antarctic, is probably related to the circumpolar vortex. In the Antarctic this is very stable and can block air inside a large volume for a long time so that it cannot heat-exchange with surrounding air at lower latitudes. Therefore the air is cooled down to such extreme temperatures. In the Arctic the circumpolar vortex is broken up because of a larger variety on the Earth's surface, topographical variations between land and sea. The air in the Arctic will therefore not be cooled down to such extremes that the ice crystals can start to form in the same amount as in the Antarctic. A reservoir of Cl atoms will not be built up on such a large scale during the winter, and probably therefore the spring decrease in the total ozone content at northern high latitudes is not observed to accelerate as in the southern counterpart.

It has been argued against the theory of anthropogenic destruction of ozone due to chlorine that little is known about the natural production of chlorine on Earth which enters the atmosphere. It is known that volcanoes create Cl in large amounts (Fig. 3.34), but the lifetime of Cl in the atmosphere is not long enough that it can possibly reach the central ozone layer and reduce it to the extent that has been observed, although it is released by rather high vertical speeds. Statistically it may appear that the number (Fig. 3.35) of active volcanoes has increased during the last century, and therefore an increased effect on the ozone layer due to volcanic activity cannot be discarded. We should note, however, that the statistics shown in

Fig. 3.35 represents the number of reported volcanic eruptions and not necessarily the number of eruptions.

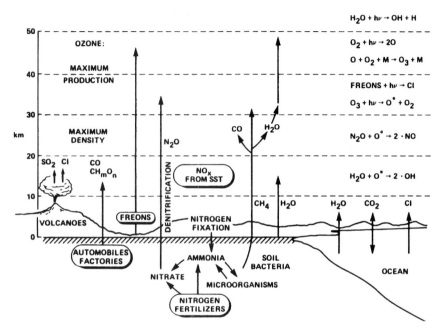

Fig. 3.34. A schematic diagram showing the sources of different gas species entering the atmosphere from the ground. Man-made sources are encircled. The photochemical reactions shown to the right indicate reactions important for production and loss of stratospheric ozone at the indicated heights. (From Shimazaki, 1987.)

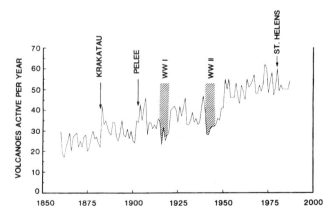

Fig. 3.35. A graph showing the variations in observed numbers of active volcanoes per year from 1860 to 1985. (From McClelland et al., 1989.)

When calculating the so-called annual running mean value of the total ozone content by deriving the mean value of 12 months and then letting it slide forward one month at a time, a very interesting result based on the observations in Oslo has been found. This is demonstrated in Fig. 3.36 where the annual running mean of the total ozone content observed in Oslo between 1978 and 1995 is shown. Two marked abrupt reductions in the total ozone content are seen; one between 1982 and 1984 and the other about 10 years later. In April 1982 the volcano El Chichon in Mexico had a very strong eruption, and in June 1991 the volcano Mt. Pinatubo in the Philippines also erupted. These events were the strongest volcanic eruptions in the period covered, and it is known that aerosol gases from these eruptions were observed in the atmosphere all over the globe for extended period of time. The amount of aerosol released from the Pinatubo eruption is estimated to be 30 Tg (30×10^9 kg) and that from El Chichon to be 12 Tg (12×10^9 kg). This should be compared to the largest volcanic eruption during the last 150 years, Krakatoa, which erupted in 1883 and released of the order of 50 Tg (50×10^9 kg). The onset of the events marked in Fig. 3.36 and in particular the El Chichon event is close to the onset of the first strong decrease of the ozone content seen in Oslo. These results strongly support the hypothesis that volcanic eruptions may have very strong effects on the global ozone layer for extended periods after the eruptions. By this way of presenting the data, however, an overall decline of about 0.26% per year can be seen in Oslo between 1978 and 1995.

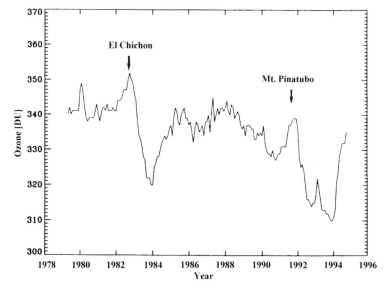

Fig. 3.36. Annual running mean values of the total ozone content as observed from the ground in Oslo between 1978 and 1995. The time of eruptions of the El Chichon and Mt. Pinatubo volcanoes are also indicated by arrows in the figure. (From Henriksen, 1994.)

For the Cl bindings in the freon gases, however, the main concern is that the lifetime in the atmosphere is so long that these gases can be transported over very large areas and accumulate in favourable conditions such as within the Antarctic circumpolar vortex.

The key question is now whether the reduction of the ozone layer above the Antarctic can be followed by an additional cooling of the stratosphere and an even stronger destruction of the ozone content, a positive feedback process, or will the reduction of the ozone content lead to an unstable stratosphere which can reduce the conditions for Cl build-up during the winter season. These problems can only be answered after many years of observations. As there also are indications that the ozone content is influenced by conditions on the Sun, the solar cycle, such effects must first be sorted out before the residues can be blamed on humankind. In order to establish these long-time trends we need good observations through several solar cycles, and these are not available at present. One of the most important tasks the international geophysical community can take on is to secure a systematic and well controlled global network for surveying geophysical parameters such as the ozone content, solar EUV radiation and other important climatic factors.

If, however, the problem really is anthropogenic, it is likely that we do not have the time to wait for the final results concerning the solar and other natural influences on the ozone, but should stop the use of any gases that can possibly be a threat to life on Earth.

3.16 POSSIBLE POLLUTION EFFECTS IN THE MESOSPHERE

It is established as a fact that the beautiful twilight sky phenomena, the noctilucent clouds (NLC), often observed at high latitudes, especially in the Scandinavian region, had not been noticed by humans until about 100 years ago. The first known observation of these has been dated to June 8, 1885, a few years after the great eruption of the Krakatoa volcano in 1883. Since then they have been the subject of much research. They have been observed from 45° to 80° of latitude, while the ideal zone for viewing them is between 53° and 57° of latitude. They are visible only in twilight and are found to be confined to thin layers 1–3 km thick at a typical altitude of 82 km.

The fact that they appear to be a new phenomenon of modern times is strongly suggestive that they are related to human activity and are in fact a result of man-made pollution of the atmosphere. Contrary to the Antarctic ozone hole which has obtained such widespread publicity in spite of the uncertainties related to its actual cause, the NLCs which clearly are a result of anthropogenic activity and may be a clearer manifestation of atmospheric pollution are hardly noticed except within some scholarly circles.

It is well-known that the amount of carbon dioxide (CO_2) released into the atmosphere has increased dramatically through the last 200 years since the Industrial Revolution. In Fig. 3.37 the increase per year in concentration of CO_2 given in particles per million since 1700 is shown. There is close to a 25% increase within less than 200 years of this gas which can have an important influence on the global temperature through a modification of the greenhouse effect mentioned in section 3.2. It has been indicated that the said increase of about 0.5 °C in the global mean temperature through this period can be related to the increase of the CO_2 content although it is far from proven since the temperature measurements 200 years ago did not give a comparable global coverage to what can be achieved today. Even today there are lacunas, especially at sea, in surface temperature measurements, although satellites have improved the global coverage significantly.

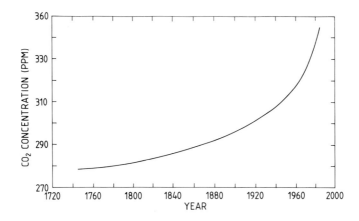

Fig. 3.37. Change of concentration of carbon dioxide in the air since 1720. Concentration is given in particles per million.

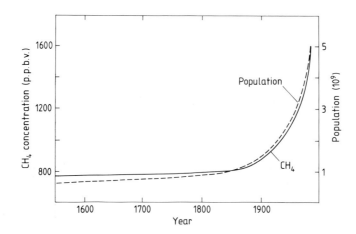

Fig. 3.38. Dashed curve shows the growth in the Earth's population since 1600 AD. Full curve shows the amount of CH_4 (methane) released in the atmosphere through the same period.

In relation to the pollution of the mesosphere there is another species that the human activity has deposited into the atmosphere at an increasing rate for the last 400 years or so, and that is methane (CH_4). Fig. 3.38 shows the total amount of methane in the atmosphere given in units of particles per billion by volume. Here we notice that within the last 400 years the methane concentration in the Earth's atmosphere has increased by 100%. Almost all this increase has taken place within the last 200 years. Methane is released in large amounts from agricultural areas, and the dramatic increase in the CH_4 concentration most certainly is related to the growing demand of agricultural products as the population increases. Also shown in Fig. 3.38 is the total number of people on Earth in the period 1600 to 2000 AD.

The lifetime of CH_4 in the troposphere is estimated to be about 12 years which is about the time it will take for the methane to reach the tropical tropopause. Another 2 to 3 years is probably needed before the CH_4 reaches the mesopause.

There are strong indications in the observation material that the NCL are due to water nucleation on either dust particles or heavy ions in the extremely low temperature of the summer mesopause. If now the methane can reach the mesopause from ground oxidation of CH_4, it will form water. Therefore the methane release from the ground is a source of water vapour in the atmosphere as high up as the mesosphere. We must here remember that the amount of water vapour in the atmosphere is as important to the global climate as the content of CO_2. CH_4 may therefore represent a larger threat to the greenhouse effect than does CO_2.

The mesospheric pollution that we are witnessing when we admire the brilliant noctilucent clouds is a strong warning that human activity is out of control. It reflects a fundamental problem related to the dramatic population growth since the Industrial Revolution. In order to feed all these people farming must intensify and the release of methane in the atmosphere must grow in consequence. It is possible that the globe can take this enormous population (12×10^9 in 2020) but the atmosphere cannot.

3.17 REFERENCES

Aikin, A. C. (1992) *Planet. Space Sci.*, **40**, 7–26.

Anderson, J. G., Brune, W. H. and Proffitt, M. H. (1989) *J. Geophys. Res.*, **94**, 11465–11479.

Dütsch, H. U. (1978) *Pure Appl. Geophys.*, **116**, 511.

Earth–Atmosphere Radiation Budget Analyses Derived from NOAA Satellite Data June 1974 – February 1978, Washington D.C., NOAA–NESS, 1979.

Fahey, D. W., Murphy, D. M., Kelly, K. K., Ko, M. K. W., Proffitt, M. H., Eubank, C. S., Ferry, G. W., Lowenstein, M. and Chan, K. R. (1989) *J. Geophys. Res.* **94**, 16665–16681.

Farman, J. C., Gardiner, B. C. and Shanklin, D. J. (1985) *Nature* **315**, 207–210.

Giraud, A. and Petit, M. (1978) *Ionospheric Techniques and Phenomena*, D. Reidel, Dordrecht, The Netherlands.

Hargreaves, J. K. (1979) *The Upper Atmosphere and Solar–Terrestrial Relations.*

An introduction to the aerospace environment. Van Nostrand Reinhold, London.

Henriksen, T. (1994) *Fra Fysikkens Verden,* 4, 110–114.

Herman, J. R. and Goldberg, R. A. (1978) *Sun. Weather and Climate,* NASA, Washington, D.C.

Hofmann, D. J., Harder, J. W., Rosen, J. M., Hereford, J. V. and Carpenter, J. R. (1989) *J. Geophys. Res.,* 94, 16527–16536.

Koike, M., Kondo, Y., Hayashi, M., Iwasaka, Y., Newman, P. A., Helten, M. and Aimedieu, P. (1991) *Geophys. Res. Lett.,* 18, 791–794.

London, J. (1985) in *Ozone in the Free Atmosphere,* Chapter 1, Van Nostrand Reinhold, London.

McClelland, L., Simkin, T., Summers, T., Nielsen, E. and Stein, T. C. (1989) *Global Volcanism 1975–1985,* AGU, Washington, D.C. and Prentice Hall, New Jersey, p. 30.

Neuendorffer, A. (1994) Private communication.

Paetzold, H. K. and Zschörner, H. (1960) in *Space Research,* Kallmann, H. (Ed.), pp. 24–36, Bijl, North-Holland, Amsterdam.

Pittoc, A. B. (1965) *Nature* 207, 182.

Ratcliffe, J. A. (1972) *An Introduction to the Ionosphere and Magnetosphere,* Cambridge.

Rees, M. H., Emery, B. A. and Roble, R. G. (1983) *J. Geophys. Res.,* 88, 6289–6300.

Roble, R. G. (1987) in *The Solar Wind and the Earth,* Akasofu, S.-I. and Kamide, Y. (Eds.), Terra Sci. Publ. Comp., Tokyo, Japan.

Roble, R. G. and Emery, B. A. (1983) *Planet. Space Sci.,* 31, 597–614.

Shimazaki, T. (1987) in *The Solar Wind and the Earth,* Akasofu, S.-I. and Kamide, Y. (Eds.), Terra Sci. Publ. Comp., Tokyo, Japan.

Tohmatsu, T. (1990) *Compendium of Aeronomy,* Terra Sci. Publ. Comp., Tokyo.

U.S. Standard Atmosphere (1976) NOAA, NASA, USAF, Washington.

Walterscheid, R. L. (1989) *J. Spacecraft and Rockets,* 26, 439–444.

3.18 EXERCISES

1. Let the normalized velocity distribution in a gas in thermal equilibrium be given by:

$$f(v) = 4\pi \left(\frac{m}{2\pi \kappa T}\right)^{3/2} v^2 \exp\left(-\frac{mv^2}{2\kappa T}\right)$$

    and derive the results in (3.29), (3.30) and 3.33).

2. Suppose that a part (α_{ar}) of the solar radiation is reflected from the atmosphere and a part (α_{aa}) of it is absorbed by the atmosphere, when we assume radiative equilibrium, derive an expression for the Earth's radiative temperature. The atmosphere is assumed to be a thin shell with negligible height compared to the Earth's radius. The Earth's albedo is α_r.

3. Assume that the Earth's atmosphere in Fig. 3.5 has a reflection coefficient α_{ar} and an absorption coefficient α_{aa}.

   (a) Assume radiative equilibrium and express the energy balance equations for the atmosphere and the Earth.

   (b) Find an expression for the radiative temperature of the Earth.

4

The Earth's magnetic field

4.1 AN HISTORICAL INTRODUCTION

It has been known for hundreds of years that the Earth possesses a magnetic field. The modern science of geomagnetism, however, can probably be dated back to the English physicist William Gilbert who in the year 1600 published his book "De Magnete" in which he stated that "Magnus magnetis ipse est globus terrestris" (The Earth itself is a great magnet). Fig. 4.1 is an illustration from Gilbert's book demonstrating how small magnets placed on the Earth's surface will orient themselves towards the poles. Not much happened in the field of geomagnetism from when his book was printed and until about 100 years later. Then the English scientist Halley compared observations of the magnetic field in London as made by Gilbert, a few others and himself through a period of 100 years or more. He found that the direction of the magnetic needle had changed from 10° east in Gilbert's time to almost 10° west around 1700 (Fig. 4.2). Halley immediately came up with an explanation for this and maintained that the Earth itself consists of two separate magnetic systems, one connected to a solid sphere in the internal regions of the Earth, and another one connected to a concentric spherical shell or the Earth's crest. If these two were rotating with different rotation speeds, a drift in the magnetic system could according to Halley be expected.

Halley and many others were fairly convinced that by a more accurate mapping of the geomagnetic field, especially at sea. it should be possible to obtain a remedy by which one could safely sail irrespective of cloudiness, in contrast to navigation by help of the Sun and the stars. He then convinced the British government to equip several naval expeditions to observe the magnetic elements as often and as accurately as possible. He also encouraged captains in the merchant navy to do the same. From these observations Halley was able to construct the first map showing the deviation of the magnetic needle from true north at many places on the globe and drew the first so-called isodeclination charts (Fig. 4.3). From then on the study of the geomagnetic field and field mapping was an important issue in the scientific and marine communities. It was not until 1838 that the magnetic

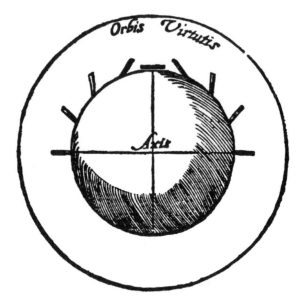

Fig. 4.1. William Gilbert's drawing based on his "terella" experiments with small magnets placed on a magnetized sphere to demonstrate that the Earth is in itself a magnet (Gilbert, 1600).

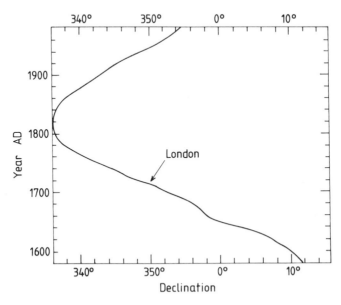

Fig. 4.2. A graph showing the variation in the magnetic declination angle in London from 1600 to 1980. (After Merrill and McElhinny, 1983.)

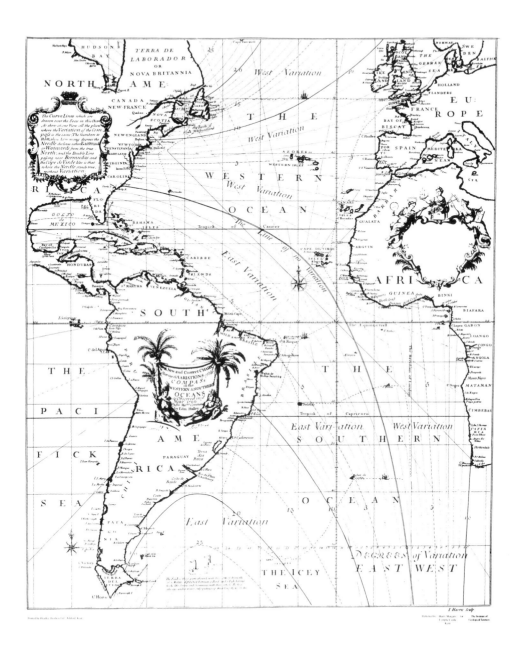

Fig. 4.3. Sir Edmond Halley's first magnetic chart from 1702. (Courtesy The Institute of Geological Sciences, Edinburgh.)

field observations were organized in such a manner that they could be continuous and referred to accurate absolute standards. Thanks to C.F. Gauss who urged different observatories to cooperate, the geomagnetic field has been measured with great accuracy and continuity ever since. Today about 200 observatories all over the globe are monitoring the Earth's magnetic field.

4.2 DESCRIPTION OF THE EARTH'S MAGNETIC FIELD

Within a few Earth radii the magnetic field of the Earth is similar to the field one would find if the Earth itself was a magnetized sphere (Fig. 4.4) or if a gigantic rod

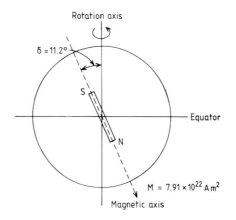

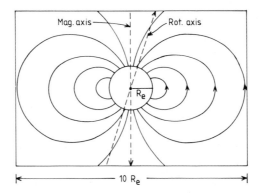

Fig. 4.4. A sketch showing (below) the magnetic field line of the Earth if it was a perfect dipole magnet in vacuum. The rotation axis is tilted with respect to the magnetic dipole axis. In order to represent the Earth's magnetic field in the best possible way by a simple dipole, it had to have a magnetic moment $M = 7.91 \times 10^{22}$ A m^2 to be situated close to the centre of the Earth and tilted by 11.2° with respect to the Earth's rotation axis (above).

Sec. 4.2]   Description of the Earth's magnetic field   131

magnet was situated somewhere close to the centre of the sphere. This imaginary rod magnet, in order to represent the real field as accurate as possible, had to be situated about 400 km from the centre of the Earth with a magnetic moment (strength) of 7.91×10^{22} A m^2. The magnetic axis would make an angle of 11.2° with the Earth axis, and the extensions of this axis would penetrate the Earth at two points; the geomagnetic poles (78.8° N, 289.1° E and 78.8° S, 109.1° E) as of 1980. The first point is at the northwestern tip of Greenland and the other is in the Antarctic. Note that these two points are different from the magnetic poles, or the points where the magnetic needle will point vertically. The north magnetic pole is situated at 75.3° N and 101.8° W. In contrast to the magnetic poles that have a practical significance, the geomagnetic poles are imaginary and serve as a model to simplify the description of the Earth's magnetic field. With such a simple model, however, the field is correct to within 30% of the real field up to a distance of 4 R_e. Closer to the Earth the error is within 10%. As a curiosity one should remember that due to our convention of magnetic poles, it is the north magnetic pole of the magnetic needle that points toward the north, therefore the magnetic north pole is in fact a south magnetic pole.

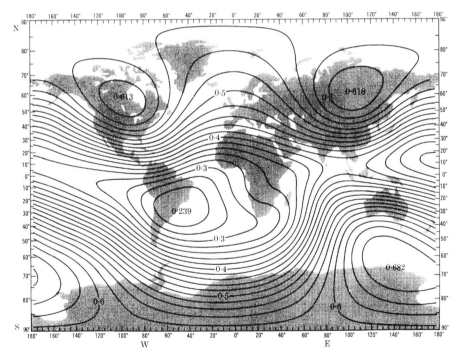

Fig. 4.5. A chart showing the isointensity lines of the global magnetic field. The magnetic poles are represented by two antipodal maximum areas. Another maximum is seen in Siberia and a distinct minimum in the Atlantic Ocean close to South America (the Atlantic anomaly). (From Cain and Cain, 1968.)

Fig. 4.5 shows a global map of the total magnetic field intensity of the Earth illustrated with the so-called isointensity lines. We notice the magnetic north and south poles west of Hudson Bay and at the northern coast of the Antarctic, respectively. Furthermore, there is an area of very strong magnetic field in Siberia. For many years this was thought to be one of the poles of another pair. The discussion of whether the globe had two or four poles was very vivid up towards the middle of the last century. Finally there is an area of unusually low magnetic field intensity just off the Atlantic coast of South America, the so-called Atlantic anomaly.

It is often very convenient to use the geomagnetic field system as an alternative reference system since many problems dealt with in geophysics are influenced or controlled by the magnetic field of the Earth. One often-used elementary geomagnetic coordinate system is spherical with its origin at the Earth's centre and with its axis along the geomagnetic axis just described. We then designate the north geomagnetic or dipole pole with the geographic coordinates $\lambda_p = 78.8°$ N and $\varphi_p = 289.1°$ E. The geomagnetic pole and the meridian plane through the geographic and geomagnetic poles define the geomagnetic coordinate system (see Fig. 4.6). The geographic coordinates (λ = latitude and φ = longitude) for a place on Earth are related to the geomagnetic coordinates (λ_m, φ_m) by the following formulas, obtained from spherical triangles, as shown in Fig. 4.6:

$$\sin \lambda_m = \sin \lambda_p \cdot \sin \lambda + \cos \lambda_p \cos \lambda \cos(\varphi_p - \varphi) \tag{4.1}$$

$$\sin \varphi_m = \frac{\cos \lambda}{\cos \lambda_m} \cdot \sin(\varphi - \varphi_p) \tag{4.2}$$

φ_m is measured positive eastward from the dipole meridian through the north geomagnetic pole. The formulas of course do not apply when $\lambda_m = \pm 90°$. For a place like Tromsø with geographical coordinates (69.66° N, 18.94° E), the geomagnetic coordinates will be (66.97°, 117.29°). The magnetic dipole declination ψ (see Fig. 4.6) is the angle between the directions to the two poles as seen from the observing station. It is measured positive eastward from geographic north and can be found by help of the spherical triangle in Fig. 4.6:

$$\sin \psi = -\frac{\cos \lambda_p}{\cos \lambda_m} \cdot \sin(\varphi - \varphi_p) \tag{4.3}$$

At Tromsø $\psi = -25.06°$ which is a westward declination.

The concept of geomagnetic time is also sometimes useful in auroral studies. It is defined by the angle between the magnetic meridians that pass through the observing site and through the Sun. For example, geomagnetic noon is the time when the Sun is on the geomagnetic meridian of the station, and geomagnetic midnight occurs when the Sun is on the opposite meridian. The corresponding geomagnetic times at an observing site are the times when the Sun reaches an azimuth $180° + \psi$ or ψ respectively. Geomagnetic midnight in Tromsø is therefore 2400 hrs − 25.06/15 hrs = 2220 hrs or 2220 Local Time.

Sec. 4.2]   **Description of the Earth's magnetic field**   133

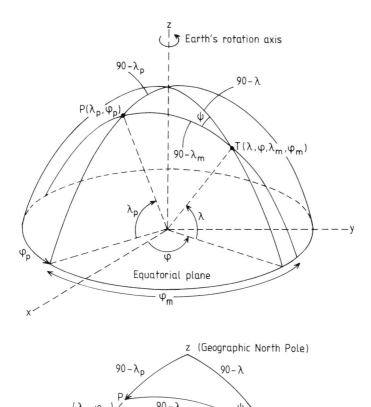

Fig. 4.6. Geometrical sketches to illustrate the spherical coordinates used in the text to derive dipole latitude (λ_m), longitude (φ_m) and time (ψ).

Several geomagnetic coordinate systems, often called corrected geomagnetic coordinate systems, exist in the literature. They differ from this simple system just described by application of a geomagnetic model that is a better representative to the true magnetic field of the Earth. Since the magnetic system is always changing, these corrected geomagnetic systems are regularly updated (see Table 4.1).

Observations of the geomagnetic field through decades indicate that the electric current flowing between the Earth's surface and space, although of considerable interest themselves, are of minor importance with respect to the origin of the geomagnetic field. Outside the Earth we can in this regard assume that the current is negligible ($\mathbf{j} = 0$) and therefore:

$$\nabla \times \mathbf{B} = 0 \qquad (4.4)$$

outside the Earth. There must then exist a magnetic potential such that:

$$\mathbf{B} = -\nabla V_M \tag{4.5}$$

Since $\nabla \cdot \mathbf{B} = 0$ everywhere and especially at the Earth's surface, we find:

$$\nabla^2 V_M = 0 \tag{4.6}$$

Solving then for V_M the magnetic field could be derived at any point in space by the following formulas in spherical polar coordinates (Fig. 4.7):

$$\begin{align}
B_\theta &= -\frac{1}{r}\frac{\partial V_M}{\partial \theta} & \text{(a)} \\
B_\varphi &= -\frac{1}{r\sin\theta}\frac{\partial V_M}{\partial \varphi} & \text{(b)} \\
B_r &= -\frac{\partial V_M}{\partial r} & \text{(c)}
\end{align} \tag{4.7}$$

The derived values of $(B_\theta, B_\varphi, B_r)$ could then be compared to the measured components of $\mathbf{B}$ in any point in space and especially to the measured elements (X, Y, Z) or (H, D, Z) (Fig. 4.8) of the magnetic field on ground. The degree of correspondence between these values could then be used to assess the value of the derived model V_M. To solve for V_M is, however, no straightforward process since the real B-field cannot be described by any analytical function.

When discussing the elements of $\mathbf{B}$, (X, Y, Z) or (H, D, Z) at a point of the Earth, it is customary to let the z-axis point downwards towards the centre of the Earth (Fig. 4.8) and the x- and y-axis towards geographic north and east, respectively. In the northern hemisphere $\mathbf{B}$ points downward slanted towards north. The (X, Y, Z) elements of $\mathbf{B}$ are then defined positive in the north, east and downward directions respectively. The angle D is the declination or the deviation with which the magnetic field is pointing with respect to true north. It is defined positive eastward. The angle I is called inclination and describes the angle the magnetic field makes with the horizontal plane, measured positive downward from this plane. H is the component of the B-field in the horizontal plane.

We notice that when (X, Y, Z) are measured, H, D and I can be derived:

$$\begin{align}
H &= \sqrt{X^2 + Y^2} & \text{(a)} \\
D &= \arctan(X/Y) & \text{(b)} \\
I &= \arctan(Z/H) & \text{(c)}
\end{align} \tag{4.8}$$

which also give the relationship between the two sets of elements depending on which is measured.

Sec. 4.2]   Description of the Earth's magnetic field   135

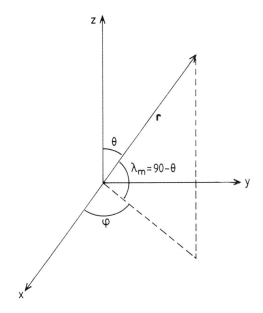

Fig. 4.7. The polar coordinate system used in the text.

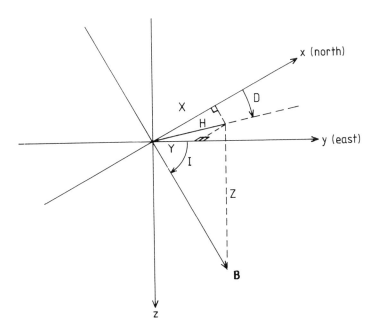

Fig. 4.8. The magnetic elements used to describe the Earth's magnetic field (X, Y, Z), (H, D, Z) or (H, D, I).

Historically the Earth's magnetic field has been measured by the effect the Earth's magnetic field has on small magnets suspended in torsion threads of quartz or balanced on tiny knife-edges. A schematic illustration of the principle for a magnetometer recording on photographic paper is shown in Fig. 4.9. Today, however, these magnetometers are being replaced by computerized electronic devices called fluxgate magnetometers. At present about 200 magnetic observatories are in operation, but many of them are under constant threat due to the expansion of the urban areas all over the globe. Sensitive as the magnetometers are, they are exposed to disturbances from many kinds.

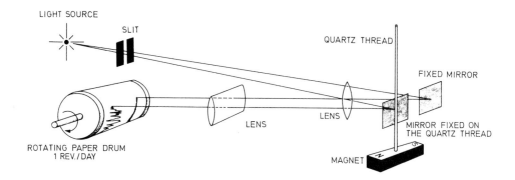

Fig. 4.9. A schematic diagram showing the principles for magnetic field recording by a suspended magnet and a rotating photographic recording paper. The classical variometer set-up.

4.3 MATHEMATICAL REPRESENTATION OF THE EARTH'S MAGNETIC SYSTEM

Since the magnetic field of the Earth is not derivable from a simple analytical function, one has to take the opposite approach to find V_M from a set of observations of **B** and make a fitting analysis of these measurements to the derivatives of a model potential. The customary way to handle this problem is to expand the magnetic potential in a series of spherical harmonics. Such a solution expressed in the polar coordinates (r, θ, φ) (Fig. 4.7) are given below:

$$V_M = \frac{4\pi}{\mu_0} R_e \sum_{n=0}^{\infty} \left(\frac{R_e}{r}\right)^{n+1} \sum_{m=0}^{n} P_n^m(\cos\theta) \cdot \qquad (4.9)$$
$$[g_n^m \cdot \cos(m\varphi) + h_n^m \cdot \sin(m\varphi)]$$

where R_e is the radius of the Earth, θ and φ are the geographic colatitude and east longitude, respectively. $P_n^m(\cos\theta)$ are normalized associated Legendre functions. g_n^m and h_n^m are the so-called Gaussian coefficients which one wishes to derive in

Sec. 4.3] **Mathematical representation of the magnetic system** 137

order to find V_M. μ_0 is the permeability in vacuum. V_M here refers to the internal sources. There is also a term related to external sources in the full expression for V_M, but as already mentioned, these are negligible and contribute less than 1% to the total field at the Earth's surface and are therefore neglected in the following.

Since the potential is not a measured quantity but the components of **B** at different places on the Earth's surface, a spherical harmonic analysis can be performed on a number of observed sets of the elements (components of **B**).

When such a large set of the elements (X, Y, Z) are collected from many stations all over the globe, these can be fitted by a method of least squares, and the coefficients g_n^m and h_n^m can be derived at the Earth's surface $(r = R_e)$. For practical reasons, however, the summation is truncated at a reasonable m. Table 4.1 lists the 8 first coefficients derived by this method for the International Geomagnetic Reference Field 1990 (IGRF 1990) together with some earlier models.

Table 4.1. Spherical harmonic expansion coefficients in nanoteslas (nT)

Model	g_1^0	g_1^1	h_1^1	g_2^0	g_2^1	g_2^2	h_2^1	h_2^2
IGRF* 90	−29775	−1851	5411	−2136	3058	1693	−2278	−380
DGRF** 85	−29873	−1905	5500	−2072	3044	1687	−2197	−306
DGRF 80	−29992	−1956	5604	−1997	3027	1663	−2129	−200
DGRF 75	−30100	−2013	5675	−1902	3010	1632	−2067	−68
DGRF 70	−30220	−2068	5737	−1781	3000	1611	−2047	25
DGRF 65	−30334	−2119	5776	−1662	2997	1594	−2016	114

*IGRF, International Reference Field (to be updated in 1995)
**DGRF, Definite Geomagnetic Reference Field

The potential due to the first term g_1^0 is the main dipole term $(n = 1)$ given by

$$V_1 = \frac{4\pi}{\mu_0} g_1^0 \cos\theta \left(\frac{R_e}{r}\right)^2 R_e \qquad (4.10)$$

g_1^0 is obviously zero since no monomagnetic pole $(n = 0)$ exists. The dipole will then be associated with a magnetic moment given by:

$$M_1 = \frac{4\pi}{\mu_0} g_1^0 R_e^3 \qquad (4.11)$$

which is oriented along and antiparallel to the Earth's rotation axis since g_1^0 is negative (Fig. 4.10). The next two terms g_1^1 and h_1^1 are also dipole terms in the plane of geographic equator. $M_2 = (4\pi/\mu_0)g_1^1 R_e^3$ points towards 180° geographic longitude ($g_1^1 < 0$), and $M_3 = (4\pi/\mu_0)h_1^1 R_e^3$ points towards 90° E longitude (Fig. 4.10). The resultant of these 3 dipoles is a dipole with a magnetic moment of

$$M_0 = \frac{4\pi}{\mu_0} H_0 R_e^3 \qquad (4.12)$$

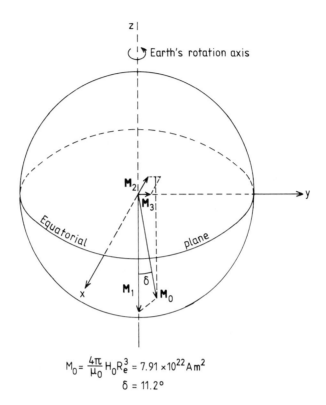

Fig. 4.10. The first 3 dipole terms $\mathbf{M}_1$, $\mathbf{M}_2$ and $\mathbf{M}_3$ in the harmonic series expression of the Earth's magnetic field represented by the coefficients g_1^0, g_1^1 and h_1^1, respectively (see Table 4.1). The resultant magnetic moments $\mathbf{M}$ ($= 7.91 \times 10^{22}$ A m^2) is making an angle δ ($= 11.2°$) with the Earth's rotation axis.

where H_0 is determined by

$$H_0 = \left[(g_1^0)^2 + (g_1^1)^2 + (h_1^1)^2\right]^{1/2} \tag{4.13}$$

Inserting from Table 4.1 (IGRF 90) we find

$$H_0 = 30319 \text{ nanotesla} = 3.0 \times 10^{-5} \text{ tesla} \tag{4.14}$$

and

$$M_0 = 7.91 \times 10^{22} \text{ A m}^2 \tag{4.15}$$

Note that $H_0 R_e^3 = 7.91 \times 10^{15}$ Wb m. We also notice that $\mathbf{M}_0$ makes an angle with the rotation axis of the Earth which is given by:

$$\tan \delta = \left[(g_1^1)^2 + (h_1^1)^2\right]^{1/2} / g_1^0 = 0.2 \tag{4.16}$$

Sec. 4.3] **Mathematical representation of the magnetic system** 139

and
$$\delta = 11.20° \tag{4.17}$$

which is the angle we mentioned at the beginning of this chapter.

With the magnetic potential derived with the proper number of coefficients demanded for the problem of interest, the empirical components of **B** can be obtained in any point of space where the assumptions made apply and **B** is given by (4.5) in any coordinate system chosen for the problem to be solved.

We notice, however, that the higher-order terms except the dipole terms, will be trifling. The magnetic field system can to a good approximation (within 10%) be represented by a dipole with magnetic moment M_0. Furthermore we notice that the angle δ between M_0 and the Earth's rotation axis is very small. We will therefore in the following neglect all other terms than the dipole moment M_0 and assume that this is antiparallel to the Earth's rotation axis. By the choice of our coordinate system we then have

$$\mathbf{M_0} = -M_0 \hat{z} \tag{4.18}$$

The dipole is in itself a solution of the Laplace equation (4.6) and the dipole potential is

$$V_M = -\frac{\mu_0}{4\pi} \mathbf{M_0} \cdot \nabla \left(\frac{1}{r}\right) = \frac{\mu_0}{4\pi} \frac{\mathbf{M_0} \cdot \mathbf{r}}{r^3} = -\frac{\mu_0 M_0}{4\pi} \frac{\cos\theta}{r^2} \tag{4.19}$$

The components of **B** can then be derived by taking the partial derivatives of V_M according to (4.7)

$$B_r = -\frac{1}{r}\frac{\partial V_M}{\partial r} = -\frac{\mu_0 M_0}{2\pi}\frac{\cos\theta}{r^3} = -\frac{\mu_0 M_0}{2\pi}\frac{\sin\lambda_m}{r^3} \quad \text{(a)} \tag{4.20}$$

$$B_\varphi = -\frac{1}{r\sin\theta}\frac{\partial V_M}{\partial \varphi} = 0 \quad \text{(b)}$$

$$B_\theta = -\frac{1}{r}\frac{\partial V_M}{\partial \theta} = -\frac{\mu_0 M_0}{4\pi}\frac{\sin\theta}{r^3} = -\frac{\mu_0 M_0}{4\pi}\frac{\cos\lambda_m}{r^3} \quad \text{(c)}$$

The magnitude of **B** is given by:

$$B(r,\lambda_m) = (B_r^2 + B_\varphi^2 + B_\theta^2)^{1/2} = \frac{\mu_0}{4\pi}\frac{M_0}{r^3}(1 + 3\sin^2\lambda_m)^{1/2} \tag{4.21}$$

Introducing (4.12) we get

$$B(r,\lambda_m) = H_0 \left(\frac{R_e}{r}\right)^3 (1 + 3\sin^2\lambda_m)^{1/2} \tag{4.22}$$

We notice that at the pole where $\lambda_m = 90°$ and $r = R_e$

$$B_p = 2H_0 \tag{4.23}$$

and at the equator where $\lambda_m = 0$

$$B_e = H_0 \tag{4.24}$$

At the Earth's surface we find for a position with magnetic latitude λ_m when we introduce the unit vector $\hat{\boldsymbol{\lambda}}_m$ where

$$\hat{\boldsymbol{\lambda}}_m = -\hat{\boldsymbol{\theta}} \tag{4.25}$$

$$\begin{aligned} \mathbf{B} &= B_r \hat{\mathbf{r}} + B_{\lambda_m} \hat{\boldsymbol{\lambda}}_m \\ &= H_0(-2\sin\lambda_m \hat{\mathbf{r}} + \cos\lambda_m \hat{\boldsymbol{\lambda}}_m) \end{aligned} \tag{4.26}$$

At Tromsø where we have $\lambda_m = 66.97°$, we find that $B_r = -57522$ nanotesla. The measured field in Tromsø, however, is -52800 nanotesla, i.e. a difference of 9%.

The inclination angle for the magnetic field at any point of the Earth would be (Fig. 4.11)

$$\tan I = -\frac{B_r}{B_\lambda} = 2\tan\lambda_m \tag{4.27}$$

In particular for Tromsø I becomes

$$I_T = 78.0° \tag{4.28}$$

while the measured angle is 77.8° at present.

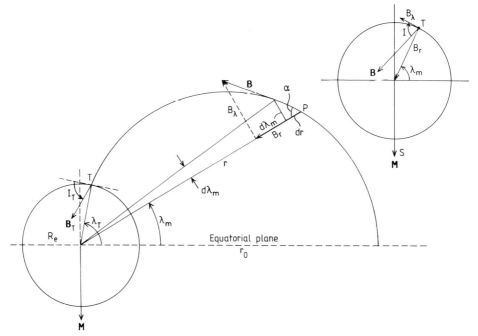

Fig. 4.11. An illustration to show the geometry of the magnetic field line to assist in deriving the geometric formula for B.

Sec. 4.3] Mathematical representation of the magnetic system

We notice from Fig. 4.11 that since **B** is parallel to the field line drawn at any point, $\tan\alpha$, where α is the angle indicated, can be expressed in two ways

$$\tan\alpha = \frac{r \cdot d\lambda_m}{dr} \tag{4.29}$$

$$\tan\alpha = \frac{B_\lambda}{B_r} = -\frac{\cos\lambda_m}{2\sin\lambda_m} \tag{4.30}$$

where $d\lambda_m$ and dr are increments in λ_m and r, respectively. There is therefore a simple relationship between dr and $d\lambda_m$ given by:

$$\frac{dr}{r} = -2\frac{\sin\lambda_m}{\cos\lambda_m} d\lambda_m \tag{4.31}$$

and the solution is

$$\ln r = 2\ln\cos\lambda_m + C \tag{4.32}$$

If we now designate $r = r_0$ where the given field line hits the equatorial plane, i.e. where $\lambda_m = 0$, then:

$$C = \ln r_0 \tag{4.33}$$

and

$$r = r_0 \cos^2\lambda_m \tag{4.34}$$

For a place on Earth where $r = R_e$ we find that the magnetic field line through that place reaches the equatorial plane at a distance

$$r_0 = \frac{R_e}{\cos^2\lambda_m} \tag{4.35}$$

In particular for Tromsø we have

$$r_0 = 6.53 \cdot R_e \tag{4.36}$$

Frequently the ratio r_0/R_e is termed L and

$$L = \cos^{-2}\lambda_m \tag{4.37}$$

and the magnetic latitude or invariant latitude λ_m is given by:

$$\lambda_m = \arccos\sqrt{\frac{1}{L}} \tag{4.38}$$

4.4 E-FIELD MAPPING ALONG CONDUCTING MAGNETIC FIELD LINES

Let us now assume that the magnetic field lines are highly conducting. An electric potential that might have formed perpendicular to the magnetic field lines in the ionosphere due to polarization effects or induction by neutral winds blowing charges across, will then be propagated to the equatorial plane. Similar electric potentials formed between field lines in the equatorial plane or anywhere along the lines, will propagate downward to the ionosphere.

Let us therefore consider two field lines (Fig. 4.12) separated in azimuth by a distance $l_{i\varphi}$ along a constant geomagnetic latitude in the ionosphere. The distance between the same field lines in the equatorial plane will be denoted $L_{m\varphi}$. When neglecting the height of the ionosphere, we find:

$$l_{i\varphi} = R_e \cdot \cos\lambda_m \cdot d\varphi \tag{4.39}$$

where $d\varphi$ is the azimuthal angle between these two field lines. In the magnetic equatorial plane we find

$$L_{m\varphi} = r_0 \cdot d\varphi = \frac{R_e}{\cos^2\lambda_m} d\varphi \tag{4.40}$$

When conserving the electric potential between the equatorial plane and the ionosphere, we therefore have

$$V_\varphi = E_{i\varphi} l_{i\varphi} = E_{m\varphi} L_{m\varphi} \tag{4.41}$$

where $E_{i\varphi}$ and $E_{m\varphi}$ are the azimuthal components of the electric field in the ionosphere and the equatorial plane, respectively. The mapping ratio for the electric field in the azimuthal direction is then given by:

$$\frac{E_{i\varphi}}{E_{m\varphi}} = \frac{L_{m\varphi}}{L_{i\varphi}} = \cos^{-3}\lambda_m = L^{3/2} \tag{4.42}$$

The electric field in the ionosphere is therefore enhanced by a factor $L^{3/2}$ with respect to the electric field in the equatorial plane, and the enhancement factor increases for higher L values or higher geomagnetic dipole latitudes. For Tromsø, where $L = 6.53$, the enhancement factor is 16.7.

For two magnetic field lines separated by an angle $d\lambda_m$ in the meridional plane the distance between these lines in the ionosphere is $l_{i\lambda}$, and the corresponding distance in the equatorial plane is $L_{m\lambda}$ (see Fig. 4.12). Since $r_0 = R_e \cos^{-2}\lambda_m$ according to (4.36), we find that

$$L_{m\lambda} = dr_0 = -\frac{2R_e}{\cos^3\lambda_m}(-\sin\lambda_m)d\lambda_m \tag{4.43}$$

or

$$L_{m\lambda} = 2R_e \frac{\sin\lambda_m}{\cos^3\lambda_m} d\lambda_m \tag{4.44}$$

Sec. 4.4]   **E-field mapping along magnetic field lines**   143

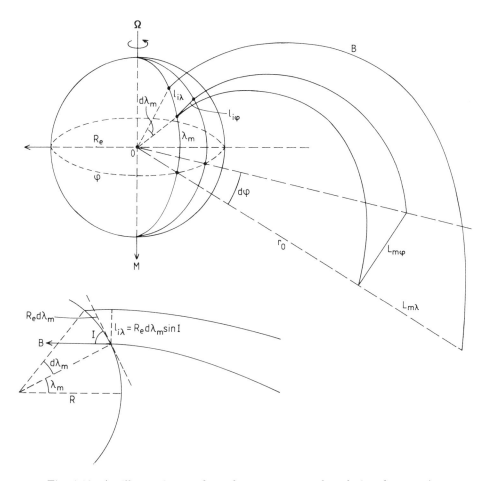

Fig. 4.12. An illustration to show the geometry used to derive the mapping ratios between the azimuthal distances $l_{i\varphi}$ and $L_{m\varphi}$ and the latitudinal distances $l_{i\lambda}$ and $L_{m\lambda}$ for neighbouring field lines in the ionosphere and equatorial plane respectively.

The corresponding distance in the ionosphere must be the line marked $l_{i\lambda}$ in Fig. 4.12 (bottom part) which is perpendicular to the field line leaving the Earth at a geomagnetic latitude λ_m. From the figure we see that

$$l_{i\lambda} = R_e \, d\lambda_m \cdot \sin I \tag{4.45}$$

where I is the inclination angle of the B-field. For the corresponding electric field components in the meridional direction $E_{i\lambda}$ and $E_{m\lambda}$ in the ionosphere and the equatorial plane, respectively, the conservation of the electric potential between the two field lines gives:

$$V_\lambda = E_{i\lambda} \cdot l_{i\lambda} = E_{m\lambda} \cdot L_{m\lambda} \tag{4.46}$$

The mapping ratio is therefore

$$\frac{E_{i\lambda}}{E_{m\lambda}} = \frac{L_{m\lambda}}{l_{i\lambda}} = \frac{1}{\cos^2 \lambda_m} \cdot \frac{1}{\cos I} \qquad (4.47)$$

From (4.27) we have

$$\tan I = \sqrt{\frac{1 - \cos^2 I}{\cos^2 I}} = 2 \tan \lambda_m \qquad (4.48)$$

Solving for $1/(\cos I)$ we get:

$$\frac{1}{\cos I} = \sqrt{4 \tan^2 \lambda_m + 1} \qquad (4.49)$$

and inserting in (4.47) we have:

$$\frac{E_{i\lambda}}{E_{m\lambda}} = \frac{1}{\cos^2 \lambda_m} \sqrt{4 \tan^2 \lambda_m + 1} = \frac{2}{\cos^3 \lambda_m} \sqrt{1 - \frac{3}{4} \cos^2 \lambda_m} \qquad (4.50)$$

By finally introducing the L value from (4.37) we obtain:

$$\frac{E_{i\lambda}}{E_{m\lambda}} = 2 \cdot L \cdot \sqrt{L - \frac{3}{4}} \qquad (4.51)$$

And for Tromsø, where $L = 6.53$, we derive the mapping ratio 31.4. A radially electric field of 1 mV/m in the equatorial plane would therefore be enhanced to 31.4 mV/m when mapped to the ionosphere above Tromsø assuming the magnetic field lines are superconducting.

Also notice that since E_λ and E_φ do not map identically, the field will be rotated when transplanted from the equatorial plane to the ionosphere or vice versa (Fig. 4.13). Let the angle between the E-field and the meridian plane in the equatorial plane and in the ionosphere be δ_m and δ_i, respectively, then:

$$\tan \delta_m = \frac{E_{m\varphi}}{E_{m\lambda}} \qquad (4.52)$$

$$\tan \delta_i = \frac{E_{i\varphi}}{E_{i\lambda}} \qquad (4.53)$$

The relationship between $\tan \delta_i$ and $\tan \delta_m$ is then given by:

$$\tan \delta_i = \tan \delta_m \cdot \sqrt{\frac{L}{4L - 3}} \qquad (4.54)$$

For Tromsø where $L = 6.53$, we find that $\tan \delta_i = \tan \delta_m \cdot 0.53$. If $\delta_m = 45°$, then $\delta_i = 28°$. The electric field will be more meridional in the ionosphere. Notice also that while the azimuthal component maintains its direction between the equatorial plane and the ionosphere, the meridional component flips over from being outward in the equatorial plane to becoming poleward in the ionosphere (Fig. 4.13).

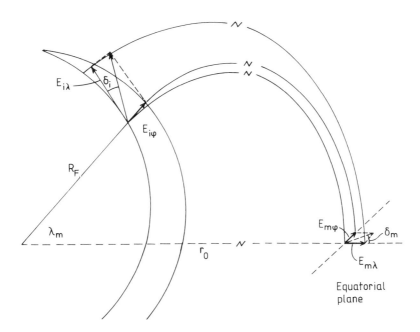

Fig. 4.13. A diagram illustrating the rotation of an ionospheric E-field due to non-uniform mapping along field lines when projected out to the equatorial plane.

4.5 SECULAR VARIATIONS IN THE EARTH'S MAGNETIC FIELD

As mentioned earlier the magnetic field system of the Earth is always changing and the series of the spherical harmonics have to be updated from year to year (Table 4.1). For the last 150 years or so we have reasonable good data which enable us to study the variation in the Earth's dipole moment through this time. Fig. 4.14 illustrates this variation where the dipole moment, M_0, derived between 1820 and 1980, is given in units of A m^2. We notice that there has been a decreasing trend in M_0 and that the decrease in fact is stronger now than it has been throughout this period. If this tendency continues, the magnetic field will disappear within about 2000 years.

As already pointed out by Halley, the magnetic field system is drifting. The speed of the drift is not constant but decreases by latitude as shown in Fig. 4.15, where the average westward drift per year is plotted versus latitude for some stations in the northern hemisphere. In Canada and Norway the drift is more than 0.5 degrees per year. From these analyses it is also possible to deduce where the geomagnetic pole would be at any time. In Fig. 4.16 the geographic longitude and latitude of the geomagnetic north pole is shown for the period between 1600 AD and 1900. While it was at almost 83° N and 40° W at Gilbert's time, it is now closer to 79° N and 70° W.

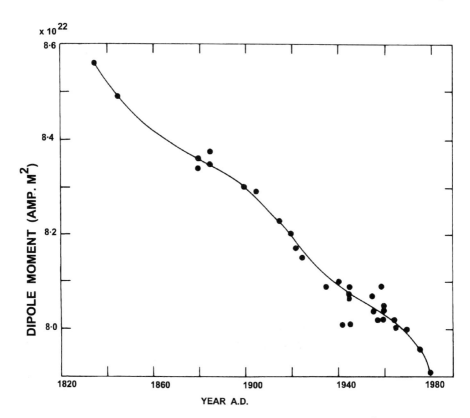

Fig. 4.14. The variation in the Earth's dipole moment (in A m^2) as function of time between 1840 and 1980 showing a gradual decline in strength. (From Merrill and McElhinny, 1983.)

It is in fact possible to study the variations in the Earth's magnetic field further back than 400 years by so-called archaeomagnetic studies. This research is based on magnetic material in pottery, burned clay, sediments and lava. Without going into detail we reproduce a figure (Fig. 4.17) showing the drift of the geomagnetic pole in the northern hemisphere as far back as 700 AD. At the present time the drift is about 2 arc minutes per year westwards.

This drift of the magnetic system has a special importance of the auroral morphology. As is well known from years of ground-based observations and especially revealed by satellite imaging from the DE–1 and Viking satellites, the aurora tends to appear in an oval-shaped pattern with its centre close to the geomagnetic pole (Fig. 4.18). Since this pole has changed position over the years, it is also likely that the auroral oval must have been situated differently in periods compared to the present average situation. This is illustrated in Fig. 4.19 where the auroral oval is drawn for 4 different periods in historic time. It is especially noted that at about 1200 AD the oval was well north of Europe. The shift of the oval towards south over northern Europe is also evident in the figure.

Sec. 4.5] Secular variations in the Earth's magnetic field 147

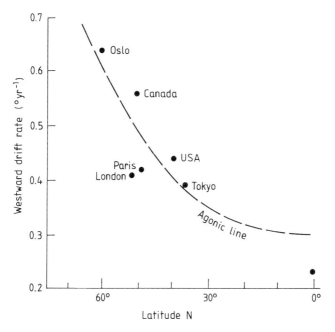

Fig. 4.15. The westward drift of the global magnetic system indicated by the rate of change (°/year) of the declination angle at different latitudes. (After Yukutake, 1967.)

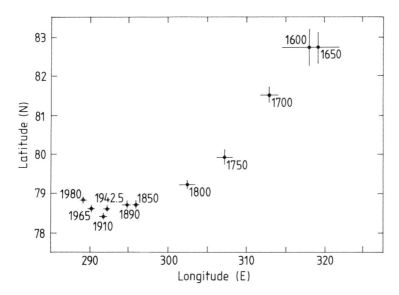

Fig. 4.16. The variation in position of the north geomagnetic pole from 1600 AD to 1980 AD showing a southwestward drift. (After Barraclough, 1974.)

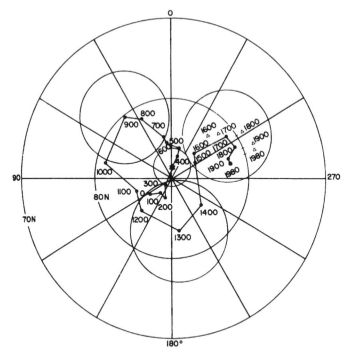

Fig. 4.17. The position of the geomagnetic north pole for the last 1300 years. (From Merrill and McElhinny, 1983.)

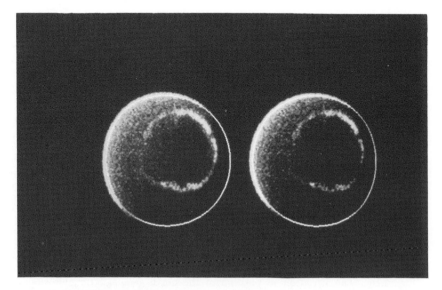

Fig. 4.18. Modern pictures of the auroral oval as derived from imaging techniques onboard DE-1 satellite. (From Frank, 1994.)

Sec. 4.5]   **Secular variations in the Earth's magnetic field**   149

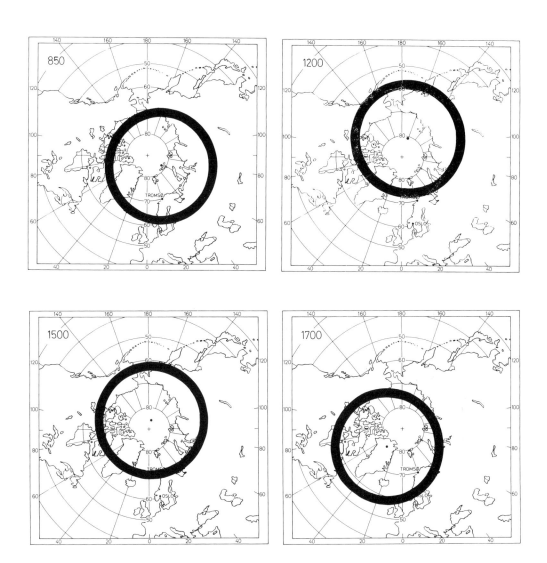

Fig. 4.19. The position of the auroral oval for 4 periods in historic time.

In a chronicle called "The King's Mirror", written in Norway about 1250 AD, the aurora mentioned then for the first time by the name "the northern light", is said to be a phenomenon often observed in Greenland. It is not mentioned that it was seen in Norway. Maybe the reason was that the centre of the oval was far to the east and north of Siberia.

The fact that the magnetic field decreases, will also have an impact on the geographical appearance of the aurora. The weaker the field, the lower in latitude the aurora will appear. For the last 3000 years or so (Fig. 4.20) the magnetic field strength has been decreasing. Therefore, more people, not only living at extreme latitudes, can occasionally enjoy the displays of the aurora. 6000 years ago the magnetic moment of the Earth was weaker than it is today, this should then be a period of aurora occurring at lower latitudes. We have indications from this early time that the aurora was observed in China at about 30° of latitude.

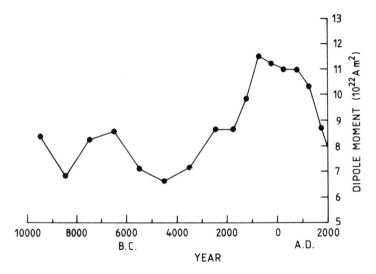

Fig. 4.20. The change of the magnetic dipole moment as deduced from archaeomagnetic data for the last 12 000 years. (After McElhinny and Senanayake, 1982.)

By taking this reduction into account and assuming a continuous drift of the geomagnetic axes, the position of the oval in the future can be predicted. One such prediction for the year 2300 AD is shown in Fig. 4.21. According to this the oval will probably be crossing the southern part of Scandinavia in a few hundred years.

It was indicated earlier that if the decrease in the magnetic moment, as we observe today, is continuing with the same trend, it might disappear within 2000 years or so. The question is, however, whether it disappears or it just changes direction. We can get an idea about this feature of the Earth's magnetic field by conferring archaeomagnetic data, some of which are shown in Fig. 4.22. The results are presented in terms of geomagnetic declination referring to the angle δ

Sec. 4.5]   **Secular variations in the Earth's magnetic field**   151

between the Earth's rotation axis and the magnetic axis. At present and for the last 1 million years the polarity of the magnetic field of the Earth has probably been the same. About 1 million years ago, however, the magnetic moment flipped over by 180° in the course of 100 000 years or less. Similar events appear to have taken place several times in the last 4 million years.

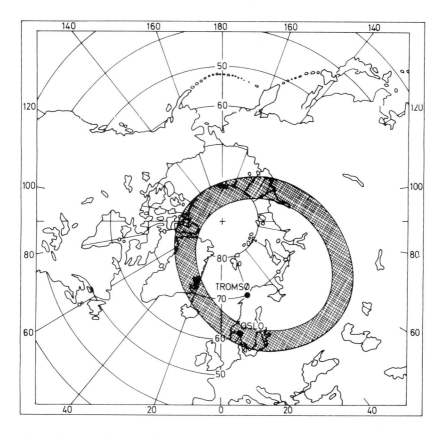

Fig. 4.21. A prediction of the position of the auroral oval in 2300 AD. (After Oguti, 1994.)

The magnetic field of the Earth acts as a shield against energetic particles radiating from the Sun. At present it mainly shields the lower latitudes of the Earth from this radiation, while the particles can penetrate more freely along the magnetic field lines at higher latitudes. By coincidence the high latitudes at present geological time are rather hostile areas, and only a few individuals of living species are able to survive there. Those who do so are probably adapted to any high-energy particle radiation that might hit the Earth. If the magnetic field disappeared for a while, all living species on the Earth would be exposed to a much higher radiation

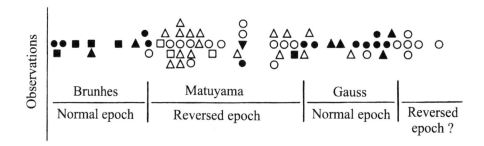
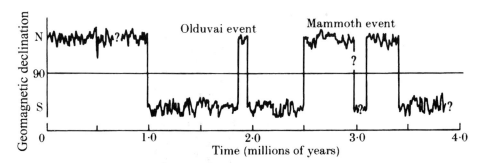

Fig. 4.22. The magnetic polarities as deduced from 64 volcanic rocks taken from ○ North America, ▽ Africa, □ Europe and △ Hawaii. The open and closed symbols represent different polarities. (From Cox et al., 1967.)

source. Since some of these species probably would not be adapted to such high radiation, it could happen that they become extinct. There have been speculations presented where it has been indicated that some of the dinosaurs actually succumbed to extraterrestrial radiation in periods when the Earth's magnetic field underwent a reversal. The time history for such reversals is shown in Fig. 4.22.

4.6 THE SOURCE OF THE MAGNETIC FIELD OF THE EARTH

It is rather useless to speculate about the past and future behaviour of the geomagnetic field unless we have a quite well founded theory for the creation of the field itself to lean on. At present it is probably fair to claim that no existing theory about the source of the magnetic field can explain the dramatic episodes that appear to have taken place in the Earth's magnetic field. It is implicitly assumed in our discussions of the magnetic potential that the source of the field is in the Earth's interior. We notice from the harmonic expansion of the geomagnetic potential that the contribution to the magnetic field from a term of order n will decrease by distance from the centre of the Earth by $r^{-(n+2)}$. The contribution from higher-order terms will therefore vanish very rapidly, for example at or just above the Earth's surface. Furthermore, we notice that the higher harmonic terms

Sec. 4.6] The source of the magnetic field of the Earth 153

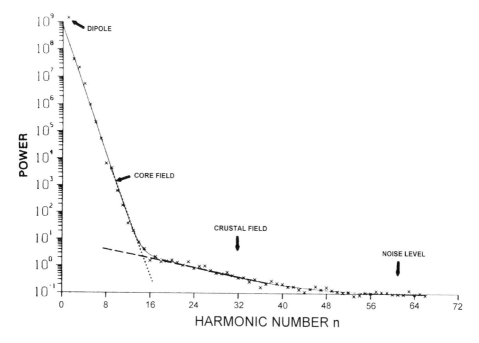

Fig. 4.23. Spectrum of the geomagnetic field as seen by the satellite Magsat at 400 km altitude. The n values are those of the spherical harmonic modes; $n = 1$ is the dipole. The break in the spectrum represents the transition from the core field to that of crustal magnetic anomalies. The scale in (nanotesla)^2 is representing the energy density of the field at 400 km altitude. (From Cain, 1987.)

represent the smallest scale sizes in the field structure. For the term of order n

$$2\pi = n\varphi_n \tag{4.55}$$

$$\varphi_n = \frac{2\pi}{n} \tag{4.56}$$

where φ_n is the angular dimension of the term of order n. Converted into Earth radii:

$$L_n = R_e \varphi_n = \frac{2\pi R_e}{n} \tag{4.57}$$

Now the power of each sinusoidal component of the magnetic field can be measured outside the Earth, for example by a satellite. In Fig. 4.23 the results from a satellite at 400 km are illustrated. The unit along the ordinate is in (nanotesla)^2. Up to $n = 13$ which corresponds to distances of $\approx 0.75\ R_e$, the field terms decrease very rapidly with n. It is believed that these terms have their sources in the core region of the Earth. (The core has a radius of 3485 km.) The terms between $n = 13$ and $n = 50$ have a very different slope from those at lower n, and these terms are

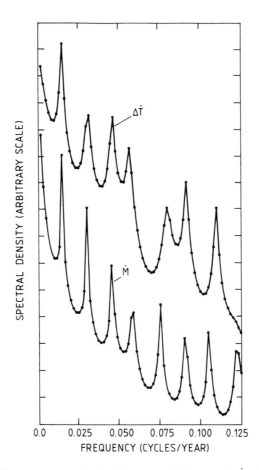

Fig. 4.24. Power spectra of the time rate of change, $\Delta \dot{T}$, for the rotation period of the Earth based on data from 1865 to 1965. This is compared with the power spectrum of the variations by time $\dot{M}$ in the geomagnetic moment for the period between 1901 and 1969. The spectral densities are given in arbitrary scales. (After Jin and Thomas, 1977.)

believed to originate in crustal anomalies. For $n = 50$ the dimensions of these anomalies are ≈ 800 km.

As the magnetic field appears to originate from internal processes of the Earth, it is expected that the rotation of the Earth itself is associated with the creation of magnetic field. A strong indication lending support to this assumption can be found in Fig. 4.24 where the power spectra of the time rate of change $\Delta \dot{T}$ for the period 1865–1965 (T meaning the rotation period of the Earth), and of the variations in the dipole geomagnetic moment ($\dot{M}$) for the period 1901–1969 are shown. There is a striking similarity between these two spectra, as if they have a common course or are causally related to one another.

Sec. 4.6] The source of the magnetic field of the Earth 155

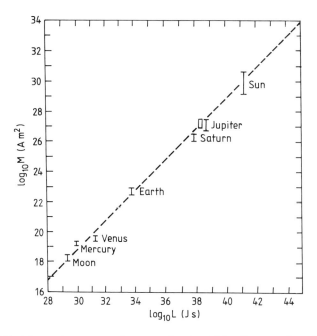

Fig. 4.25. Magnetic moment for several celestial bodies as a function of the corresponding angular momenta.

Table 4.2. The ratio of magnetic moment M in units A m^2 to angular momentum L in units of J s for some celestial bodies

	M/L
The Earth	1.15×10^{-11} C/kg
The Sun	0.067–2.7×10^{-11} C/kg
78 Virginis	0.1–41×10^{-11} C/kg
The Moon	$-.48$–1.3×10^{-11} C/kg
Mercury	1.6–3.3×10^{-11} C/kg
Venus	0.16–0.35×10^{-11} C/kg
Jupiter	0.12–0.57×10^{-11} C/kg
Saturn	0.13–0.35×10^{-11} C/kg
Hercules X–1	0.28–1.9×10^{-11} C/kg

Another observed evidence in support of a connection between the rotation of the globe and its magnetic moment can be Bode's law for the magnetism of celestial bodies. This is illustrated in Fig. 4.25 where the magnetic moments (in A m^2) for several planets in the solar system are shown versus their respective angular moments L (is J s). The logarithms of the two parameters appear to be

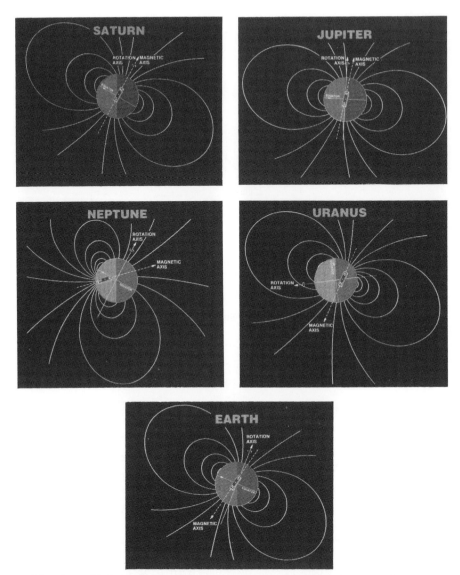

Fig. 4.26. Models of the magnetic systems for the large planets indicating the relationship between the rotation axis and the magnetic axis for each planet. Also illustrated is the corresponding model for the Earth.

linearly related within 15 orders of magnitude. Table 4.2 shows the ratios between the magnetic moment M and the angular moment for the bodies shown in Fig. 4.25.

Fig. 4.26 shows the magnetic field systems for the outermost planets together with the geomagnetic system of the Earth. We notice that for the Earth $\mathbf{M} \cdot \mathbf{L} < 0$, while for the other planets $\mathbf{M} \cdot \mathbf{L} > 0$. The rotation direction and the magnetic

Sec. 4.7] The unipolar inductor 157

moment therefore do not appear to be related in the same sense for all bodies. Furthermore, if the magnetic moment was strictly related to the direction of the angular momentum, the Earth would have had to reverse its rotation about the polar axis on several occasions in prehistoric time. There is no indication that so has happened.

Since the interior of the Earth is extremely hot and part of it is supposed to be a fluid of highly conducting plasma, the Earth's rotation is supposed to create rotation cells in the interior which will give rise to currents and magnetic fields.

4.7 THE UNIPOLAR INDUCTOR

One attempt to visualize the processes that can create the Earth's magnetic field is by the so-called unipolar inductor (Fig. 4.27a). The rotating, highly

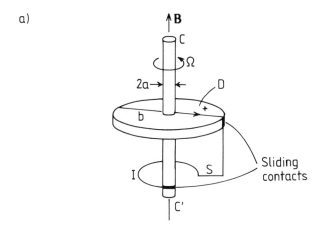

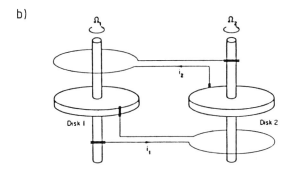

Fig. 4.27. Illustrations of (a) one unipolar inductor rotating with angular velocity Ω in the presence of a magnetic field **B**. (b) A coupled pair of two unipolar inductors rotating with different angular velocities Ω_1 and Ω_2. (From Rikitake, 1958).

conducting plasma is represented by a disc spinning around an external magnetic field, let us say with its origin in the interplanetary space. A current feedback loop is illustrated which, when encircling the rotation axis, can enhance or decrease the external field.

The current will be a function of the rotation speed of the disc and the external magnetic field. Such a model can only give rise to oscillations around a constant background magnetic field but not to a complete turn of its direction by 180°.

We notice that it is again impossible to turn the current around from plus to minus, but very large oscillations around a mean value can occur.

In order to achieve a complete turn of the current, one can link together two or more such inductors with different rotation speeds and feedback current loops (Fig. 4.27b).

It is fair to say, however, that to establish a reliable theory for the geomagnetic field or any magnetic field associated with celestial bodies is among the most difficult problems of cosmic geophysics.

4.8 MOTION OF A CHARGED PARTICLE IN A MAGNETIC FIELD

4.8.1 General introduction of the guiding centre

The equation of motion for a charged particle with mass m and charge q moving in space with a magnetic field **B**, an electric field **E** as well as external non-magnetic force **F** can be expressed as:

$$\frac{d}{dt}\left(m\frac{d\mathbf{r}}{dt}\right) = q\left(\frac{d\mathbf{r}}{dt} \times \mathbf{B} + \mathbf{E}\right) + \mathbf{F} \qquad (4.58)$$

The general solution of this equation represented by the position vector $\mathbf{r}(t)$ of the particle as function of time and, depending upon the initial conditions applied to the problem, is extremely complicated. Within some limitations of the field geometry, the strength of the external forces and the energy of the particles, the solutions can take on a periodic character which can simplify the problem to some extent.

There are 3 periodicities which stand out:

1. *Cyclotron motion:* The particle will gyrate in a plane perpendicular to the magnetic field.

2. *Mirror reflection motion:* The particle will move back and forth along a magnetic field line.

3. *Drift motion:* The particle will move around a closed surface described by a series of magnetic field lines.

The first motion is always the first to occur and it has the highest frequency.

The reflection motion is often several orders of magnitude lower in terms of frequency. Any magnetic field where such motions occur, is called a trapping magnetic field.

Sec. 4.8]   **Motion of a charged particle in a magnetic field**   159

The drift frequency is again several orders of magnitude less than the reflection frequency. A trapping magnetic field where this drift also occurs, is called a stable trapping field.

The cyclotron motion exists if there at any time can be found a reference system in which the motion is circular in a plane perpendicular to the magnetic field. If such a reference system can be found, then the so-called guiding centre approximation is satisfied, where the guiding centre is the centre for the approximate circular velocity. The radius of this circle is called the Larmor radius or gyroradius r_C, and the period representing this motion is the cyclotron or gyro period τ_C.

The instantaneous velocity of this guiding centre is by definition the velocity of the Guiding Centre System (GCS) as seen from the inertial coordinate system at O (Fig. 4.28). The position of the particle can therefore be described by the position vector of $\mathbf{R}_C$ of the guiding centre, the Larmor radius vector $\mathbf{r}_C$ and the phase angle φ. If such a system, where the motion is periodic, cannot be found, there can be no cyclotron motion. This is the case for very high energetic cosmic radiation.

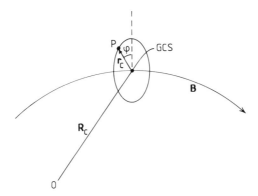

Fig. 4.28. The guiding centre for a particle spiralling around a magnetic field **B**. O is the centre in the inertial coordinate system and $\mathbf{R}_C$ is the position vector of the guiding centre. In the Guiding Centre System (GCS) the particle makes a circular motion with gyroradius r_C and phase angle φ.

The instantaneous velocity $\mathbf{V}_G$ of GCS can be divided into two terms with respect to the local magnetic field.

$$\mathbf{V}_G = \mathbf{V}_{\parallel} + \mathbf{V}_{\perp} \quad (4.59)$$

The parallel velocity $\mathbf{V}_{\parallel}$ must be equal to the velocity of the particle along the magnetic field $\mathbf{v}_{\parallel}$ so that

$$\mathbf{V}_{\parallel} = \mathbf{v}_{\parallel} \quad (4.60)$$

The perpendicular velocity $\mathbf{V}_\perp$ of GCS is different from $\mathbf{v}_\perp$, the perpendicular velocity of the particle. $\mathbf{V}_\perp$ is often called the drift velocity since it represents the drift of the guiding centre with respect to the original field line. We will denote this drift velocity by $\mathbf{V}_D$, and therefore

$$\mathbf{V}_D = \mathbf{V}_\perp \tag{4.61}$$

For a particle moving with the perpendicular velocity $\mathbf{v}_\perp$ in the inertial system, the perpendicular velocity $\mathbf{v}_\perp^*$ in GCS will be

$$\mathbf{v}_\perp^* = \mathbf{v}_\perp - \mathbf{V}_D \tag{4.62}$$

The parallel velocity of the particle in GCS, however, is simply zero so that

$$\mathbf{v}_\parallel^* = 0 \tag{4.63}$$

In a non-relativistic case when the magnetic field in the inertial system is $\mathbf{B}$, the magnetic field in GCS is

$$\mathbf{B}^* = \mathbf{B} \tag{4.64}$$

and the electric field in GCS is

$$\mathbf{E}^* = \mathbf{E} + \mathbf{V}_D \times \mathbf{B} \tag{4.65}$$

if $\mathbf{E}$ is the field in the inertial system.

4.8.2 Motion in a static and uniform magnetic field

For a particle moving in a uniform static magnetic field without being influenced by other external forces, the equation of motion may be reduced to

$$m \frac{d\mathbf{v}}{dt} = q \mathbf{v} \times \mathbf{B} \tag{4.66}$$

for a non-relativistic motion. We notice here that since $d\mathbf{v}/dt$ is perpendicular to $\mathbf{v}$, the kinetic energy of the particle is constant, and also the magnitude of $\mathbf{v}$.

The angle between $\mathbf{v}$ and $\mathbf{B}$, called the pitch angle and denoted by α, is defined by:

$$\mathbf{B} \cdot \mathbf{v} = Bv \cos \alpha = B v_\parallel \tag{4.67}$$

and therefore

$$\cos \alpha = \frac{v_\parallel}{v} \tag{4.68}$$

Since we also have that

$$|\mathbf{v} \times \mathbf{B}| = vB \sin \alpha = B v_\perp \tag{4.69}$$

we find that

$$\sin \alpha = \frac{v_\perp}{v} \tag{4.70}$$

Sec. 4.8]   **Motion of a charged particle in a magnetic field**   161

For the parallel acceleration $\mathbf{a}_\| = (d/dt)/\mathbf{v}_\|$ we have

$$m\,\mathbf{a}_\| = q(\mathbf{v} \times \mathbf{B})_\| = 0 \tag{4.71}$$

and therefore is

$$\mathbf{a}_\| = 0 \quad \text{and} \quad \mathbf{v}_\| = \text{const.}$$

Since $\mathbf{v}_\|$ is constant, the motion must be linear along $\mathbf{B}$. We have also concluded that the magnitude v of the velocity is constant. Therefore we have that

$$v_\perp = \text{const.} \tag{4.72}$$

Finally

$$\alpha = \text{const.} \tag{4.73}$$

and the pitch angle is conserved.

The perpendicular acceleration is now given by:

$$\mathbf{a}_\perp = \frac{q}{m}(\mathbf{v} \times \mathbf{B})_\perp = \frac{q}{m}(\mathbf{v}_\perp \times \mathbf{B}) \tag{4.74}$$

and its magnitude is:

$$a_\perp = \frac{qv_\perp B}{m} \tag{4.75}$$

since $\mathbf{v}_\perp$ is perpendicular to $\mathbf{B}$.

Since all parameters to the right are constants, $a_\perp$ is also constant. The motion perpendicular to $\mathbf{B}$ is therefore a circle, and $\mathbf{a}_\perp$ is simply the centripetal acceleration.

In this situation where $\mathbf{B}$ is static, GCS will move along $\mathbf{B}$ with a constant velocity:

$$\mathbf{V} = \mathbf{V}_\| = \mathbf{v}_\| \tag{4.76}$$

and

$$\mathbf{V}_D = 0 \tag{4.77}$$

In GCS the particle will move in a circle with radius r_C given by

$$r_C = \frac{v_\perp^2}{a_\perp} = \frac{mv_\perp^*}{qB^*} = \frac{p_\perp^*}{qB} \tag{4.78}$$

since $v_\perp = v_\perp^*$.

It is important to note that due to the charge, $a_\perp$ will have different signs for positive and negative charges. Seen in a direction antiparallel to $\mathbf{B}$, the positive particles rotate clockwise and the negative particles anticlockwise (Fig. 4.29a).

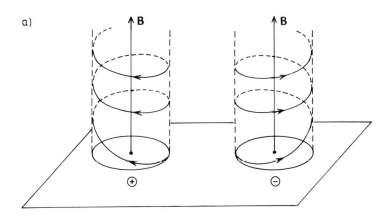

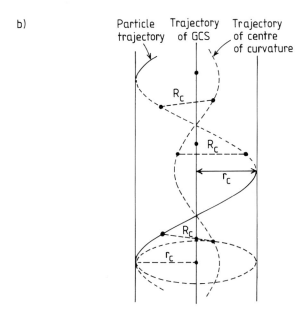

Fig. 4.29. (a) The spiralling motion of a positive and a negative particle around a static and homogeneous magnetic field **B**. The direction of rotation is opposite for the two charges being clockwise for the positive ions and anticlockwise for the electrons and negative ions when seen against the direction of **B**. The direction of the currents carried by the different charges, however, will be the same, namely clockwise. (b) The spiralling motion of a charge in a static and homogeneous **B**-field is tracing out the surface of a cylinder. The radius of curvature of the spiral trajectory R_C is always larger than the gyroradius r_C. Only if there is no velocity component along **B** will they be equal.

Sec. 4.8] Motion of a charged particle in a magnetic field

The gyroperiod is defined as

$$\tau_C = \frac{2\pi r_C}{v_\perp^*} = \frac{2\pi m}{qB} \qquad (4.79)$$

and the corresponding gyrofrequency

$$\omega_C = \frac{2\pi}{\tau_C} = \frac{qB}{m} \qquad (4.80)$$

For $\alpha = 90°$, $v_\parallel = 0$ and the motion of the particle is always circular and perpendicular to $\mathbf{B}$. For $\alpha = 0$ there is only a motion along $\mathbf{B}$ and no cyclotron motion.

In general, the motion will be along a spiral winding around $\mathbf{B}$ and the gyroradius r_C will not be equal to the curvature radius R_C of the trajectory. R_C is always greater than r_C since r_C is the radius of the cylinder around which the spiral is winding (Fig. 4.29b).

4.8.3 Zeroth-order drift

We will now consider a particle moving in a static uniform B-field again, but in addition assume an external force $\mathbf{F}$ acting on the particle. Let then $\mathbf{F}$ be decomposed along and perpendicular to $\mathbf{B}$, then

$$\mathbf{F} = \mathbf{F}_\parallel + \mathbf{F}_\perp \qquad (4.81)$$

The equations of motion along these two directions are then:

$$m\frac{d\mathbf{v}_\parallel}{dt} = \mathbf{F}_\parallel \qquad (4.82)$$

$$m\frac{d\mathbf{v}_\perp}{dt} = \mathbf{F}_\perp + q\,\mathbf{v}_\perp \times \mathbf{B} \qquad (4.83)$$

The first equation shows that the particle is accelerated along $\mathbf{B}$ as long as $\mathbf{F}_\parallel \neq 0$. Let therefore $\mathbf{F}_\parallel = 0$ so that this motion has a constant velocity. We will now determine the drift velocity, $\mathbf{V}_D$, of GCS in which the velocity is a circle. In this system the external force $\mathbf{F}_\perp$ has to be balanced by the induced electric field

$$q\mathbf{E}_\perp^* = q\,\mathbf{V}_D \times \mathbf{B} \qquad (4.84)$$

such that

$$q\mathbf{E}_\perp^* + \mathbf{F}_\perp = q\,\mathbf{V}_D \times \mathbf{B} + \mathbf{F}_\perp = 0 \qquad (4.85)$$

or

$$\mathbf{V}_D \times \mathbf{B} = -\frac{1}{q}\mathbf{F}_\perp \qquad (4.86)$$

By multiplying with $\mathbf{B}$ and noticing that $\mathbf{V}_D$ is perpendicular to $\mathbf{B}$ we have

$$\mathbf{V}_D = \frac{\mathbf{F}_\perp \times \mathbf{B}}{qB^2} \qquad (4.87)$$

Since $\mathbf{F}_\parallel$ does not enter into the expression for $\mathbf{V}_D$, this expression is also correct even if $\mathbf{F}_\parallel \neq 0$. The motion of the particle seen from the inertial reference system will be composed of a constant motion along $\mathbf{B}$ together with a cyclotron motion in GCS and a velocity perpendicular to $\mathbf{B}$ due to $\mathbf{V}_D$.

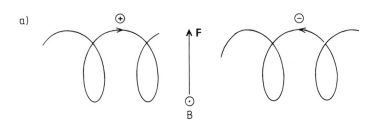

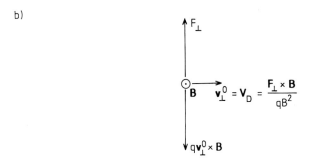

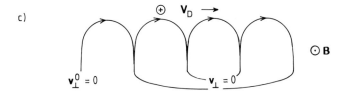

Fig. 4.30. (a) For charges moving in a static and homogeneous magnetic field under the influence of a constant external force $\mathbf{F}$ perpendicular to $\mathbf{B}$, the charges will drift perpendicular to both $\mathbf{F}$ and $\mathbf{B}$ as they gyrate around $\mathbf{B}$. The trajectories will therefore form cycloids. The drifts of positive and negative charges will be opposite to each other. (b) Depending on the initial conditions of the charges the cycloids may change form. If $\mathbf{v}_\perp^0 = \mathbf{V}_D$ at $t = 0$, then $\mathbf{v}_\perp^{*0} = 0$ and there is no cyclotron motion in GCS; the charge will move in a straight line. (c) If the perpendicular velocity $\mathbf{v}_\perp^0$ is zero at $t = 0$, then the cycloid will be open.

Sec. 4.8]   Motion of a charged particle in a magnetic field

If $\mathbf{F}_\parallel = 0$ and $\mathbf{v}_\parallel = 0$, then the motion in a plane perpendicular to $\mathbf{B}$ will be a cycloid (Fig. 4.30a). Since $\mathbf{V}_D$ is always perpendicular to $\mathbf{B}$ as well as $\mathbf{F}$, the force will not result in any work, and the energy remains constant. The cycloidal motion in GCS will depend on the initial condition for the particle in the inertial reference system. If $\mathbf{v}_\perp^0$ is the initial velocity in this system, then the initial velocity in GCS will be:

$$\mathbf{v}_\perp^{*0} = \mathbf{v}_\perp^0 - \mathbf{V}_D \qquad (4.88)$$

If $\mathbf{v}_\perp^0 = \mathbf{V}_D$, then $\mathbf{v}_\perp^{*0} = 0$, and there is no cyclotron motion in GCS. The Lorentz force $q\mathbf{v}_\perp^0 \times \mathbf{B}$ exactly cancels the external force $\mathbf{F}_\perp$. The particle motion is therefore a straight line (Fig. 4.30b).

If, on the other hand, the particle is at rest in the inertial system, then $\mathbf{v}_\perp^0 = 0$ and

$$\mathbf{v}_\perp^{*0} = -\mathbf{V}_D \qquad (4.89)$$

The trajectory of the particle now becomes an open cycloid (Fig. 4.30c).

Let the external force $\mathbf{F}$ be derived from a potential U, then

$$\mathbf{F} = -\nabla U \qquad (4.90)$$

and the drift velocity is given by

$$\mathbf{V}_D = -\frac{\nabla U \times \mathbf{B}}{qB^2} \qquad (4.91)$$

If now $\mathbf{F}_\parallel = 0$ and since the field lines for $\mathbf{F}$ will be forming potential surfaces, the drift velocity will be tangential to these surfaces. A particle (Fig. 4.31a) having a 90° pitch angle will therefore follow an equipotential line along a plane perpendicular to $\mathbf{B}$ if the radius of curvature for the equipotential lines are much larger than the gyroradius.

For $\mathbf{F}$ being the force of gravity

$$\mathbf{F} = m\mathbf{g} \qquad (4.92)$$

the drift velocity becomes

$$\mathbf{V}_D = \frac{m\mathbf{g} \times \mathbf{B}}{qB^2} \qquad (4.93)$$

For example, at magnetic equator (Fig. 4.31b), where $\mathbf{B}$ is approximately horizontal and $\mathbf{g}$ is vertical downward, there will be an eastward motion of the positive ions and a westward motion of the electrons leading to a charge separation and a build-up of a polarization field from east to west. There will therefore be a pile-up of positive charges on the evening side (east) and of negative charges on the morning side (west). These charges must be neutralized by charges moving along the magnetic field lines from the two hemispheres (Fig. 4.31c), and since the electrons have a larger mobility, there will be a drizzle of electrons into the atmosphere on the morning side and an outflow out of the atmosphere on the evening side.

At high latitudes where **g** becomes close to parallel (or antiparallel) to the magnetic field lines, this drift is negligible.

If the external potential is an electrostatic potential, then:

$$\mathbf{E} = -\nabla \phi \tag{4.94}$$

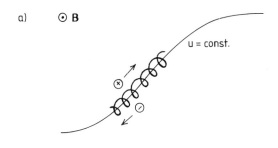

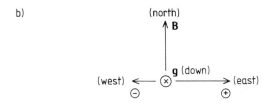

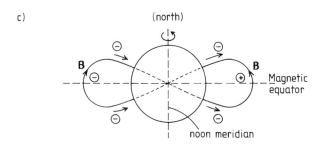

Fig. 4.31. (a) The external force can be derived from a potential U, and then the charge will spiral around equipotential lines in a plane perpendicular to **B**. (b) At equator the acceleration of gravity is perpendicular to the Earth's magnetic field, and the drift motion will be eastward for positive charges and westward for negative charges. (c) The charge separation in (b) will accumulate negative charges on the morning side and positive charges on the evening side. The charges can be neutralized by field-aligned current flows to and from the two hemispheres, thus resulting in a drizzle of electrons to the lower ionosphere in the morning side and an outflow of electrons from the atmosphere in the evening side.

and since
$$\mathbf{F} = q\mathbf{E} = -q\nabla\phi \tag{4.95}$$
the drift velocity becomes
$$\mathbf{V}_D = \frac{\mathbf{E} \times \mathbf{B}}{B^2} = -\frac{\nabla\phi \times \mathbf{B}}{B^2} \tag{4.96}$$

This drift is therefore independent of the sign of the charge such that negatively and positively charged particles are drifting along the same direction with the same velocity (Fig. 4.32a). If $\mathbf{E}_\parallel = 0$, particles with $\alpha = 90°$ will drift along equipotential lines (Fig. 4.32b) in a uniform magnetic field if the change in the electric potential is small across one gyroradius.

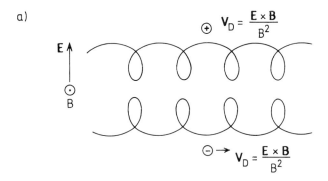

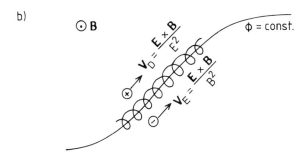

Fig. 4.32. (a) Positive and negative charges drift in the same direction with the same velocity when they are under the influence of perpendicular and homogeneous, static electric and magnetic fields. (b) When the electric field is derivable from a potential, the charges will drift along equipotential surfaces with the same speed perpendicular to the B-field. Observing such drifting particles therefore can be used as tracers of the electrostatic potential lines.

4.8.4 First-order drift motions

If the magnetic field is no longer homogeneous but changes in space, then we can write

$$\nabla B = \nabla_{\parallel} B + \nabla_{\perp} B \qquad (4.97)$$

and first-order drift motion will occur.

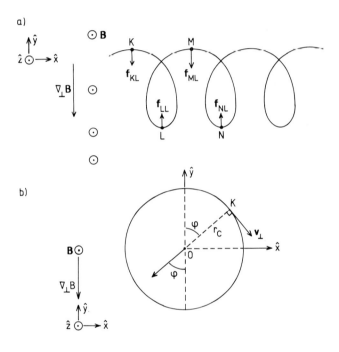

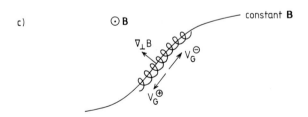

Fig. 4.33. (a) A charge moving in a magnetic field **B** having a gradient $\nabla_{\perp} B$ perpendicular to **B**, will experience a drift perpendicular to both **B** and $\nabla_{\perp} B$. The drift occurs because the **B**-field increases along $\nabla_{\perp} B$, and the radius of gyration will decrease along the $\nabla_{\perp} B$ direction. (b) Schematic illustration to explain the geometry associated with the gradient drift. (c) The gradient drift is opposite for positive and negative charges and it follows contours of constant **B**, perpendicular to both **B** and $\nabla_{\perp} B$.

Sec. 4.8] Motion of a charged particle in a magnetic field

If a particle with $\alpha = 90°$ enters an area where the magnetic field lines are almost straight lines, but where they have a gradient perpendicular to $\mathbf{B}$ ($\nabla_\perp B \neq 0$) (Fig. 4.33a), then the Larmor radius will become small where B is large and vice versa since r_C according to (4.78) is inversely proportional to B. The Lorentz force at point K in Fig. 4.33a given by

$$f_{KL} = qv_\perp B_K \tag{4.98}$$

seen from the inertial reference frame is smaller than the Lorentz force at L in Fig. 4.33a given by

$$f_{LL} = qv_\perp B_L \tag{4.99}$$

since B_K is less than B_L and $v_\perp$ is constant as long as no external force is acting on the particle. There will therefore be a net mean Lorentz force directed from L to K or opposite to $\nabla_\perp B$ during one full revolution, KLM.

Let us consider this situation more in detail. The magnetic field is almost a straight line along the z-axis (Fig. 4.33b) but has a perpendicular gradient

$$\nabla_\perp B = -\frac{\partial B}{\partial y}\hat{\mathbf{y}} \tag{4.100}$$

The particle in K makes a gyration around B with an approximate gyroradius r_C. If the magnetic strength in origo is B_0, then, due to the vertical gradient, the magnetic field strength in K becomes:

$$\begin{aligned} B_K &= B_0 + \delta B = B_0 - \frac{\partial B}{\partial y}\hat{\mathbf{y}} \cdot d\mathbf{y} \\ &= B_0 - \frac{\partial B}{\partial y} r_C \cos\varphi \end{aligned} \tag{4.101}$$

where φ is the phase angle measured positive clockwise from $\hat{\mathbf{y}}$, and r_C is the radius of the gyrocircle.

The strength of the Lorentz force in K becomes:

$$f_{KL} = |q\mathbf{v}_\perp \times \mathbf{B}_K| = qv_\perp \left(B_0 - \frac{\partial B}{\partial y} r_C \cos\varphi\right) \tag{4.102}$$

where $\mathbf{v}_\perp$ is the velocity of the particle perpendicular to $\mathbf{B}$. The component of this force in the y-direction is:

$$\begin{aligned} f_y &= +f_{KL}\cos(180+\varphi) \\ &= qv_\perp \left(B_0 - \frac{\partial B}{\partial y} r_C \cos\varphi\right)\cos\varphi \end{aligned} \tag{4.103}$$

During one cycle the net force in the y-direction will be:

$$\begin{aligned} F_G &= \frac{1}{2\pi}\int_0^{2\pi} f_y \, d\varphi \\ &= \frac{1}{2\pi}\int_0^{2\pi}\left[-qv_\perp\left(B_0 - \frac{\partial B}{\partial y} r_C \cos\varphi\right)\cos\varphi\right]d\varphi \end{aligned} \tag{4.104}$$

This force is directed along the positive y-axis, i.e. opposite to the gradient in B. As there are no net forces along the x- and z-axes, the net gradient force on the particle during one revolution is:

$$\mathbf{F}_G = \frac{1}{2} q v_\perp r_C \frac{\partial B}{\partial y} \hat{\mathbf{y}} \quad (4.105)$$

There is therefore an associated drift due to the gradient in B given by:

$$\mathbf{V}_G = \frac{\mathbf{F}_G \times \mathbf{B}}{qB^2} = \frac{v_\perp r_C}{2} \frac{1}{B} \frac{\partial B}{\partial y} \hat{\mathbf{x}} \quad (4.106)$$

By introducing the expression for r_C from (4.78) we get:

$$\mathbf{V}_G = \frac{m v_\perp^2}{2qB^2} \frac{\partial B}{\partial y} \hat{\mathbf{x}} \quad (4.107)$$

A positive particle will drift along the positive x-axis, while a negative particle drift in the opposite direction.

We also notice that

$$\mathbf{V}_G = \frac{1}{2} \frac{m v_\perp^2}{qB^2} \frac{\partial B}{\partial y} (\hat{\mathbf{y}} \times \hat{\mathbf{z}}) \quad (4.108)$$

$$= \frac{1}{2} \frac{m v_\perp^2}{qB^2} \hat{\mathbf{B}} \times \nabla_\perp B$$

since $\nabla_\perp B = -(\partial B / \partial y) \hat{\mathbf{y}}$ and $\hat{\mathbf{z}} = \hat{\mathbf{B}}$ is the unit vector along $\mathbf{B}$. It is, of course, assumed that the change δB, in B, over a gyroradius is small compared to B_0, i.e.:

$$\delta B = \nabla_\perp B \cdot r_C \ll B_0$$

This gradient drift $\mathbf{V}_G$ (Fig. 4.33c) is perpendicular both to $\mathbf{B}$ and $\nabla_\perp B$ and therefore will a particle with $\alpha = 90°$ drift in a plane perpendicular to $\mathbf{B}$ along trajectories where $\mathbf{B}$ is constant.

4.8.5 Curvature drift

For a particle gyrating along a field line which has a radius of curvature R_S where

$$R_S \gg r_C$$

there will be a centrifugal force acting on the particle. If we now assume that $\alpha = 0$ and that the gyro centre follows this curved field line, then the particle experiences the centripetal force

$$\mathbf{F}_S = + \frac{m v_\parallel^2}{R_S^2} \mathbf{R}_S \quad (4.109)$$

Sec. 4.8] Motion of a charged particle in a magnetic field

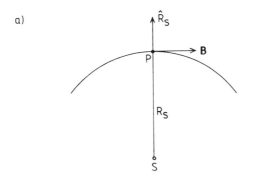

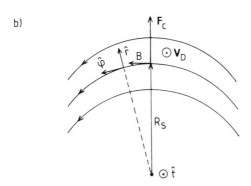

Fig. 4.34. (a) A curved magnetic field line will impose a centrifugal acceleration to a charge spiralling along this line. The radius of curvature is R_S, perpendicular to **B** in the point P. (b) Geometry to explain the effect of curvature drift. The **B**-field is directed along $\hat{\varphi}$ and has a radius of curvature R_S. The centrifugal force will impose a drift directed out of the paper which will force the particle away from the original field line.

where $v_\parallel$ is the velocity of the particle in GCS along the field line, and $\mathbf{R}_S$ is the position vector for the particle as seen from the centre of curvature (Fig. 4.34a). This force will again lead to a drift

$$\mathbf{V}_C = \frac{\mathbf{F}_S \times \mathbf{B}}{qB^2} = \frac{mv_\parallel^2}{qB^2 R_S^2} \mathbf{R}_S \times \mathbf{B} \qquad (4.110)$$

which is called the curvature drift. We will now try to find an expression for $\mathbf{R}_S/R_S^2$ expressed with the gradient in **B**. From Maxwell's equation (4.4) we find in cylindrical coordinates (r, φ, z) (Fig. 4.34b)

$$(\nabla \times \mathbf{B})_z = \frac{1}{r}\frac{\partial}{\partial r}(rB_\varphi) = 0 \qquad (4.111)$$

This gives:
$$B_\varphi \propto \frac{1}{r} \tag{4.112}$$

and thus:
$$B \propto \frac{1}{R_S} \tag{4.113}$$

Therefore we can have:
$$\nabla_\perp B \propto -\frac{1}{R_S^2}\hat{\mathbf{R}}_S \tag{4.114}$$

or
$$\frac{\nabla_\perp B}{B} = -\frac{\mathbf{R}_S}{R_S^2} \tag{4.115}$$

and finally we find
$$\mathbf{V}_C = \frac{mv_\parallel^2}{qB^2}\hat{\mathbf{B}} \times \nabla_\perp B \tag{4.116}$$

Adding together $\mathbf{V}_G$ and $\mathbf{V}_C$ now gives.
$$\mathbf{V}_G + \mathbf{V}_C = \left(\frac{1}{2}\frac{mv_\perp^2}{qB^2} + \frac{mv_\parallel^2}{qB^2}\right)\hat{\mathbf{B}} \times \nabla_\perp B \tag{4.117}$$
$$= \frac{m}{2qB^2}\left(v_\perp^2 + 2v_\parallel^2\right)\hat{\mathbf{B}} \times \nabla_\perp B$$

By introducing the pitch angle α
$$\mathbf{V}_{CG} = \frac{mv^2}{2qB^2}(1 + \cos^2\alpha)\hat{\mathbf{B}} \times \nabla_\perp B \tag{4.118}$$

4.8.6 Second-order drift
If the drift velocity $\mathbf{V}_D$ changes by time, then GCS will be accelerated and an inertial force $-m\dot{\mathbf{V}}_D$ will occur on the particle. The guiding centre system therefore will experience a drift

$$\mathbf{V}_s = -\frac{m}{qB^2}(\dot{\mathbf{V}}_D \times \mathbf{B}) \tag{4.119}$$

If the change in V_D during one gyro period is very small, then the gyro centre approximation holds. For example, a time varying $\mathbf{E}$-field will induce a second-order drift. Assume the electric field $\mathbf{E}_\perp$ is homogeneous and perpendicular to $\mathbf{B}$, then

$$\dot{\mathbf{V}}_D = \frac{\dot{\mathbf{E}}_\perp \times \mathbf{B}}{B^2} \tag{4.120}$$

Sec. 4.8]   **Motion of a charged particle in a magnetic field**   173

and the associated drift is

$$\mathbf{V}_P = -\frac{m}{qB^2}(\dot{\mathbf{E}}_\perp \times \mathbf{B}) \times \frac{\mathbf{B}}{B^2} = \frac{m}{qB^2}\dot{\mathbf{E}}_\perp \qquad (4.121)$$

This drift is often called the polarization drift.

4.8.7 The first adiabatic invariant – the magnetic moment

A charged particle moving in a circle around a B-field is equivalent to a current loop whose magnetic field is antiparallel to the external magnetic field independent of the charge (Fig. 4.35).

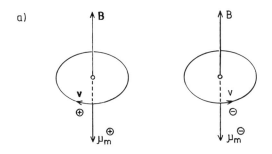

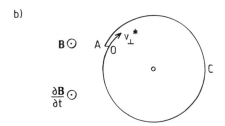

Fig. 4.35. (a) A charge gyrating around a static and homogeneous **B**-field creates a current in a direction which is independent of the sign of the charge and which is associated with a magnetic moment μ_m antiparallel to **B**, rendering the medium diamagnetic. (b) A diagram to illustate the geometry related to the gyration motion when the intensity of the magnetic field changes in time $\partial B/\partial t \neq 0$.

The current magnitude is

$$I = \frac{|q|}{\tau_C} = \frac{q^2 B}{2\pi m} \qquad (4.122)$$

where τ_C is the gyration period. The gyro circle covers an area given by

$$A_C = \pi r_C^2 \qquad (4.123)$$

The magnetic moment is therefore

$$\mu_m = I \cdot A_C = \frac{q^2 B}{2\pi m} \cdot \pi r_C^2 = \frac{q^2}{2\pi m} \phi_C \qquad (4.124)$$

where ϕ_C is the magnetic flux through the gyro circle. Introducing the expression for r_C in GCS from (4.78)

$$\mu_m = \frac{q^2 B}{2\pi m} \cdot \pi \cdot \left(\frac{mv_\perp^*}{qB}\right)^2 = \frac{W_\perp^*}{B} \qquad (4.125)$$

This gives the relationship between the magnetic moment μ_m and the kinetic energy $W_\perp^*$ perpendicular to B in the GCS reference system.

Since $\mathbf{B}$ is changing by time, there is an electric field $\mathbf{E}^*$ induced in the GCS during one gyration which is given by

$$\nabla \times \mathbf{E}^* = -\frac{\partial \mathbf{B}}{\partial t} \qquad (4.126)$$

The time variation in $\mathbf{B}$ can either happen by a genuine time variation or that GCS passes through a spatially changing $\mathbf{B}$ which will be observed as a time variation in GCS. We will now assume that the time variation is very slow such that

$$\frac{\partial B}{\partial t} \cdot \tau_C \ll B \qquad (4.127)$$

This then implies that the gyration trajectory is close to a circle with radius r_C. The work done by the electric field $\mathbf{E}^*$ during one gyration is (Fig. 4.35b):

$$\Delta W_\perp^* = -|q| \int_{ACO} \mathbf{E}^* \cdot d\mathbf{l} \approx -|q| \oint \mathbf{E}^* \cdot d\mathbf{l} \qquad (4.128)$$
$$= -|q| \iint_{A_C} \nabla \times \mathbf{E}^* \, d\mathbf{s} = +|q| \iint_{A_C} \frac{\partial \mathbf{B}}{\partial t} \, d\mathbf{s}$$

Here $d\mathbf{s}$ is an infinitesimal area within the circular area $\mathbf{A}_C$. Since $\mathbf{A}_C$ is parallel with $\mathbf{B}$ and $\partial \mathbf{B}/\partial t$ we get:

$$W_\perp^* = |q|\mathbf{A}_C \cdot \frac{\partial \mathbf{B}}{\partial t} = |q|\frac{\partial}{\partial t} \phi_C \qquad (4.129)$$

where

$$\phi_C = \mathbf{A}_C \cdot \mathbf{B} = A_C B \qquad (4.130)$$

is the magnetic flux through the circular area $A_C = \pi r_C^2$. Now we also have that

$$\Delta W_\perp^* = \tau_C \cdot \frac{\partial W_\perp^*}{\partial t} = \frac{2\pi m}{|q|B} \frac{\partial W_\perp^*}{\partial t} = |q| \cdot \pi r_C^2 \frac{\partial B}{\partial t} = \pi |q| \left(\frac{p_\perp^*}{qB}\right)^2 \frac{\partial B}{\partial t} \qquad (4.131)$$

Since we have for the kinetic energy due to the perpendicular motion:

$$W_\perp^* = \frac{(p_\perp^*)^2}{2m} \qquad (4.132)$$

we derive:
$$\frac{1}{W_\perp^*}\frac{\partial W_\perp^*}{\partial t} = \frac{1}{B}\frac{\partial B}{\partial t} \tag{4.133}$$

and finally find that:
$$\mu_m = \frac{W_\perp^*}{B} = \text{const.} \tag{4.134}$$

In vector form
$$\boldsymbol{\mu}_m = -\frac{W_\perp^*}{B^2}\mathbf{B} = -\frac{1}{2}\frac{mv_\perp^2}{B^2}\mathbf{B} \tag{4.135}$$

and the particle is associated with a constant magnetic moment antiparallel to the external magnetic field independent of the charge as long as $|\dot{B}|\cdot\tau_C \ll B$. μ_m is also often called the first adiabatic invariant.

As long as μ_m is conserved, $W_\perp$ will increase when B does, and this is referred to as an adiabatic acceleration. The increase in B in GCS can occur as already stated by true time variation or by particles drifting into a space with a stronger field. In both cases the particle senses a time-varying field and ascribes the energy variation to a non-vanishing $\nabla\times\mathbf{E}$ as in a betatron. Therefore this is also often referred to as betatron acceleration.

For a gyro veolcity $v_\perp^*$ much larger than $\mathbf{V}_D$ we have $\mathbf{v}_\perp^* = \mathbf{v}_\perp$, and therefore $v_\perp = v\sin\alpha$. The magnetic moment is then given by:
$$\mu_m = \frac{p_\perp^2}{2mB} = \frac{p^2\sin^2\alpha}{2mB} = \frac{W\sin^2\alpha}{B} \tag{4.136}$$

where W is the kinetic energy. Since μ_m as well as W are constants, we then have
$$\frac{\sin^2\alpha}{B} = \text{const.} \tag{4.137}$$

4.8.8 The magnetic mirror
For a particle moving with negligible drift motion ($\mathbf{v}_0 \approx 0$) along a magnetic field line we have found:
$$\frac{\sin^2\alpha(s)}{B(s)} = \frac{\sin^2\alpha_0}{B_0} \tag{4.138}$$

where $\alpha(s)$ and $B(s)$ are the pitch angle and magnetic field strength at a position s along the field line, while α_0 and B_0 are the corresponding values at a fixed reference point on the same field line. If a particle is to move with a pitch angle α_0 into a magnetic field where its strength is B_0, then the particle will spiral along this field line with a parallel velocity at position s given by:
$$v_\parallel(s) = v\cos\alpha(s) = \left[1 - \frac{B(s)}{B_0}\sin^2\alpha_0\right]^{1/2}\cdot v \tag{4.139}$$

The vertical component at position s is:

$$v_\perp(s) = v \sin \alpha(s) = v \sqrt{\frac{B(s)}{B_0}} \sin \alpha_0 \qquad (4.140)$$

If the particle now moves toward an increasing field strength, then $B(s)$ will increase until the particle reaches a critical point where the magnetic field strength is given by:

$$B_m = \frac{B_0}{\sin^2 \alpha_0} \qquad (4.141)$$

The particle velocity parallel to B will be zero at this point, and the pitch angle is $\alpha_m = 90°$. A particle with a smaller pitch angle at this point, however, will penetrate further down the field line. Introducing B_m we also have:

$$v_\parallel(s) = \left[1 - \frac{B(s)}{B_m}\right]^{1/2} \cdot v \qquad (4.142)$$

Now we consider a magnetic field which is pointing primarily in the z-direction and whose magnitude varies in the same direction. If the field is axisymmetric with $B_\varphi = 0$ and $\partial/\partial\varphi = 0$, then, since the field converges or diverges, there must be a radial component (Fig. 4.36). The task is now to show that this configuration can give a trapping of the particles in a magnetic field. From Maxwell's equation

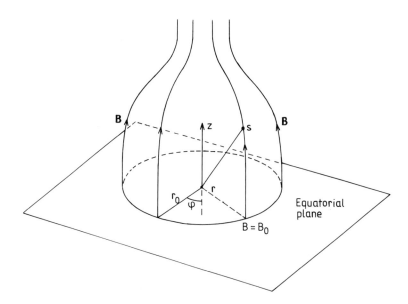

Fig. 4.36. Half of the section of a magnetic bottle above the equatorial plane. The field lines are axisymmetric ($\partial/\partial\varphi = 0$ and $B_\varphi = 0$).

Sec. 4.8] Motion of a charged particle in a magnetic field

in polar coordinates

$$\nabla \cdot \mathbf{B} = \frac{1}{r}\frac{\partial}{\partial r}(rB_r) + \frac{\partial B_z}{\partial z} = 0 \tag{4.143}$$

and

$$rB_r = -\int_0^r r\frac{\partial B_z}{\partial z}dr \simeq -\frac{1}{2}r^2\left[\frac{\partial B_z}{\partial z}\right]_{r=0} \tag{4.144}$$

if $\partial B_z/\partial z$ is given at $r=0$ and it does not vary much with r. We then obtain

$$B_r = -\frac{1}{2}r\left[\frac{\partial B_z}{\partial z}\right]_{r=0} \tag{4.145}$$

This variation of B with r causes a gradient drift of the guiding centre about the axis of symmetry, but there is no radial grad-B drift since $\partial B/\partial\varphi = 0$. The component of the Lorentz force along the z-axis now becomes

$$F_z = -qv_\varphi B_r \tag{4.146}$$

and by introducing the expression for B_r just derived

$$F_z = \frac{1}{2}qv_\varphi r\frac{\partial B_z}{\partial z} \tag{4.147}$$

Averaging this force over one gyration period gives

$$\bar{F}_z = +\frac{1}{2}qv_\varphi r_C\frac{\partial B_z}{\partial z} \tag{4.148}$$

Since $v_\varphi = \mp v_\perp$ depending on whether the charge is positive or negative, we can also express this force by:

$$\bar{F}_z = -\frac{1}{2}\frac{mv_\perp^2}{B}\frac{\partial B_z}{\partial z} = -\mu_m\frac{\partial B_z}{\partial z} \tag{4.149}$$

where we have introduced the expression for r_C (equation (4.78)) and the magnetic moment μ_m (equation (4.125)). This is the z-component of a more general expression for a force acting on a diamagnetic particle

$$\mathbf{F}_\parallel = -\mu_m\nabla_\parallel B \tag{4.150}$$

This parallel force is always acting against the particle travelling along a converging magnetic field geometry. Eventually a particle may be reflected by this force. The trapping, however, is not perfect since a particle with $\mu_m = 0$ will not feel any force along the field line. A particle with small enough pitch angle at the midplane where $B = B_0$ will also escape if the maximum field strength B_m is not large enough. The smallest angle a particle can have at the midplane in order to be reflected is therefore

$$\sin^2\alpha_0 = \frac{B_0}{B_m} = R_m \tag{4.151}$$

For any lower angle than this B_m will be too small and the particle escapes. R_m is often called the mirror ratio.

4.8.9 The second adiabatic invariant

A particle trapped between two magnetic mirrors bounces between them in a periodic motion at the bounce period. A constant of this motion, called the second adiabatic invariant, is defined as

$$J = \oint mv_\parallel ds \tag{4.152}$$

where ds is an element of the path length (of the guiding centre) along the field line. If the magnetic field lines change so that the separation of the mirror points changes, the integration path also changes. If the integration path is shortened, then $v_\parallel$ has to increase and conclusively $W_\parallel$, the kinetic energy due to the motion parallel to **B**. This is then called Fermi acceleration. In the magnetosphere this can happen by a true time variation in **B** or by the particle drifting into a spatially changing **B**. Most often the second alternative is at play.

Since we have found that (equation (4.142))

$$v_\parallel = v\cos\alpha = v\sqrt{1 - \frac{B(s)}{B_m}} \tag{4.153}$$

we also have that the second adiabatic invariant can be expressed by

$$J = \oint mv\sqrt{1 - \frac{B(s)}{B_m}}\, ds \tag{4.154}$$

If now $B_m \gg B(s)$ over most of the path, we see that the conservation of J implies that $\oint ds \approx$ constant, i.e. the mirror separation measured along the field line must be conserved. Such a conservation obviously implies that the variations in $B(s)$ during a complete bounce period must be small compared to B_m.

4.9 STØRMER'S CALCULATION OF PARTICLE MOTIONS IN MAGNETIC FIELDS

Let **B** be the magnetic field from a dipole situated at the centre of a Cartesian coordinate system such that the positive z-axis is parallel to the magnetic moment $\mathbf{M_0}$ (Fig. 4.37). Then we have:

$$V_M = \frac{\mathbf{M_0} \cdot \mathbf{r}}{r^3} = \frac{M_0 z}{r^3} \tag{4.155}$$

Since

$$r^2 = x^2 + y^2 + z^2 \tag{4.156}$$

Sec. 4.9] Størmer's calculation of particle motions

and

$$\mathbf{B} = -\frac{\partial V_M}{\partial x}\hat{\mathbf{x}} - \frac{\partial V_M}{\partial y}\hat{\mathbf{y}} - \frac{\partial V_M}{\partial z}\hat{\mathbf{z}} \qquad (4.157)$$

we find **B** on component form in a Cartesian coordinate system:

$$B_x = M_0 \frac{3xz}{r^5} \quad \text{(a)} \qquad (4.158)$$

$$B_y = M_0 \frac{3yz}{r^5} \quad \text{(b)}$$

$$B_z = M_0 \frac{3z^2 - r^2}{r^5} \quad \text{(c)}$$

The equation of motion for a charged (q) particle of mass m moving in this field **B** is found from (4.66) on component form:

$$\frac{d^2x}{dt^2} = \frac{qM_0}{m}\left(\frac{dy}{dt}\frac{3z^2 - r^2}{r^5} - \frac{dz}{dt}\frac{3yz}{r^5}\right) \quad \text{(a)} \qquad (4.159)$$

$$\frac{d^2y}{dt^2} = \frac{qM_0}{m}\left(\frac{dz}{dt}\frac{3xz}{r^5} - \frac{dx}{dt}\frac{3z^2 - r^2}{r^5}\right) \quad \text{(b)}$$

$$\frac{d^2z}{dt^2} = \frac{qM_0}{m}\left(\frac{dx}{dt}\frac{3yz}{r^5} - \frac{dy}{dt}\frac{3xz}{r^5}\right) \quad \text{(c)}$$

By changing variable from time t to path length $s = v \cdot t$ we get

$$\frac{d^2x}{ds^2} = \frac{qM_0}{mv}\left(\frac{dy}{ds}(3z^2 - r^2) - \frac{dz}{ds}3yz\right)\frac{1}{r^5} \quad \text{(a)} \qquad (4.160)$$

$$\frac{d^2y}{ds^2} = \frac{qM_0}{mv}\left(\frac{dz}{ds}3xz - \frac{dx}{ds}(3z^2 - r^2)\right)\frac{1}{r^5} \quad \text{(b)}$$

$$\frac{d^2z}{ds^2} = \frac{qM_0}{mv}\left(\frac{dx}{ds}3yz - \frac{dy}{ds}3xz\right)\frac{1}{r^5} \quad \text{(c)}$$

We notice that if q changes sign, then by changing sign on x simultaneously the structure of the equations are retained. Therefore, if the solution for one type of charge is found, the solution for the opposite charge is given by a change in sign of x. We then need to consider only the solution for positive q.

We now introduce cylindrical coodinates (R, φ, z) (Fig. 4.37)

$$x = R\cos\varphi \quad \text{(a)} \qquad (4.161)$$
$$y = R\sin\varphi \quad \text{(b)}$$
$$z = z \quad \text{(c)}$$

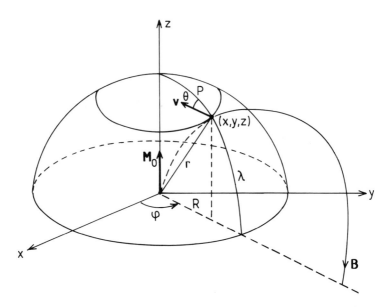

Fig. 4.37. The coordinate system used to illustrate Størmer's geometry. $\mathbf{M_0}$ is the dipole magnetic moment along positive z-axis. $\mathbf{B}$ is the magnetic field line through a point $P(x, y, z)$. $\mathbf{V}$ is the velocity of the particle which has a component $\mathbf{v_\perp}$ in the azimuthal direction. λ is the latitude angle from the equatorial plane. R is the distance to the projection of P in the equatorial plane and φ the azimuthal angle.

With these relations we form:

$$x \frac{dy}{ds} - y \frac{dx}{ds} = R^2 \frac{d\varphi}{ds} \tag{4.162}$$

and by a second derivation of (4.162)

$$x \frac{d^2y}{ds^2} - y \frac{d^2x}{ds^2} = \frac{d}{ds}\left(R^2 \frac{d\varphi}{ds}\right) \tag{4.163}$$

When introducing the Størmer length defined by:

$$C_{st}^2 = -\frac{M_0 q}{mv} \tag{4.164}$$

we can also derive:

$$\frac{d}{ds}\left(R^2 \frac{d\varphi}{ds}\right) = C_{st}^2 \frac{d}{ds}\left(\frac{R^2}{r^3}\right) \tag{4.165}$$

This equation can be easily integrated to yield

$$R^2 \frac{d\varphi}{ds} = C_{st}^2 \frac{R^2}{r^3} + C \tag{4.166}$$

We also have that:

$$R^2 \frac{d\varphi}{ds} = R^2 \left(\frac{d\varphi}{v\, dt} \right) = \frac{R}{v} v_\perp \qquad (4.167)$$

where $v_\perp = R(d\varphi/dt)$ is the component of **v** perpendicular to the magnetic meridional plane and therefore

$$\frac{v_\perp}{v} = \pm \sin \theta \qquad (4.168)$$

θ is actually the angle between the velocity vector and the meridian plane. Now substituting $R^2(d\varphi/ds)$ into (4.166) we get

$$\pm \sin \theta = C_{st}'^2 \frac{R}{r^3} + \frac{C}{R} \qquad (4.169)$$

If we now express the lengths R and r as well as the constant C is Størmer lengths, then

$$\begin{aligned} R &= \mathcal{R}\, C_{st} & \text{(a)} \\ r &= \rho\, C_{st} & \text{(b)} \\ C &= 2\gamma\, C_{st} & \text{(c)} \end{aligned} \qquad (4.170)$$

where $\mathcal{R}$, ρ and γ now are dimensionless numbers. Then we obtain

$$\pm \sin \theta = \frac{\mathcal{R}}{\rho^3} + \frac{2\gamma}{\mathcal{R}} \qquad (4.171)$$

Since $-1 \leq \sin \theta \leq 1$, there is only a limited number of $\mathcal{R}$, ρ and γ which can satisfy the equation, namely

$$-1 \leq \frac{\mathcal{R}}{\rho^3} + \frac{2\gamma}{\mathcal{R}} \leq 1$$

Now introducing the latitude angle λ (Fig. 4.37) we have

$$\mathcal{R} = \rho \cos \lambda \qquad (4.172)$$

and by introducing

$$k = \pm \sin \theta \qquad (4.173)$$

we obtain from (4.171)

$$k\rho^2 = \cos \lambda + \frac{2\gamma \rho}{\cos \lambda} \qquad (4.174)$$

or by multiplying by $\cos \lambda$ and rearranging

$$k\rho^2 \cos \lambda - 2\gamma \rho - \cos^2 \lambda = 0 \qquad (4.175)$$

This equation has the solution

$$\rho = \frac{\gamma \pm \sqrt{\gamma^2 + k\cos^3\lambda}}{k\cos\lambda} = \frac{-\cos^2\lambda}{\gamma \mp \sqrt{\gamma^2 + k\cos^3\lambda}} \tag{4.176}$$

Or if expressed in terms of $\mathcal{R}$

$$\mathcal{R} = \frac{\gamma}{k} \pm \sqrt{\left(\frac{\gamma}{k}\right)^2 + \frac{\cos^3\lambda}{k}} \tag{4.177}$$

This solution is symmetric around the xy-plane, therefore we only consider λ between 0° and 90°. Since $\gamma \in [-\infty, \infty]$, we first consider the negative values of γ and introduce

$$\gamma = -\gamma_1 \tag{4.178}$$

where $\gamma_1 \in [1, \infty]$. We can have two types of solutions depending on k.

(a) $k > 0$

$$\rho_1 = \frac{\cos^2\lambda}{\gamma_1 + (\gamma_1^2 + k\cos^3\lambda)^{1/2}} \tag{4.179}$$

since ρ_1 must be positive. If λ increases from 0° to 90°, ρ_1 will decrease from

$$\rho_{1,\max} = \frac{1}{\gamma_1 + (\gamma_1^2 + k)^{1/2}} \tag{4.180}$$

to

$$\rho_{1,\min} = 0 \tag{4.181}$$

The curve is an oval passing through origo and approaching asymptotically

$$\rho_{1,A} = \frac{\cos^2\lambda}{2\gamma_1} \tag{4.182}$$

when $k \to 0$ which means that when a particle enters the dipole field parallel to the meridional plane. The particle will then follow a field line. The solutions are shown schematically in the $\mathcal{R}z$ plane in Fig. 4.38.

(b) $k < 0$

We then introduce

$$k = -k_1 \tag{4.183}$$

and obtain

$$\rho_2 = \frac{\cos^2\lambda}{\gamma_1 + (\gamma_1^2 - k_1\cos^3\lambda)^{1/2}} \tag{4.184}$$

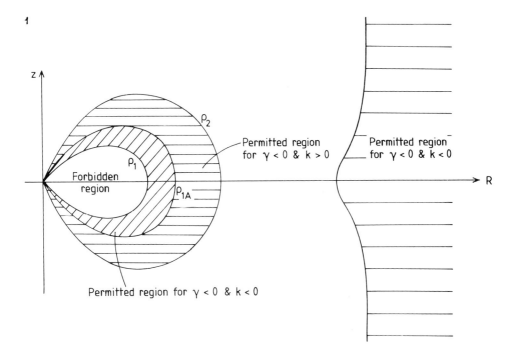

Fig. 4.38. Forbidden and permitted regions for charged particles depending on their initial conditions γ ($\gamma < 0$) and their azimuthal velocity component $\mathbf{v}_\perp$ in the azimuthal direction. Particles from one region cannot reach the other.

for the upper sign, and

$$\rho_3 = \frac{\cos^2 \lambda}{\gamma_1 - (\gamma_1^2 - k_1 \cos^2 \lambda)^{1/2}} \qquad (4.185)$$

for the lower sign. Here ρ_3 is always larger than ρ_2. If λ increases from $0°$ to $90°$, then ρ_2 will decrease from

$$\rho_{2,\max} = \frac{1}{\gamma_1 + (\gamma_1^2 - k^2)^{1/2}} \qquad (4.186)$$

to

$$\rho_{2,\min} = 0 \qquad (4.187)$$

which is an oval outside ρ_1. This oval approaches asymptotically

$$\rho_{2,A} = \frac{\cos^2 \lambda}{2\gamma_1} \qquad (4.188)$$

when $k = 0$, i.e. when the particle enters parallel to the meridian plane again it follows a field line. This oval, however, is always inside ρ_2.

If λ increases from $0°$ to $90°$, then ρ_3 will increase from

$$\rho_{3,\min} = \frac{1}{\gamma_1 - (\gamma_1^2 - k_1)^{1/2}} \tag{4.189}$$

to

$$\rho_{3,\max} = \infty \tag{4.190}$$

The solutions for ρ_2 and ρ_3 are shown schematically in Fig. 4.38.

We have now shown that particles with a negative γ can move into different regions, one inner region as well as one outer region. The particles from one region cannot, however, reach the other. In the case of $0 > \gamma > -1$, there are several solutions to be considered depending on k and the relation between k and γ_1. One can show that particles can reach toward the dipole poles from infinity through a limited region. This is illustrated schematically in Fig. 4.39a. For $\gamma = 0$ there will be only one real solution, namely

$$\rho = \sqrt{\frac{\cos \lambda}{k}} \tag{4.191}$$

which is describing the surface of a torus which is a forbidden area. Outside this torus, however, particles can reach (Fig. 4.39b). For $\gamma > 0$ there is only one real solution again expressed by

$$\rho = \frac{\gamma + (\gamma^2 + k \cos^3 \lambda)^{1/2}}{k \cos \lambda} \tag{4.192}$$

This region never reaches the Earth, but for small values of γ it will get closer towards the z-axis (Fig. 4.39c).

For very high velocities, $v \approx c$, where c is the speed of light, the Størmer length can be smaller than the Earth's radius, and the particles will reach the Earth at any point. This will be the situation if the mass of the particle is larger than m_1 given by:

$$m_1 = \left| \frac{qM}{R_e^2 c} \right| \tag{4.193}$$

where R_e is the radius of the Earth. A particle with an energy

$$mc^2 \geq m_1 c^2$$

will therefore reach the Earth's surface while for particles where $mc^2 \leq m_1 c^2$ the magnetic field will act as a filter. The particles with the highest energy can reach the Earth at any place, while the particles with less energy will be forced toward the poles.

Sec. 4.9]   Størmer's calculation of particle motions   185

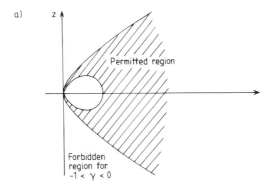

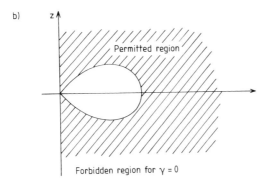

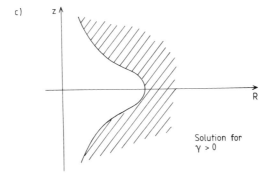

Fig. 4.39. (a) Forbidden and permitted regions for particles with $-1 < \gamma < 0$. These particles cannot reach the torus region but can advance toward the pole within a limited region at high latitudes. These particles could be a potential source for auroral particles. (b) Forbidden and permitted regions for $\gamma = 0$. These particles could reach the whole polar cap region down to a latitude determined by the forbidden region. (c) The permitted and forbidden regions for $\gamma = 0$. The permitted region gets closer to the z-axis as γ decreases for which panel (b) represents the limit when $\gamma = 0$.

4.10 THE RADIATION BELT

Fig. 4.40 shows how the Størmer length changes for different energies of the electrons and the protons.

The magnetic rigidity defined by

$$R_m = r_C \cdot B = \frac{mv}{q} \qquad (4.194)$$

is often used as a measure for the ability of the particles to penetrate a magnetic field. We notice that

$$|C_{st}^2| = \frac{M_0}{R_m} \qquad (4.195)$$

Particles with a high magnetic rigidity will be associated with a small Størmer length and vice versa, as shown in Fig. 4.40. From these calculations performed by Størmer before 1920 it was clear that there are regions outside the Earth where particles cannot reach from the outside under steady-state conditions. These regions were called "Størmer-trons" or forbidden regions. They were first observed by the first satellites in the late 1950s and are now called *Van Allen Belts* after him who first realized the existence of these regions from the response of instruments on the Explorer satellite in 1958.

When cosmic radiation reaches the atmosphere and collides with the nuclei of atomic nitrogen and oxygen, neutrons with energy between 1 MeV and 1 GeV may be formed. These particles can move freely independent of the magnetic field and

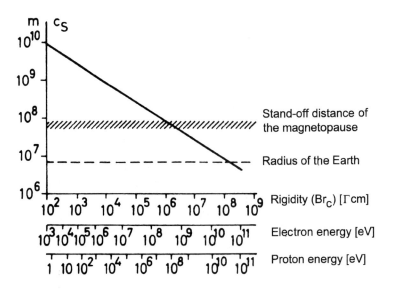

Fig. 4.40. Diagram showing the Størmer length C_{st} as function of proton and electron energy as well as magnetic rigidity $R_m = r_C \cdot B$. (From Fälthammar, 1973.)

can leave the atmosphere again. They have, however, a lifetime of about 1000 s before they decay into an electron, a proton and a neutrino. If this disintegration happens while the neutron still is within the inner part of the magnetic field, the charged particles may become trapped. The continuous influx of cosmic rays therefore represents a source of trapped particles in the forbidden region, which otherwise would be unreachable by charged particles. It is, however, not the only source.

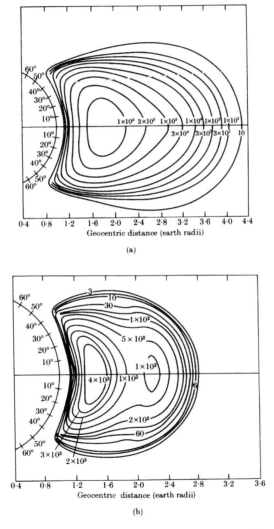

Fig. 4.41. The distribution of protons in the radiation zones. (a) Protons with energy greater than 4 MeV. (b) Protons with energy greater than 50 MeV. The numbers on the curves are fluxes in units of $cm^{-2}\ s^{-1}$. The angles indicated are latitude. (From Freden, 1969.)

In the case of a disturbance in the magnetic field, however, particles that otherwise would not reach these forbidden regions, may do so and become trapped when the field returns to the steady state. Therefore these zones also are called the trapping zones or the radiation belts.

Fig. 4.41 shows some typical observations of the proton fluxes in units of cm^{-2} s^{-1} for proton energies of 4 MeV and 50 MeV, respectively. The 4 MeV protons have a maximum zone typically at 1.8 R_e (Fig. 4.41a), while the 50 MeV protons have two maxima, one at about 1.4 R_e (Fig. 4.41b) and one at about 2.2 R_e.

Unfortunately, the devastations in our environment due to man's activity on Earth reaches far beyond the lower atmosphere, the ozone layer and maybe the upper stratosphere. Some of the most detestable experiments in the name of science performed by mankind were the atomic bomb tests in the atmosphere in the 1960s. In spite of world-wide protests against these tests, they were continued for some time irrespective of the polluting effects they had on the atmosphere. One piece of evidence among the many used in the actions to try to stop the demolition of the environment was the fact that the effects of the debris of the bombs could be traced for long times in the radiation belts. Fig. 4.42 shows the results of satellite

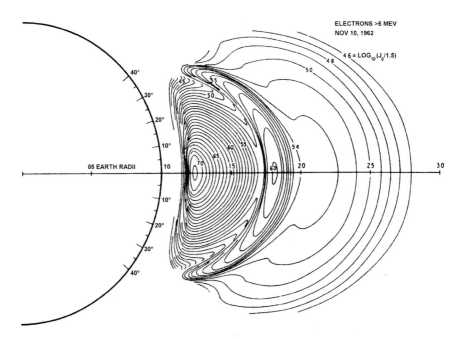

Fig. 4.42. Contours of constant intensity of fluxes of electrons with energy higher than 5 MeV. The observations are made Nov. 10, 1962. The peak around 1.4 R_e is due to USA's high altitude bomb test Starfish of July 9, 1962, and the peak around 1.75 R_e is probably due to a USSR high altitude atomic bomb test October–November 1962. (From Röderer, 1970.)

observations of the flux of electrons with energy above 5 MeV as obtained between 1.2 and 3 R_e above ground. At about 1.4 R_e there is a maximum identified as the debris of the Starfish explosion July 9, 1962, made by the USA, and the peak at about 1.75 R_e is probably due to tests in USSR between October and November 1962. The observations were made on November 10, 1962.

4.11 REFERENCES

Barraclough, D. R. (1974) *Geophys. J. Roy. Astron. Soc.*, **36**, 497–513.
Cain, J. C. (1987) in *The Solar Wind and the Earth*, Akasofu, S. I. and Kamide, Y. (Eds.), Terra Sci. Publ. Comp., Tokyo.
Cain, J. C. and Cain, S. J. (1968) *IGRF (10/68)*, Goddard Space Flight Center Report X–612–68–501.
Cox, A., Dalrymple, G. B. and Doell, R. R. (1967) *Scient. Am.*, **216**, 44–54.
Frank, L. (1994) Private communication
Freden, S. C. (1969) *Space Sci. Rev.*, **9**, 198–242.
Fälthammar, C.-G. (1973) in *Cosmical Geophysics,* Egeland, A., Holter, Ø. and Omholt, A. (Eds.) pp. 121–142, Universitetsforlaget, Oslo.
Gilbert, W. *De Magnete*. Translated by P. Fleur Mottelay, Dover, New York, 1958.
Jin, R.-S. and Thomas, D. M. (1977) *J. Geophys. Res.*, **82**, 828–834.
McElhinny, M.W. and Senanayake, W. E. (1982) *J. Geomag. Geoelect.*, **34**, 39–51.
Merrill, R. T. and McElhinny, M. (1983) *International Geophysics Series*, **32**, Academic Press.
Oguti, T. (1994) Private communication.
Rikitake, T. (1958) *Proc. Camb. Phil. Soc. math. phys. Sci.*, **54**, 89–105.
Röderer, J. G. (1970) *Dynamics of Geomagnetically Trapped Radiation*, Springer-Verlag, New York.
Yukutake, T. (1967) *J. Geomag. Geoelect.*, **19**, 103.

4.12 EXERCISES

1. (a) Prove that
   $$\mathbf{M}_0 \cdot \nabla \left(\frac{1}{r}\right) = \frac{\mathbf{M}_0 \cdot \mathbf{r}}{r^3} = -\frac{M_0 \cos\theta}{r^3}$$
   when $\mathbf{M}_0 = -M_0 \hat{z}$ in the coordinate system shown in Fig. 4.7.

   (b) Prove by using the same coordinate system the identity
   $$\nabla \left(\mathbf{M}_0 \cdot \nabla \left(\frac{1}{r}\right)\right) = 3(\mathbf{M}_0 \cdot \mathbf{r})\frac{\mathbf{r}}{r^5} - \frac{\mathbf{M}_0}{r^3}$$
   Use this to derive equations (4.20a,b,c).

   (c) Prove by applying a Cartesian coordinate system equations (4.158a,b,c).

2. Assume the Earth has an ideal dipole as outlined in Fig. 4.4.

(a) Derive the geomagnetic latitude of your site.

(b) Find the magnetic field strength at your site.

(c) Find the distance from the Earth's centre to the point where the magnetic field line through your site crosses the magnetic equatorial plane.

(d) Derive the L value at your site.

3. The auroral oval is at present to a good approximation equal to a circle around the geomagnetic dipole axis at 67° dipole latitude. Assume that the magnetic moment is reduced to half its present value and determine then the geomagnetic latitude of the auroral oval.

4. (a) Prove by applying (4.66) that the kinetic energy of the particle is conserved.

   (b) Explain also why $v_\perp$ is conserved.

5. Prove the relationship given by (4.165).

6. Prove the relationship given by (4.195).

5
The ionosphere

5.1 THE PRODUCTION OF IONIZATION BY SOLAR RADIATION

The existence of an ionized layer in the upper atmosphere was probably already appreciated at the beginning of this century when Marconi demonstrated that radio waves could propagate across large distances beyond the free horizon, as if they were guided between a conducting layer and the ground. A conducting layer in the upper atmosphere was also inferred as early as 1880 when Stewart, from studies of diurnal oscillations in the Earth's magnetic field, indicated that these variations might be caused by tidal oscillations in the upper atmosphere. This was later substantiated by Schuster in 1908 based on similar studies of geomagnetic field variations.

It was, however, not until 1924 that the existence of an ionized conducting layer in the upper atmosphere was fully proven, when Appleton and Barnett in England and Breit and Tuve in the USA demonstrated the existence of the ionosphere by studying the reflections from the atmosphere of radio waves.

The ionosphere can be viewed as a variable shell of ionization (plasma) surrounding the Earth. The range of variability in vertical profiles of electron density up to altitudes of 800 km are shown in Fig. 5.1. Typically there is a maximum around 300 km, this can, however, vary between 200 and 600 km. Below this maximum there often appear bumps in the profile, sometimes distinct enough to form secondary peaks such as during auroral events when they can form rather sharp layers (Fig. 5.2).

Historically the ionosphere has been divided into layers, the earliest detected was the E-layer, named so due to reflection of electric fields. For alphabetical order we have since got D- and F-layers below and above the E-layer, respectively. Today it is more common to speak about regions as the distinctions between layers are not all that clear, except maybe for some special events in the E-region such as sporadic and auroral E-layers.

Below the E-region, between 60 and 90 km, is the D-region which is very variable with lesser electron density. Sometimes the lower D-region may even be referred to as the C-region, alluding to its complicated chemistry.

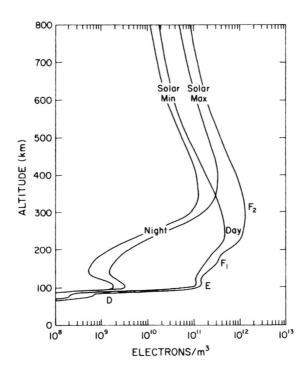

Fig. 5.1. Typical midlatitude ionospheric electron density profiles for sunspot maximum and minimum conditions at daytime and night-time. The different altitude regions in the ionosphere are labelled with appropriate nomenclature. (From Richmond, 1987.)

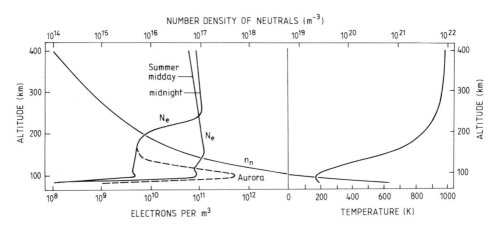

Fig. 5.2. Electron density profiles representing average daytime and night-time conditions at high latitudes. Also indicated is a profile for auroral conditions. The background neutral density profile together with an average neutral atmosphere temperature profile are also schematically illustrated.

Sec. 5.1]   **The production of ionization by solar radiation**   193

The electron density in all these regions, however, varies by time of day, season, solar cycle and level of magnetospheric or solar wind disturbance. At the uppermost altitudes the ionosphere merges with the magnetosphere of the Earth.

Fig. 5.3 shows some typical electron density profiles at 3 different latitudes for varying level of solar activity. The F-region peak is found to be stronger for increasing solar activity and the height of the peak is also increasing. The electron density is higher at all stations during winter than during summer above say 200 km.

In Fig. 5.4 is shown a global contour plot of the electron density at F-region maximum at equinox. Although the electron density has a bulge at midday at lower latitudes, there is in fact a minimum above geomagnetic equator. This so-called equatorial anomaly is related to transport phenomena along the geomagnetic field lines.

In Fig. 5.5 an alternative illustration of the same phenomenon is shown where the electron densities at different heights are plotted versus latitude. A clear minimum is seen at most altitudes close to the magnetic equator while maxima

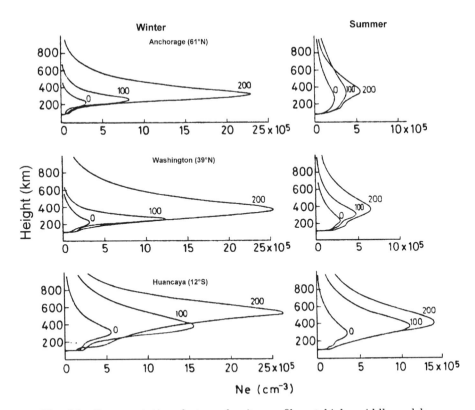

Fig. 5.3. Representative electron density profiles at high, middle and low latitudes for three levels of solar activity, here indicated by the solar sunspot number ($R = 0, 100, 200$). (From Wright, 1962, and Hargreaves, 1979.)

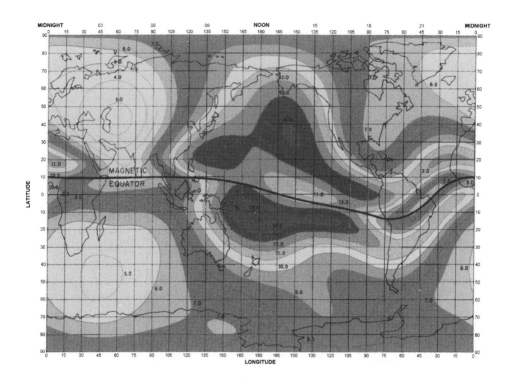

Fig. 5.4. Worldwide map of average ionospheric critical frequency for northern winter conditions at midnight Greenwich Mean Time. (From Richmond, 1987.)

occur at about 20° from the equator in both hemispheres. The explanation is found in electrodynamic lifting at magnetic equator. As the plasma is lifted up due to a horizontal electric field produced by dynamic action in the E-region, the plasma will drift due to a combination of thermal diffusion and electrodynamic drift along the magnetic field lines to higher latitudes. The phenomenon has been described as a "fountain effect".

At high latitudes variations in the electron densities can be very dramatic due to ionization by sporadic auroral particle events. This is illustrated in Fig. 5.6 where 3 different electron density profiles observed during such events are shown. At 100 km the electron density varies by about a factor of 5 within 20 seconds.

As indicated in Fig. 5.2 the ionosphere is situated in an altitude region of rather low neutral particle density, $< 10^{20}$ m^{-3}, i.e. less than one part in a million of the particle density at ground. Even so, the number of charged particles or the degree of ionization is less than one in a million below say 150 km. Above the E-region, however, the degree of ionization increases, and above 400 km more than one particle per thousand is ionized.

Sec. 5.1]   The production of ionization by solar radiation   195

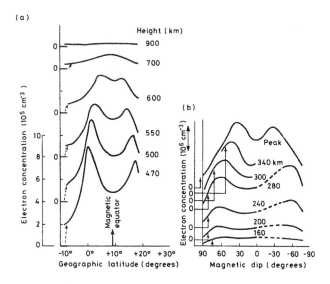

Fig. 5.5. Latitudinal variation of electron density across the equator at various altitudes. (a) from topside sounding, (b) from bottomside sounding. (From Croom et al., 1959, Eccles and King, 1969, and Hargreaves, 1979.)

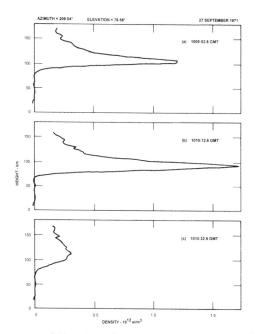

Fig. 5.6. Sequence of E-region electron concentration profiles obtained on Sept. 27, 1971, using the Chatanika Alaska Radar. (From Baron et al., 1971.)

The ion composition changes in accordance with the change of molecular constituents in the neutral atmosphere. As illustrated in Fig. 5.7, NO^+ and O_2^+ are the dominant ions below say 150 km, above which atomic oxygen ions, O^+, are more abundant. Above 300 km, H^+ is a more abundant ion species than either O_2^+ or NO^+. O^+ may dominate even up to 600 km and higher. This will, however, depend strongly on magnetospheric and solar conditions. For strong disturbances in the auroral region heavy ions can dominate up to 200 km or above.

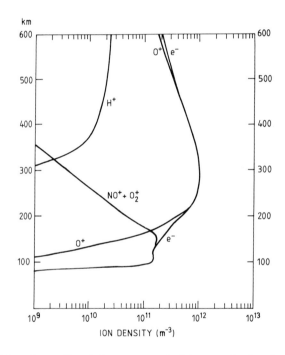

Fig. 5.7. Altitude profile of the most typical ion species in the ionosphere between 100 and 600 km, together with the corresponding electron density profile. (After Richmond, 1987.)

The temperature of the ionosphere is essentially controlled by the absorption of solar UV radiation in the thermosphere. Since this radiation is able to ionize molecules and atoms, free electrons as well as ions are formed which carry some excess energy with them that is obtained during the ionization process. The electrons have a larger mobility and heat conductivity so the temperature of the electrons becomes usually higher than for the ions. The ions are heavier and interact by collisions more strongly with the neutral gas. Because of the similar masses between the colliding species, much of the excess energy of the ions is transferred to the neutral gas. The temperature of the ions is therefore smaller than for the electrons, but higher than for the neutrals. A typical model of temperature profiles for the neutrals, ions and electrons are shown in Fig. 5.8.

Sec. 5.1]   The production of ionization by solar radiation   197

The ion and electron temperature vary strongly with time of day, season and level of disturbance. Illustrated in Fig. 5.9 are isotherm plots for the electrons and the ions according to various local time and height as observed at midlatitude stations. At all heights, from 200 km to 600 km, the electron and ion temperatures increase abruptly at sunrise and decrease almost as rapidly at sunset. The variations in the electron temperature, however, is more rapid than in the ion temperature.

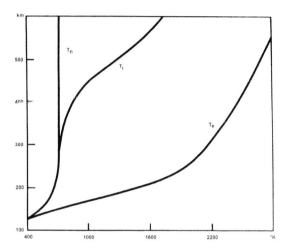

Fig. 5.8. Representative altitude profiles of the ion electron and neutral temperatures between 100 and 600 km. (After Giraud and Petit, 1978.)

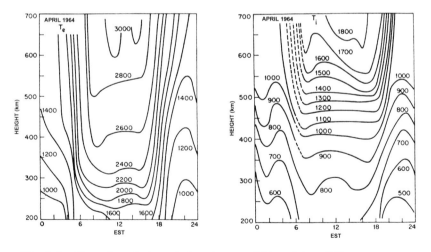

Fig. 5.9. Diurnal variation of the electron (T_e) and ion (T_i) temperatures between 200 and 700 km observed at midlatitude by an incoherent scatter radar. (From Evans, 1967.)

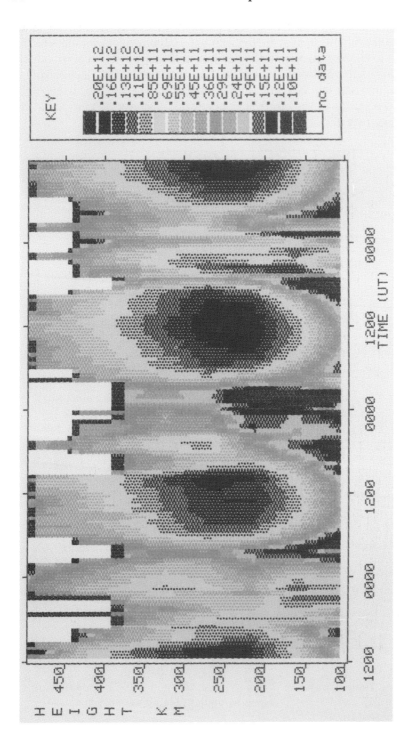

Fig. 5.10. The electron density observed between 100 and 500 km altitude by the European Incoherent Scatter Radar (EISCAT) as function of time for three consecutive days in March 1985. The scale is given in colour code to the right in units per cubic metre. (Courtesy C. Hall, 1989.)

Sec. 5.1] The production of ionization by solar radiation 199

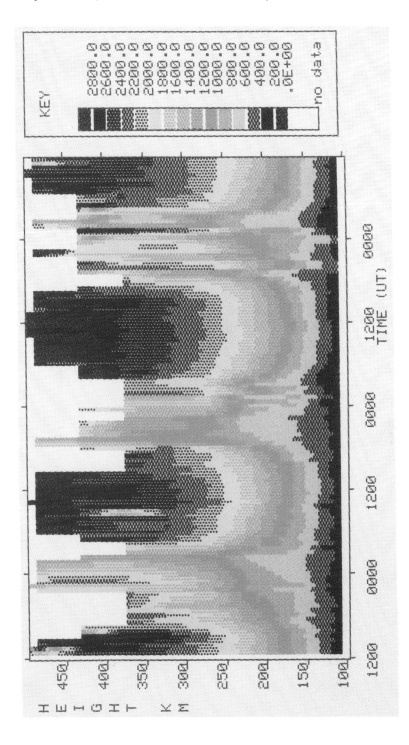

Fig. 5.11. Same format as for Fig. 5.10 but showing the electron temperature for the same days. The scale is given in colour code to the right in kelvins. (Courtesy C. Hall, 1989.)

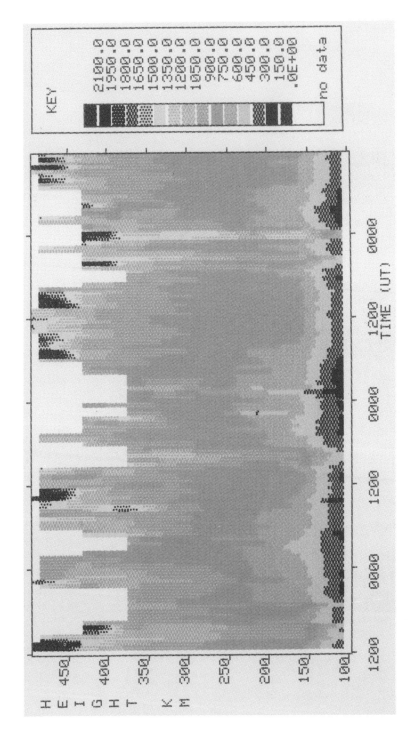

Fig. 5.12. Same format as for Fig. 5.10 but showing the ion temperature for the same days. The scale is given in colour code to the right in kelvins. (Courtesy C. Hall, 1989.)

One of the most powerful tools to study the conditions of the ionosphere from ground is the incoherent scatter radar. One such radar system is situated in the auroral zone close to Tromsø, Norway. That is the European Incoherent Scatter radar (EISCAT). Figs. 5.10, 5.11 and 5.12 show some typical observations of electron densities and temperatures, and ion temperatures, respectively, observed between 100 km and 500 km above ground for 3 consecutive days in March 1985. The electron densities (Fig. 5.10) are strongly enhanced during daytime between 150 and 400 km and reduced especially below 250 km at midnight on March 20 (the middle day). These variations are typical for the ionization due to solar radiation. In the late evening on March 19 (the first day) and during the midnight hours between March 21 and 22 there are strong enhancements in the electron densities reaching down to 100 km and lower. These enhancements are due to ionization caused by precipitating particles or auroral particles.

The electron temperatures shown in Fig. 5.11 in the same format as the electron densities in Fig. 5.12 display a similar behaviour with strong enhancements at daytime and low temperatures or cooling at night-time. Strong temperature enhancements are also seen above 150 km during the events of auroral precipitation discussed above. These temperature enhancements are due to interactions between the beam of auroral particles and the background plasma in the ionosphere. The auroral particle beam is found to enhance the electron temperature in the F-region and the electron density in the E-region.

Finally the ion temperature between 100 and 500 km is displayed in Fig. 5.12 for the same days as the two previous figures. Only small variations are found in the ion temperature in this height region. This is due to the fact that the ions are much more strongly coupled to the neutral atmosphere by collisions and therefore the temperature of the ions will be more dominated by the neutral temperature which is not changing so strongly at auroral latitudes at equinox. The event observed close to midnight on March 21, however, is related to the event of precipitating auroral particles. This event is associated with a large enhancement in the ion velocities, which creates strong ion heating due to collisions with the neutrals. It is characteristic for such events that the ion temperature is enhanced through the whole ionosphere between 100 and 500 km since the ion velocities are increased in the whole height interval.

Before we go into more detail about the different aspects of ionospheric composition, temperature and dynamics, we will discuss how the ionization in the upper atmosphere is produced by solar radiation.

5.2 THE IONIZATION PROFILE OF THE UPPER ATMOSPHERE

We will assume that the target atmosphere which the solar radiation is penetrating through, is a horizontally stratified medium, and that this medium obeys the equation of state for an ideal gas. For a static atmospheric gas in hydrostatic equilibrium the particle density (m^{-3}) will then decrease by altitude as (equation (3.48)):

$$n(z) = n_0 \exp(-z/H) \tag{5.1}$$

where H is the scale height and n_0 the density at some reference height $z = 0$.

The incoming solar radiation at wavelength λ will have an intensity $I(\lambda, z)$ at altitude z. The intensity is measured in eV/m² sec. There will be a cross-section $\sigma(\lambda)$ (m²) for ionizing the neutral particles in the atmosphere by radiation at wavelength λ. The solar radiation at wavelength λ will therefore ionize a number $n \cdot \sigma \cdot I$ of the neutral particles per m³ and second.

If now this radiation with intensity $I(\lambda)$ has passed a distance s through the atmosphere, the intensity will be reduced by dI if it passes an additonal infinitesimal distance ds (Fig. 5.13) through a slab of the atmosphere. This reduction has to be proportional to the intensity of the radiation, the cross-section for ionization, the number of targets that can be ionized and the distance ds as follows:

$$dI = -n \cdot \sigma \cdot I \cdot ds \tag{5.2}$$

Let us assume that for every unit energy absorbed of the radiation, there will be formed a number C electrons. Then the production of electrons per m³ and second can be expressed as:

$$q = C \cdot \sigma \cdot n \cdot I = -C \cdot \frac{dI}{ds} \tag{5.3}$$

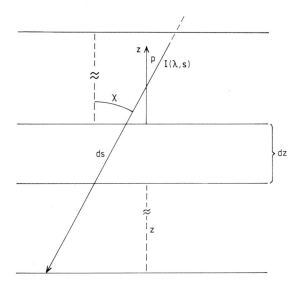

Fig. 5.13. An illustration of the geometry related to the solar irradiation (I) when penetrating a slab of thickness dz in the Earth's atmosphere at a zenith angle χ. The distance to the source and to the ground are s and z, respectively.

Sec. 5.2]   The ionization profile of the upper atmosphere   203

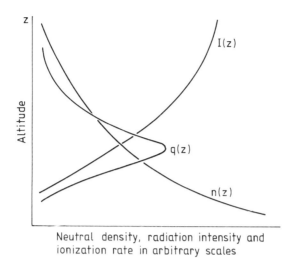

Fig. 5.14. Solar irradiation of intensity I decreases due to absorption downward in the Earth's atmosphere where the density increases toward the Earth. A maximum ionization rate q therefore occurs at the altitude where the cross-over points of the two curves $I(z)$ and $n(z)$ are situated.

C is called the ionization efficiency. We now realize that since n increases and I decreases as we go down in the atmosphere, the product $n \cdot I$ must reach a maximum somewhere (Fig. 5.14), and this maximum is found where:

$$\frac{dq}{ds} = 0 \tag{5.4}$$

or where

$$C \cdot \sigma \cdot \left(I \cdot \frac{dn}{ds} + n \cdot \frac{dI}{ds} \right) = 0 \tag{5.5}$$

since C and σ are constants. By introducing the index m for the maximum we get:

$$\frac{1}{n_m} \cdot \left(\frac{dn}{ds} \right)_m + \frac{1}{I_m} \left(\frac{dI}{ds} \right)_m = 0 \tag{5.6}$$

If the radiation falls in towards the atmosphere by an angle χ with respect to the zenith (Fig. 5.13), we see that

$$ds = -\frac{dz}{\cos \chi} \tag{5.7}$$

and therefore

$$\frac{1}{n} \frac{dn}{ds} = -\frac{1}{n} \frac{dn}{dz} \cos \chi \tag{5.8}$$

For an atmosphere in hydrostatic equilibrium

$$\frac{1}{n}\frac{dn}{dz} = -\frac{1}{H} \tag{5.9}$$

where H is the scale height. And we get:

$$\frac{1}{n}\frac{dn}{ds} = \frac{\cos \chi}{H} \tag{5.10}$$

at any distance s. In particular at the production maximum we have

$$\frac{1}{n_m}\left(\frac{dn}{ds}\right)_m = \frac{\cos \chi}{H} \tag{5.11}$$

From (5.2) we find at this maximum

$$\frac{1}{I_m}\left(\frac{dI}{ds}\right)_m = -\sigma \cdot n_m \tag{5.12}$$

Inserting (5.11) and (5.12) into (5.6) we get

$$\frac{\cos \chi}{H} - \sigma \cdot n_m = 0 \tag{5.13}$$

and then finally

$$\sigma \cdot H \cdot n_m \cdot \sec \chi = 1 \tag{5.14}$$

We know from earlier (equation (3.49)) that for an atmosphere with a constant scale height

$$n_0 \cdot H = \mathcal{N}$$

where $\mathcal{N}$ is the total number per unit area between the reference height and infinity. Especially at maximum ionization therefore

$$H \cdot n_m = \mathcal{N}_m \tag{5.15}$$

where $\mathcal{N}_m$ is the number of neutral particles above a unit area at the height z_m in the atmosphere. Therefore

$$\sigma \cdot \mathcal{N}_m \cdot \sec \chi = 1 \tag{5.16}$$

We notice from (5.2) and (5.7) that the intensity of the radiation is given by:

$$\frac{1}{I}\frac{dI}{ds} = -\frac{1}{I}\frac{dI}{dz}\cos \chi = -\sigma \cdot n = -\sigma \cdot n_0 \exp\left(-\frac{z}{H}\right) \tag{5.17}$$

The ionization profile of the upper atmosphere

and

$$\frac{dI}{I} = +\sigma \cdot n_0 \exp\left(-\frac{z}{H}\right) \sec \chi \, dz \tag{5.18}$$

For $z = \infty$, $I = I_\infty$, that is at the source of the radiation

$$\int_{I_\infty}^{I} \frac{dI}{I} = \sigma \cdot n_0 \cdot \sec \chi \int_{\infty}^{z} \exp\left(-\frac{z}{H}\right) dz \tag{5.19}$$

and

$$\ln \frac{I}{I_\infty} = -\sigma \cdot n \cdot H \cdot \sec \chi \tag{5.20}$$

At the height of maximum ionization therefore:

$$\ln \frac{I_m}{I_\infty} = -\sigma \cdot n_m \cdot H \cdot \sec \chi = -1 \tag{5.21}$$

and

$$I_m = I_\infty / e \tag{5.22}$$

The intensity of the radiation has therefore decreased to $1/e$ at the height of the ion production maximum. In general, however,

$$I = I_\infty \exp(-\sigma \cdot n \cdot H \cdot \sec \chi) = I_\infty \exp(-\tau) \tag{5.23}$$

where $\tau = \sigma \cdot n \cdot H \cdot \sec \chi$ is the "optical depth", and for the height at maximum ionization we have for this optical depth:

$$\tau_m = 1 \tag{5.24}$$

Now neither the ionization cross-section nor the scale height is constant by altitude, therefore a more correct definition of the optical depth would be:

$$\tau_\lambda(z) = -\int_{\infty}^{z} \sigma_\lambda(z') \cdot n(z') \cdot dz' \sec \chi \tag{5.25}$$

Since each wavelength in the solar spectrum will be absorbed differently at different heights in the atmosphere, each wavelength will have its own optical depth. $\sec \chi$ will always be larger than 1 for a solar zenith angle between 0 and 90°. Unit optical depth will therefore be reached at the lowest altitude for an overhead Sun. This is illustrated in Fig. 5.15, where the height in the atmosphere is displayed where the different wavelengths below 3000 Å (300 nm), have passed one optical depth from the source. As already shown, this is equivalent to the height the intensity of each wavelength has been reduced to $1/e$ times the intensity at the impact of the uppermost atmosphere.

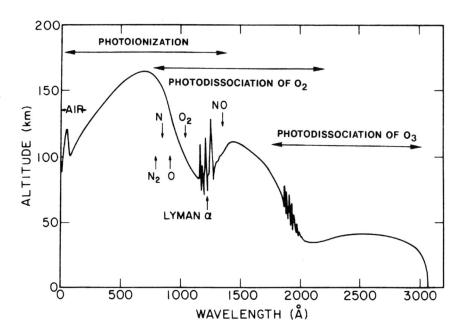

Fig. 5.15. The altitude of unit optical depth in the wavelength region below 3000 Å (300 nm). This corresponds to the level where maximum energy is dissipated at each wavelength. The arrows indicate at what wavelength region the most typical ionization and dissociation processes take place. (From Giraud and Petit, 1978.)

We notice in particular that there are a few spectral lines that are absorbed very strongly with respect to some neighbour lines, and some lines, especially Ly α at 1215 Å, are reaching deeper down in the atmosphere than lines close by in the spectrum. The Ly α line, by the way, penetrates all the way down to the D-region where it ionizes NO.

The different constituents in the atmosphere have different ionization potentials. In Table 5.1 we have listed some of the most abundant species, and their ionization and dissociation potentials in eV. These potentials are also converted to wavelengths if the ionization or dissociation is due to a radiation quantum according to

$$V_p = h\nu = h\frac{c}{\lambda} \tag{5.26}$$

where V_p is the characteristic potential. Only quanta with energies larger than V_p or wavelengths shorter than hc/V_p can ionize or dissociate the given species. From Table 5.1 we therefore notice that radiation in the X-ray region (1–170 Å) and in the extreme ultraviolet regime (170–1750 Å) will be especially significant to the ionization process in the atmosphere.

Typical ionization cross-sections in the EUV regime are of the order of 10^{-17}–10^{-18} cm^2 (10^{-21}–10^{-22} m^2). The variation of the cross-section for ionization of O_2 for wavelengths below 2000 Å (200 nm) is given in Fig. 5.16.

Table 5.1. Ionization and dissociation threshold energies and wavelengths

Species	Ionization		Dissociation	
	V_p (eV)	λ (Å)	V_p (eV)	λ (Å)
N_2	15.58	796	9.76	1270
O_2	12.08	1026	5.12	2422
O	13.61	911		
N	14.54	853		
NO	9.25	1340	6.51	1905
H	13.59	912		
He	24.58	504		

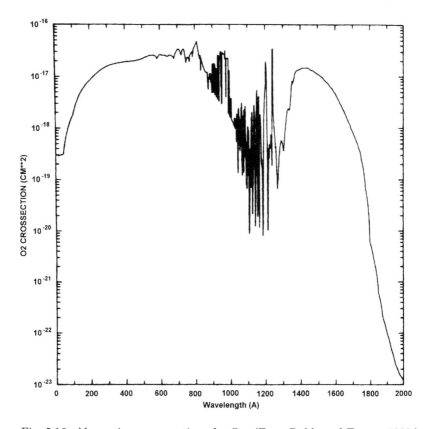

Fig. 5.16. Absorption cross-sections for O_2. (From Roble and Emery, 1983.)

Table 5.2. Some typical values of parameters below 120 km in the atmosphere

Altitude h (km)	Temperature t (K)	Scale height H (m)	Concentration n (m^{-3})	Column density nH (m^{-2})
0	288	8.40(03)*	2.55(25)*	2.14(29)*
5	256	7.50(03)	1.53(25)	1.15(29)
10	223	6.50(03)	8.61(24)	5.60(28)
15	217	6.40(03)	4.04(24)	2.59(28)
20	217	6.40(03)	1.85(24)	1.18(28)
25	222	6.50(03)	8.33(23)	5.41(27)
30	227	6.60(03)	3.83(23)	2.53(27)
35	237	6.90(03)	1.74(23)	1.20(27)
40	250	7.30(03)	6.67(22)	4.87(26)
45	264	7.70(03)	4.12(22)	3.17(26)
50	271	7.90(03)	2.14(22)	1.69(26)
55	261	7.60(03)	1.19(22)	9.04(25)
60	247	7.20(03)	6.45(21)	4.64(25)
65	233	6.80(03)	3.42(21)	2.33(25)
70	220	6.40(03)	1.71(21)	1.09(25)
75	208	6.10(03)	8.36(20)	5.10(24)
80	198	5.80(03)	4.03(20)	2.34(24)
85	189	5.50(03)	1.72(20)	9.46(23)
90	187	5.50(03)	6.98(19)	3.84(23)
95	188	5.50(03)	2.93(19)	1.61(23)
100	195	5.70(03)	1.19(19)	6.78(22)
105	209	6.10(03)	5.20(18)	3.17(22)
110	240	7.00(03)	2.14(18)	1.50(22)
115	300	8.80(03)	9.66(17)	8.50(21)
120	360	1.05(04)	5.03(17)	5.28(21)

*Read 8.40(03) for example as 8.40×10^3.

For an overhead Sun we therefore typically find that unit optical depth for EUV radiation in the Earth's atmosphere corresponds to:

$$\frac{1}{\sigma} = n \cdot H = 10^{22} \text{ m}^{-2} \qquad (5.27)$$

From Table 5.2 we notice that $n \cdot H \approx 10^{22}$ m^{-2} at about 110 km which explains why so much of the solar EUV emission is absorbed above 100 km.

Photodissociation, however, has a smaller cross-section of the order of 10^{-24} m^2. From Table 5.2 we notice that $n \cdot H \approx 10^{24}$ m^{-2} will be close to 80 km in the mesosphere. The collisional cross-section for O_2 is reduced to 10^{-27} m^2 in the Herzberg continuum (2026–2424 Å). This is the reason why these wavelengths can penetrate so deeply into the stratosphere where $n \cdot H = 10^{27}$ m^{-2} and where O_2

can dissociate to form O atoms which again can take part in the creation of ozone by the three-body reaction mentioned earlier.

For atomic species the ionization efficiency C is unity so that all the energy of the radiation goes into producing ion–electron pairs, for molecules, however, $C < 1$.

In general, the energy of the ionizing radiation $h\nu$ will be much larger than V_p, and the excess energy will then be left partly by the ejected photoelectron as kinetic energy E_e and partly as energy of the excited ion E_λ^* so that

$$h\nu = V_p + E_\lambda^* + E_e \tag{5.28}$$

If E_e is higher than V_p, the photoelectron can yield secondary or higher-order ionization during the course of its energy degradation due to further collisions.

According to laboratory experiments it is found that the mean energy $\bar{\varepsilon}$ lost per impact ionization is almost constant as long as $E_e \gg V_p$. The total number of free electrons produced per photoionization is then given by

$$N = 1 + \frac{h\nu - \bar{V}_p}{\bar{\varepsilon}} \tag{5.29}$$

where $\bar{V}_p$ is the mean ionization potential of the target atoms and molecules. It is customary to set $\bar{V}_p = 15$ eV and $\bar{\varepsilon} = 34$ eV in air. $\bar{\varepsilon}$ is larger than $\bar{V}_p$ since a photoelectron can loose its kinetic energy without ionization in the collisional process. Table 5.3 gives values of the average energy $\bar{\varepsilon}$ in eV for an ion-pair production of different species.

Table 5.3

Target species	Average energy for an ion-pair production $\bar{\varepsilon}$ in eV
N_2	35
O_2	32
O	27
H_2	36
He	45
Air	34

5.3 THE CHAPMAN IONIZATION PROFILE

The ion production rate at maximum is according to (5.3), (5.14) and (5.22) given by

$$\begin{aligned} q_m &= C \cdot \sigma \cdot n_m \cdot I_m \\ &= C \cdot \sigma \cdot n_m \cdot \frac{I_\infty}{e} = \frac{C \cdot I_\infty}{e \cdot H} \cos\chi \end{aligned} \tag{5.30}$$

For an overhead Sun we find ($\chi = 0$):

$$q_{m,0} = \frac{C \cdot I_\infty}{e \cdot H} \tag{5.31}$$

and therefore

$$q_m = q_{m,0} \cos \chi \tag{5.32}$$

The production at maximum can never be larger than for an overhead Sun, and it decreases as $\cos \chi$ for increasing zenith angle χ.

For an atmosphere with a constant scale height we also find from the expression for unit optical depth at the height of maximum production (equations (5.23) and (5.24))

$$\tau_\lambda = \sigma \cdot n_0 \exp(-z_m/H) \cdot H \cdot \sec \chi = 1 \tag{5.33}$$

and

$$\exp(z_m/H) = \sigma \cdot n_0 \cdot H \cdot \sec \chi \tag{5.34}$$

Now again, for an overhead Sun

$$\exp(z_{m,0}/H) = \sigma \cdot n_0 \cdot H \tag{5.35}$$

Therefore

$$\exp(z_m/H) = \exp(z_{m,0}/H) \cdot \sec \chi \tag{5.36}$$

and

$$\frac{z_m}{H} = \frac{z_{m,0}}{H} + \ln \sec \chi \tag{5.37}$$

The height of the maximum ion production therefore increases as the zenith angle decreases, and the lowest height ($z_{0,m}$) it can reach is for overhead Sun. Combining (5.20) and (5.21) we find:

$$\ln \frac{I}{I_m} = -(n - n_m)\sigma \cdot H \cdot \sec \chi = -\left(\frac{n}{n_m} - 1\right) \tag{5.38}$$

and finally

$$\frac{I}{I_m} = \exp\left[1 - \frac{n}{n_m}\right] \tag{5.39}$$

This shows the relationship between the intensity of the radiation at a given neutral density relative to the intensity and neutral density at the maximum of ionization. For the ion production rate we then get the following relationship:

$$\frac{q}{q_m} = \frac{C \cdot n \cdot \sigma \cdot I}{C \cdot n_m \cdot \sigma \cdot I_m} = \frac{n}{n_m} \exp\left[1 - \frac{n}{n_m}\right] \tag{5.40}$$

As we have from the barometric law (equation (3.54)) in an isothermal atmosphere

$$\frac{n}{n_m} = \exp\left(-\frac{z-z_m}{H}\right) \tag{5.41}$$

we find that the production q at height z can be expressed as

$$q = q_m \cdot \exp\left(-\frac{z-z_m}{H}\right) \cdot \exp\left[1 - \exp\left(-\frac{z-z_m}{H}\right)\right]$$

or

$$q = q_m \exp\left[1 - \frac{z-z_m}{H} - \exp\left(-\frac{z-z_m}{H}\right)\right] \tag{5.42}$$

Substituting the normalized height reduced to the height of maximum ionization

$$y = \frac{z-z_m}{H} \tag{5.43}$$

$$q = q_m \exp\left[1 - y - \exp(-y)\right]$$

We notice that independently of the zenith angle χ the ionization profile will maintain its form.

Inserting (5.37) into (5.42) we get:

$$\begin{aligned} q &= q_m \exp\left[1 - \frac{z-z_{m,0}}{H} + \ln\sec\chi - \exp\left(-\frac{z-z_{m,0}}{H} + \ln\sec\chi\right)\right] \quad (5.44) \\ &= q_m \sec\chi \exp\left[1 - \frac{z-z_{m,0}}{H} - \sec\chi \exp\left(-\frac{z-z_{m,0}}{H}\right)\right] \end{aligned}$$

By introducing (5.32) into (5.44) we get:

$$q = q_{m,0} \cdot \exp\left[1 - \frac{z-z_{m,0}}{H} - \sec\chi \cdot \exp\left(-\frac{z-z_{m,0}}{H}\right)\right] \tag{5.45}$$

By introducing the normalized height $x = (z-z_{m,0})/H$ reduced to the height of maximum ionization for overhead Sun, we find

$$q = q_{m,0} \exp(1 - x - \sec\chi \cdot \exp(-x)) \tag{5.46}$$

For very large x ($x \gg 0$ or $z \gg z_{m,0}$) the profile takes the form

$$q \approx q_{m,0} \exp(-z/H) \tag{5.47}$$

Well above the maximum in the ionization profile the profile itself decays by altitude as the density of the target atmosphere. This holds for $x > 2$, that is at a distance more than two scale heights above $z_{m,0}$. This relation arises because at those heights the intensity of the radiation is only weakly reduced, and the rate of production is essentially proportional to the density of the target gas.

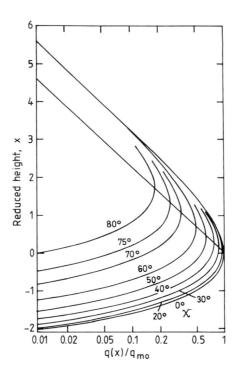

Fig. 5.17. Chapman production profiles for different solar zenith angles. (After Van Zandt and Knecht, 1964.)

For very small x ($x \ll 0$ or $z \ll z_{m,0}$) the production rate takes the form

$$q = q_{m,0} \exp\left[-\sec\chi \exp\left(-\frac{z - z_{m,0}}{H}\right)\right] \tag{5.48}$$

and the profile decreases very rapidly with height below the altitude of peak production.

Fig. 5.17 illustrates the ionization profiles for different zenith angles between 0° and 80°. The altitude parameter used is the reduced height x. The profiles all have the same form, the peak production rate decreases and the height of the maximum increases when the zenith angle increases in agreement with the mathematical treatment shown above.

For a real target atmosphere and a real solar spectrum it is a rather time consuming effort to derive the final ion production profile. The upper atmosphere is composed of different gases with different scale heights, and the solar radiation spectrum consists of a myriad of lines and bands with different intensities. The approach is therefore to assume a neutral atmosphere model with height distribution of some of the major gases in concern, and then for a finite number of wavelengths derive the individual production profiles. Examples of such a

Sec. 5.4] **The recombination process** 213

procedure is shown in Fig. 5.18. We notice that the shorter the wavelength, the deeper down into the atmosphere is the corresponding production maximum, and that wavelengths in the X-ray regime (1–170 Å) can ionize below 100 km. The individual ion density profiles being a result of the ionization of the composed radiation spectrum, are shown in the upper panel in Fig. 5.18 together with the total ionization profile. The production of the O^+ ions are found to be dominant above 200 km, and of the O_2^+ ions below 130 km according to these model calculations.

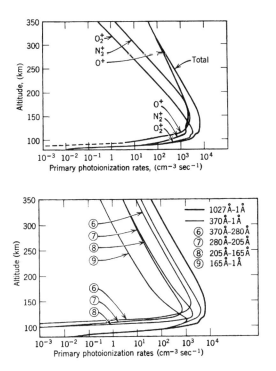

Fig. 5.18. Calculated ionization rates in the E- and F-regions. (From Hinteregger et al., 1965.)

5.4 THE RECOMBINATION PROCESS

The ionization thus being formed will immediately start to take part in chemical reactions to re-establish the equilibrium condition before the ionization took place. Some of the ionization will also be carried away by transport mechanisms such as diffusion, neutral winds and electric fields.

The continuity equation for an ion species formed can therefore be written as

$$\frac{\partial n_i}{\partial t} = q_i - l_i - \nabla \cdot (\mathbf{v}_i n_i) \qquad (5.49)$$

where n_i is the density of the ion species, q_i is the production, l_i is the loss by chemical and photochemical processes, and the last term is the divergence due to transport phenomena through a volume of interest (the convection term). In the following section we will discuss the loss term l_i for different regions of the ionosphere and study the consequences of this. Later on we will study the transport phenomena pertinent to the ionized species.

Let us start with neglecting the transport term and assume photochemical equilibrium

$$q_i = l_i \qquad (5.50)$$

Let us further assume that the electrons (e) recombine directly with the positive ions (X^+) to form a neutral species:

$$X^+ + e \longrightarrow X + h\nu \qquad (5.51)$$

with the possibility of an accompanying radiation of a photon, so-called radiative recombination. Another possibility for reducing the number of free electrons is by dissociative recombination of a molecule XY^+ as follows:

$$XY^+ + e \longrightarrow X + Y \qquad (5.52)$$

In fact, the latter is more likely to take place in the upper atmosphere because it results in two particles whereas the first reaction results in only one and a photon. The needs for conservation of energy and momentum are not so easily satisfied in the radiative recombination process, and therefore dissociative recombination is more likely. Calculated magnitudes of the dissociative recombination rate is of the order of 10^{-13} m^3 s^{-1}, while the rate for radiative recombination is much smaller and of the order of 10^{-18} m^3 s^{-1}.

Whether or not it is the radiative recombination process or the dissociative recombination process that is dominant, we will in both cases for simplicity assume charge neutrality so that the number density of positive ions $[X^+]$ or $[XY^+]$ are equal to the number density n_e of free electrons.

$$[X^+] = n_e \quad \text{or} \quad [XY^+] = n_e$$

The loss rate will be proportional to the product of the electron and ion densities, and the proportionality factor α is called the recombination coefficient, thus

$$l_i = \alpha n_e^2 \qquad (5.53)$$

Since the production is given by (5.46), we find for photochemical equilibrium

$$l_i = q_i = q_{m,0} \exp[1 - x - \sec\chi \exp(-x)] = \alpha n_e^2 \qquad (5.54)$$

The electron density at a given height z is therefore given by

$$n_e(z) = \sqrt{\frac{q_{m,0}}{\alpha}} \exp\left[\frac{1}{2}(1 - x - \sec\chi \exp(-x))\right] \qquad (5.55)$$

where x is the normalized and reduced height parameter as defined above. When neglecting height variations in α, we find that the electron density has a maximum when

$$e^{-x} = \cos \chi \tag{5.56}$$

showing that the electron density at maximum is given by

$$n_m = \sqrt{\frac{q_{m,0}}{\alpha}} \cdot (\cos \chi)^{1/2} = n_{m,0} (\cos \chi)^{1/2} \tag{5.57}$$

The maximum in the electron density therefore varies as $(\cos \chi)^{1/2}$ with the zenith angle. This is often called a Chapman α-profile and is representative for the E-region.

If, on the other hand, the electrons are lost by attachment to a molecule

$$M + e \longrightarrow M^- \tag{5.58}$$

then the loss rate is proportional to n_e and can be expressed as:

$$l_i = \beta n_e \tag{5.59}$$

where β is proportional to $[M]$, the density of the neutral molecule M. In equilibrium then

$$q_i = l_i = \beta n_e \tag{5.60}$$

and the electron density profile is given by

$$n_e = \frac{q_{m,0}}{\beta} \exp(1 - x - \sec \chi \exp(-x)) \tag{5.61}$$

If we again neglect height variations in β, the maximum electron density is found when $\exp(-x) = \cos \chi$ and therefore

$$n_m = \frac{q_{m,0}}{\beta} \cos \chi = n_{m,0} \cos \chi \tag{5.62}$$

The maximum density varies with the zenith angle as $\cos \chi$. This is often called a Chapman β-profile.

The negative ions formed in the attachment process are usually detached by other reactions so that the electron loss by attachment can be neglected in the ionospheric E-region and above. In the lower ionosphere, however, it will be of importance.

In the real ionosphere atomic and molecular ions coexist with atoms and molecules and different reactions can take place at the same time. We will now study the importance of these processes at the different regions and in particular in the transition between the E- and F-regions.

From Fig. 5.7 we notice that below 150 km molecular ions such as O_2^+ and NO^+ are the dominant species. Above 150 km, however, the O^+ ion is becoming more and more abundant and will dominate above, say, 200 km.

It is therefore evident that in the E-region dissociative recombination with molecular ions will be the most important loss mechanism for the electrons, while in the topside ionosphere loss mechanisms related to ionic oxygen atoms must dominate. In between these regions in the so-called F1-region, one or the other loss mechanism may dominate depending on the relative values of the reaction rate coefficients.

5.5 THE O^+ DOMINANT IONOSPHERE

In the F-region, say above 150 km, the dominant ions formed by solar irradiance are the O^+ ions. In the following we therefore will assume that these ions are the only ion species resulting from solar radiation and that the other ions are produced as a result of chemical reactions.

The atomic oxygen ions when produced, can be lost by several reactions. One possibility is radiative recombination with an electron as follows:

$$O^+ + e \xrightarrow{\alpha_r} O + h\nu \tag{5.63}$$

Here we denote α_r (Table 5.4) as the radiative recombination coefficient for this type of process. This process is, however, as already mentioned, rather slow for sake the energy and momentum balance which has to be maintained. A more rapid loss process for the O^+ ions is through a chain of reactions. First O^+ recombines with the abundant N_2 molecule and forms a NO^+ ion with a rate coefficient k_1 (Table 5.4) as follows:

$$O^+ + N_2 \xrightarrow{k_1} NO^+ + N \tag{5.64}$$

The NO^+ ion recombines with an electron and dissociates:

$$NO^+ + e \xrightarrow{\alpha_1} N + O \tag{5.65}$$

where we denote α_1 (Table 5.4) as the relevant dissociative recombination coefficient.

Another loss process for O^+ ions is through a similar chain of reactions with O_2^+ ions. First the O^+ ion charge exchanges with an O_2 molecule

$$O^+ + O_2 \xrightarrow{k_2} O_2^+ + O \tag{5.66}$$

with a rate coefficient for the process given by k_2 (Table 5.4). The O_2^+ ion formed recombines with an electron and dissociates as follows:

$$O_2^+ + e \xrightarrow{\alpha_2} O + O \tag{5.67}$$

The dissociative recombination rate for this process is denoted α_2 (Table 5.4).

In a quasi-chemical photoequilibrium condition the production of atomic oxygen ions must be equal to the loss of these ions

$$q(O^+) = (k_1 \cdot [N_2] + k_2 \cdot [O_2]) \cdot [O^+] = l(O^+) \tag{5.68}$$

Table 5.4. Typical reaction rates relevant for ionospheric processes as indicated

Processes	Reaction rates (m³/sec)
Dissociative recombination	
$NO^+ + e \longrightarrow N + O$	$\alpha_1 = 2.1 \times 10^{-13} \, (T_e/300)^{-0.85}$
$O_2^+ + e \longrightarrow O + O$	$\alpha_2 = 1.9 \times 10^{-13} \, (T_e/300)^{-0.5}$
$N_2^+ + e \longrightarrow N + N$	$\alpha_3 = 1.8 \times 10^{-13} \, (T_e/300)^{-0.39}$
Rearrangement	
$O^+ + N_2 \longrightarrow NO^+ + N$	$k_1 = 2 \times 10^{-18}$
$O^+ + O_2 \longrightarrow O_2^+ + O$	$k_2 = 2 \times 10^{-17} \, (T_r/300)^{-0.4}$
$O_2^+ + NO \longrightarrow NO^+ + O_2$	$k_3 = 4.4 \times 10^{-16}$
$O_2^+ + N_2 \longrightarrow NO^+ + NO$	$k_4 = 5 \times 10^{-22}$
$N_2^+ + O \longrightarrow NO^+ + NO$	$k_5 = 1.4 \times 10^{-16} \, (T_r/300)^{-0.44}$
$N_2^+ + O_2 \longrightarrow N_2 + O_2^+$	$k_6 = 5 \times 10^{-17} \, (T_r/300)^{-0.8}$
Radiative recombination	
$O^+ + e \longrightarrow O + h\nu$	$\alpha_r = 7.8 \times 10^{-14} \, (T_e/300)^{-0.5}$
Electron attachment	
$O_2 + M + e \longrightarrow O_2^- + M$	$\beta_1 = 10^{-43} \, (300/T_e) e^{-600/T_e} \, [N_2]$
	$\beta_2 = 10^{-41} \, [O_2]$
$O + e \longrightarrow O^- + h\nu$	$\beta_r = 10^{-21}$
Electron detachment	
$O_2^- + h\nu \longrightarrow O_2 + e$	$\rho = 0.33 \, \text{sec}^{-1}$
$O_2^- + N_2 \longrightarrow O_2 + N_2 + e$	$\gamma_1 = 2 \times 10^{-18} \, (T_r/300)^{1.5} \exp(5 \times 10^{-3}/T_r)$
$O_2^- + O_2 \longrightarrow O_2 + O_2 + e$	$\gamma_2 = 3 \times 10^{-16} \, (T_r/300)^{0.5} \exp(6 \times 10^{-3}/T_r)$
$O_2^- + O \longrightarrow O_3 + e$	$\gamma_3 = 5 \times 10^{-16}$
$O^- + O \longrightarrow O_2 + e$	$\gamma_4 = 3 \times 10^{-16}$
Ion–ion recombination	
$O_2^+ + O_2^- \longrightarrow O_2 + O_2$	$\alpha_{i1} = 2 \times 10^{-13}$
$O_2^+ + O_2^- + M \longrightarrow O_2 + O_2 + M$	$\alpha_{i2} = 3 \times 10^{-31} \, (T/300)^{-2.5} \, [M]$

when we neglect radiative recombination. For the molecular oxygen ions we have a similar equilibrium condition when neglecting the production of O_2^+ due to solar radiation

$$q(O_2^+) = k_2[O_2] \cdot [O^+] = \alpha_2[O_2^+] \cdot n_e = l(O_2^+) \tag{5.69}$$

and for the NO^+ ions when no solar production is assumed

$$q(NO^+) = k_1[O^+] \cdot [N_2] = \alpha_1[NO^+] \cdot n_e = l(NO^+) \tag{5.70}$$

Since the number of positive and negative charges must be equal

$$n_e = [O^+] + [O_2^+] + [NO^+] \tag{5.71}$$

we have:

$$\begin{aligned} n_e &= [O^+] + \frac{k_1}{\alpha_1}\frac{[N_2]}{n_e}[O^+] + \frac{k_2}{\alpha_2}\frac{[O_2]}{n_e}[O^+] \\ &= \left(1 + \frac{k_1}{\alpha_1}\frac{[N_2]}{n_e} + \frac{k_2}{\alpha_2}\frac{[O_2]}{n_e}\right)[O^+] \end{aligned} \tag{5.72}$$

and finally

$$q(O^+) = \left(\frac{k_1[N_2] + k_2[O_2]}{1 + \frac{k_1}{\alpha_1}\frac{[N_2]}{n_e} + \frac{k_2}{\alpha_2}\frac{[O_2]}{n_e}}\right) n_e \tag{5.73}$$

We therefore express the production as follows:

$$q(O^+) = \beta' \cdot n_e \tag{5.74}$$

where β' is the loss rate given by

$$\beta' = \frac{k_1[N_2] + k_2[O_2]}{1 + \frac{k_1}{\alpha_1}\frac{[N_2]}{n_e} + \frac{k_2}{\alpha_2}\frac{[O_2]}{n_e}} \tag{5.75}$$

The values for k_1 and k_2 are of the order of 2×10^{-18} m^3 sec (see Table 5.4), while α_1 and α_2 are of the order of 10^{-13} m^3 sec, and we see that for altitudes above 250 km or so where $[N_2] < 10^{15}$ m^{-3} and $[O_2] \approx 10^{14}$ m^3, $\alpha_1 \cdot n_e$ and $\alpha_2 \cdot n_e$ will be much larger than $k_1[N_2]$ and $k_2[O_2]$ respectively, for an electron density of 10^{11} m^{-3} or less. It is therefore a good approximation to set

$$\beta = k_1[N_2] + k_2[O_2] \tag{5.76}$$

and since N_2 and O_2 have similar scale heights, we have approximately:

$$\beta = \beta_0 \exp\left[-\frac{z - z_0}{H}\right] \tag{5.77}$$

where β_0 is the effective loss rate at reference height z_0 and H is the scale height.

If we move lower down in the ionosphere to the lower F-region and the E-region, the reactions involving charge rearrangement between O^+ and O_2 and N_2 will become more abundant because of the increase in the molecular neutral density and therefore the complete expression for β' must be maintained, and:

$$n_e = \frac{q(O^+)}{\beta'} = q(O^+) \cdot \frac{1 + \left(\frac{k_1}{\alpha_1}[N_2] + \frac{k_2}{\alpha_2}[O_2]\right)\frac{1}{n_e}}{\beta} \tag{5.78}$$

Sec. 5.5] The O⁺ dominant ionosphere

We can then form

$$n_e = q(O^+) \left[\frac{1}{\beta} + \frac{1}{\alpha_{\text{eff}} n_e} \right] \tag{5.79}$$

where

$$\frac{1}{\alpha_{\text{eff}}} = \left[\frac{k_1[N_2]}{\alpha_1} + \frac{k_2[O_2]}{\alpha_2} \right] \frac{1}{\beta} \tag{5.80}$$

Since β decreases by height and $\alpha_{\text{eff}} \cdot n_e$ will increase by height under the F-region maximum, there can be a transition height z_t at which $n_e = n_{et}$ and where

$$\beta_t = \alpha_{\text{eff}} n_{et} \tag{5.81}$$

Above this height the electron density will vary as q/β. Below this height, however, the electron density varies as $\sqrt{q/\alpha_{\text{eff}}}$.

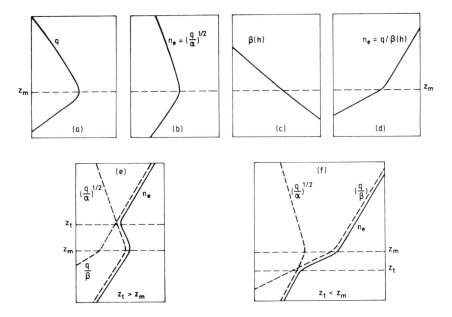

Fig 5.19. Schematic diagram showing the formation of the F_1-layer. (a) The ionization profile q has a maximum at z_m. (b) For a constant recombination constant α by height will the electron density profile have a maximum at z_m. (c) The electron loss rate β decays exponentially with height. (d) For an electron loss process dominated by β the electron density profile n_e would have a knee at z_m. (e) The electron loss due to recombination (α) and attachment (β) are equal at an altitude z_t. If the production maximum at z_m is below z_t, the electron density profile will have a maximum at z_m and a minimum at z_t. (f) If the production maximum at z_m is above z_t, the electron density will have a ledge between z_t and z_m. (After Ratcliffe, 1972.)

If the latter situation applied at all heights, the electron density profile would take the form shown in Fig. 5.19b, where a single maximum would occur at height z_m of the maximum of the production (Fig. 5.19a). If the first situation applied, then β would decrease by height as shown in Fig. 5.19c, and a knee would occur in the electron density profile at the height z_m of the production maximum (Fig. 5.19d). The decay of β by altitude would enhance the electron density above the peak due to a slower loss rate.

If now the transition point z_t is above the maximum of production, there will still be a maximum in the electron density profile at the peak of production (Fig. 5.19e), since the loss rate here is still dominated by dissociative recombination. If, however, the transition height z_t is situated below the maximum of production at z_m, there would be a ledge in the electron density profile at z_t and a knee at z_m, since the β-loss will dominate over the α-loss at z_m.

We have seen that for distances larger than 2 scale heights above the ionization maximum, the ion production profile decays by altitude with the scale height of the background atmosphere, i.e., with the scale height of the oxygen atoms (H_O) since these are the major ionization target above the F-region peak. We then have

$$q(z) \approx \exp\left(-\frac{z}{H_O}\right) \tag{5.82}$$

Since β is proportional to $k_1[N_2] + k_2[O_2]$, it decays by altitude as the density of N_2 and O_2. As N_2 is the dominant species of these two (Fig. 3.2), we assume:

$$\beta(z) \approx \exp\left(-\frac{z}{H_{N_2}}\right) \tag{5.83}$$

where H_{N_2} is the scale height of the N_2 species. The electron density profile at the topside ionosphere therefore will vary according to the following expression:

$$\begin{aligned} n_e(z) &= \frac{q(z)}{\beta(z)} \propto \exp\left(-\frac{z}{H_O}\right) \cdot \exp\left(\frac{z}{H_{N_2}}\right) \\ &= \exp\left[\left(\frac{H_O}{H_{N_2}} - 1\right)\frac{z}{H_O}\right] \end{aligned} \tag{5.84}$$

Since

$$\frac{H_O}{H_{N_2}} = \frac{7}{4} \tag{5.85}$$

we find

$$n_e(z) \propto \exp\left(\frac{3z}{4H_O}\right) \tag{5.86}$$

The electron density profile above the F-region peak will increase by a scale height which is $1.33 \cdot H_O$ if the ionosphere is in photochemical equilibrium. If our assumptions were appropriate for all regions, the electron density would therefore increase to infinity above the altitude z_t.

5.6 AMBIPOLAR DIFFUSION

There has to be more to it: the ionosphere is simply not in photochemical equilibrium above the F-region peak.

We have to turn our attention to diffusion to see what that can do to the problem.

Let us assume charge neutrality ($n_e = n_i = n$), that the neutral air is at rest ($\mathbf{u} = 0$) and that no current is present. The last assumption means that $\mathbf{v}_i = \mathbf{v}_e = \mathbf{v}$, where $\mathbf{v}_i$ and $\mathbf{v}_e$ are the ion and electron velocities respectively. Since the mass of the ions is so much greater than the mass of the electrons, a charge separation will occur due to the gravity field which creates an electric field. This field, however, will represent a strong force on the electrons compared to gravity and the charge separation will be blocked. Especially for the vertical motion we will have

$$w_i = w_e = w \tag{5.87}$$

where w_i and w_e are the vertical velocity components for the ions and the electrons respectively. We will at present assume no magnetic field and an isothermal plasma in thermal balance with the neutrals

$$T_e = T_i = T_n = T \tag{5.88}$$

where T_i, T_e and T_n are the ion, electron and neutral temperatures respectively. The momentum equations in the vertical direction for the ions and electrons now become:

$$n_i m_i \frac{\partial w_i}{\partial t} = -\frac{\partial p_i}{\partial z} - n_i m_i g + n_i e E - n_i m_i \nu_i w_i = 0 \tag{5.89}$$

$$n_e m_e \frac{\partial w_e}{\partial t} = -\frac{\partial p_e}{\partial z} - n_e m_e g - n_e e E - n_e m_e \nu_e w_e = 0 \tag{5.90}$$

when neglecting any collision between electrons and ions. By adding these two equations and leaving the collision terms on the left-hand side:

$$n(m_i \nu_i + m_e \nu_e) w = -\frac{\partial}{\partial z}(p_i + p_e) - n(m_i + m_e) g \tag{5.91}$$

Assuming that the ideal gas law applies for the plasma and that T is constant by height:

$$n(m_i \nu_i + m_e \nu_e) w = -2\kappa T \cdot \frac{\partial n}{\partial z} - n(m_i + m_e) g \tag{5.92}$$

Taking into account that $m_i \gg m_e$ and solving for the vertical particle flux nw

$$nw = -\frac{2\kappa T}{(m_i \nu_i + m_e \nu_e)} \left(\frac{\partial n}{\partial z} + \frac{n m_i g}{2\kappa T} \right) \tag{5.93}$$

We notice that in this equation the gravity term has exactly the same form as in the equation of vertical diffusion of the neutral gas (equation (3.119)) except that the mass has been changed from m to $\frac{1}{2}m_i$. The scale height of the plasma is therefore given as:

$$H_p = \frac{2\kappa T}{m_i g} = 2H \tag{5.94}$$

where H is the scale height of the neutral atmosphere. This is true if the neutrals and the ions have the same molecular mass, as they have if O^+ is moving in an atomic oxygen dominated atmosphere or if H^+ is moving in a hydrogen dominated one. We have noticed that the effect of the polarization electric field is to render the plasma to a gas with a mean molecular mass equal to $\frac{1}{2}m_i$.

The diffusion coefficient of the plasma will be given by:

$$D_p = \frac{2\kappa T}{m_i \nu_i + m_e \nu_e} \approx 2D \tag{5.95}$$

when $m_i \nu_i \gg m_e \nu_e$, and D is the diffusion coefficient of the neutral gas (equation (3.114)). D_p is often called the *ambipolar diffusion coefficient* referring to the polarization field between the ions and electrons which forces the plasma to move as a whole.

The equation of diffusion now will be

$$\frac{\partial n}{\partial t} = -\frac{\partial}{\partial z}(nw) = \frac{\partial}{\partial z}\left\{ D_p \left(\frac{\partial n}{\partial z} + \frac{n}{H_p} \right) \right\} \tag{5.96}$$

Equilibrium can be achieved if

$$\frac{\partial n}{\partial z} = -\frac{n}{H_p} \tag{5.97}$$

or

$$n = n_0 \exp\left(-\frac{z - z_0}{H_p}\right) \tag{5.98}$$

where n_0 is the plasma density at a reference height z_0; for example at the F-region maximum. The equilibrium profile of the electron density will therefore be exponential and decay by altitude with a scale height which is twice as great as the scale height of the neutral atmosphere.

Since the diffusion coefficient for $m_i \nu_i \gg m_e \nu_e$ is given by:

$$D_p = \frac{2\kappa T}{m_i \nu_i} \tag{5.99}$$

it will increase exponentially by altitude as

$$D_p = D_0 \exp\left(\frac{z}{H}\right) \tag{5.100}$$

since ν_i decreases proportionally to the neutral density, and the diffusion will become more important at greater heights. The electron density profile will, however, decrease by altitude above the F-region peak due to the effect of the diffusion.

We notice, however, that if $T_e \neq T_i$ and $T_e = c \cdot T_i$, where c is a constant, the plasma scale height would be

$$H_p = (1+c)H \tag{5.101}$$

and the plasma diffusion coefficient would become ($m_i\nu_i \gg m_e\nu_e$):

$$D_p = (1+c)D \tag{5.102}$$

The shape of the profile and the effect of diffusion therefore is strongly dependent on the electron–ion temperature ratio. This can indeed be very large (Fig. 5.8).

5.7 MULTICOMPONENT TOPSIDE IONOSPHERE

In the topside ionosphere where ions are expected to be in diffusive equilibrium, there are observed different species like O^+, H^+, He^+, N^+ etc. The electric field that occurs will result in a force that is the same for individual species, and it will combine with the gravity force to produce different effective masses for individual species. For some the upward electric field force may be larger than the downward force, and it may appear that some ions have a negative apparent mass and a negative distribution height. Let m_j and n_j be the mass and density of the jth ion species. The momentum equation when collisions are neglected for these ions in diffusive equilibrium then becomes:

$$-\kappa T_i \frac{\partial n_j}{\partial z} - n_j m_j g + n_j eE = 0 \tag{5.103}$$

when it is assumed that $w_j = w_n$, where w_n is the vertical velocity of the neutrals and T_i is the ion temperature equal for all ion species. The electron momentum equation, when collisions are neglected, becomes:

$$-\kappa T_e \frac{\partial n_e}{\partial z} - n_e eE = 0 \tag{5.104}$$

where we have neglected the gravity force on the electrons. T_e is the electron temperature. We now add all equations for all ion species and the electrons

$$-\kappa T_i \frac{\partial}{\partial z} \Sigma n_j - \kappa T_e \frac{\partial n_e}{\partial z} - \Sigma n_j m_j g + (\Sigma n_j - n_e)eE = 0 \tag{5.105}$$

Electrical neutrality requires

$$\Sigma n_j - n_e = 0 \tag{5.106}$$

and by setting the mean mass for the ions $\Sigma n_j m_j / n_e$ equal to m_+ we get:

$$-\kappa (T_i + T_e) \frac{\partial n_e}{\partial z} - n_e m_+ g = 0 \tag{5.107}$$

Let $T_e = c \cdot T_i$, where c is a constant, we finally have:

$$\frac{1}{n_e} \frac{\partial n_e}{\partial z} = \frac{-m_+ g}{\kappa(1+c)T_i} \tag{5.108}$$

which gives the distribution height for the electrons. When inserting (5.104) we find that

$$\frac{eE}{c\kappa T_i} = \frac{m_+ g}{(1+c)\kappa T_i} \tag{5.109}$$

and

$$eE = \frac{m_+ g c}{1+c} \tag{5.110}$$

The electric field is therefore a function of the electron–ion temperature ratio c, and the mean mass of the ions. For the jth ion species we then have

$$\frac{1}{n_j}\frac{\partial n_j}{\partial z} = -\frac{m_j g}{\kappa T_i} + \frac{m_+ g c}{(1+c)\kappa T_i} = -\frac{1}{H_j}\left(1 - \frac{c}{1+c}\frac{m_+}{m_j}\right) \tag{5.111}$$

where $H_j = \kappa T_i / m_j g$. This describes the vertical distribution of the jth ion species, and we see that in diffusive equilibrium this species will have a distribution height different from its scale height

$$\delta_j = H_j \left(1 - \frac{c}{1+c}\frac{m_+}{m_j}\right)^{-1} \tag{5.112}$$

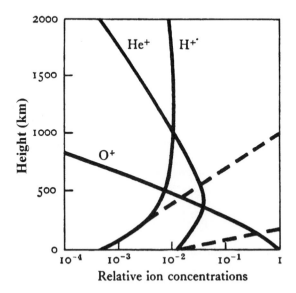

Fig. 5.20. Relative ion concentration profiles below 2000 km altitude for a mixture of H^+, He^+, O^+ ions and electrons in diffusive equilibrium for a temperature equal to 1200 K. The dashed lines correspond to conditions where photochemical equilibrium prevails. (From Ratcliffe, 1972.)

Sec. 5.8]   **Diffusion in the presence of a magnetic field**   225

which depends on the mass of the species with respect to the mean ion mass as well as the temperature ratio $c = T_e/T_i$. For $m_j < [c/(1+c)]m_+$ therefore the distribution height will be negative. In an ionosphere where $T_e > T_i$, the density of the lightest ions will increase by height.

We should notice that δ_j is dependent on the relative number of the different ion species as well as the temperature ratio c which both change by height. Therefore the distribution height is not a constant and the height variation is not a true exponential. An example of the height distribution of some ion species in the topside ionosphere when c is assumed equal to 1, is given in Fig. 5.20.

5.8 DIFFUSION IN THE PRESENCE OF A MAGNETIC FIELD

The high conductivity along a magnetic field line forces the plasma to move along the field and to some extent restrain the true vertical motion. The diffusion equation along the magnetic field line would therefore read as follows:

$$nv_\| = -\frac{2\kappa T}{m_+\nu_i}\left(\frac{\partial n}{\partial l} + \frac{nm_+ g_\|}{2\kappa T}\right) \tag{5.113}$$

where l is measured along B (Fig. 5.21).

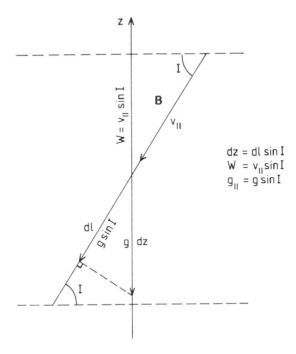

Fig. 5.21. The geometry for ionospheric diffusive motion in the presence of a magnetic field making an angle I with the horizontal plane.

Since $dl = dz/\sin I$ and $g_\| = g \sin I$, when I is the angle between the field and the horizontal plane, we find:

$$nv_\| = -\frac{2\kappa T}{m_+ \nu_i}\left(\frac{\partial n}{\partial z}\sin I + \frac{nm_+ g}{2\kappa T}\sin I\right) \tag{5.114}$$

$$= -D_p \sin I \left(\frac{\partial n}{\partial z} + \frac{n}{H_p}\right)$$

Furthermore

$$w = v_\| \sin I \tag{5.115}$$

and we have

$$nw = -\sin^2 I \frac{2\kappa T}{m_+ \nu_i}\left(\frac{\partial n}{\partial z} + \frac{nm_+ g}{2\kappa T}\right) = -\sin^2 I\, D_p \left(\frac{\partial n}{\partial z} + \frac{n}{H_p}\right) \tag{5.116}$$

and the vertical diffusion flux increases in magnitude as I increases, which is the situation when moving to higher latitudes. At the magnetic equator, however, this ambipolar diffusion would come to a stop and other forces, as we shall see later, must be included to complete the picture.

Let us return to the continuity equation (5.49) where we only allow for vertical motion of the plasma and the loss rate is equal to the linear loss (βn), in the presence of a magnetic field:

$$\frac{\partial n}{\partial t} = q_i - \beta n - \frac{\partial}{\partial z}nw \tag{5.117}$$

$$= q_i - \beta n + D_p \sin^2 I \left\{\frac{\partial^2 n}{\partial z^2} + \left(\frac{1}{H} + \frac{1}{H_p}\right)\frac{\partial n}{\partial z} + \frac{n}{H \cdot H_p}\right\}$$

when $D_p = D_0 \exp(z/H)$. Since this is a linear quadratic differential equation, it can be solved analytically by separating the variables. We will only discuss the steady-state solution, but before we do that, we must investigate the boundary conditions and the asymptotic solutions.

1. At the lowest heights where ν_i is very large and $D_p \approx 0$ for a steady state, we then obtain

$$q_i \approx \beta n \tag{5.118}$$

which is the well-known Chapman β relation.

2. At the upper height where $q_i \approx 0$ and $\beta \approx 0$ we obtain the steady-state solution:

$$\frac{d^2 n}{dz^2} + \left(\frac{1}{H} + \frac{1}{H_p}\right)\frac{dn}{dz} + \frac{n}{H \cdot H_p} = 0 \tag{5.119}$$

This equation has a solution which can be given by:

$$n = C_1 \exp\left(-\frac{z}{H}\right) + C_2 \exp\left(-\frac{z}{H_p}\right) \tag{5.120}$$

Since the mass flow at the upper heights must be zero, then

$$\lim_{z \to \infty} (nw) = 0 \tag{5.121}$$

This implies that

$$\lim_{z \to \infty} (nw) = \lim_{z \to \infty} \left[-D_p \left(\frac{dn}{dz} + \frac{n}{H_p}\right)\right] = 0 \tag{5.122}$$

By inserting for n from (5.120) and allowing for different ion and electron temperatures we get

$$\lim_{z \to \infty} (nw) = \lim_{z \to \infty} \left[(1+c)D_0 C_1 \left(\frac{1}{H} - \frac{1}{H_p}\right)\right] = 0 \tag{5.123}$$

and obviously C_1 must be zero in order to fulfil the condition.

At the upper boundary, therefore

$$n \approx C_2 \exp\left(-\frac{z}{H_p}\right) \tag{5.124}$$

Since q_i decreases more slowly with altitude than β does, the electron density must grow with altitude at the lower heights. On the other hand, we have just found that the density must decrease by height at the upper regions. There must then be a maximum somewhere in between. At this peak in the plasma density the time constant against recombination is given by

$$\tau_R = \frac{1}{\beta} \tag{5.125}$$

The time constant for diffusion, given by the time it takes for the vertical motion to cover one scale height in the atmosphere, is given by:

$$\tau_d = \frac{H}{W} \tag{5.126}$$

where W is characteristic vertical velocity. At this maximum $dn/dz = 0$ at a steady state and therefore

$$W = \frac{\sin^2 I \, D_p}{H_p} \tag{5.127}$$

The diffusion time constant is then given by:

$$\tau_d = \frac{H \cdot H_p}{\sin^2 I \, D_p} \tag{5.128}$$

Since we now assume that $\tau_d = \tau_R$ we find that

$$\frac{D_p}{\beta} = \frac{H \cdot H_p}{\sin^2 I} \tag{5.129}$$

Since $D = D_0 \exp(+z/H)$ and $\beta = \beta_0 \exp(-z/H_{N_2})$, we find that the height for the plasma density maximum occurs at:

$$z_{\max} = \frac{H_{N_2} H}{H + H_{N_2}} \ln \frac{H \cdot H_p \cdot \beta_0}{D_0 \sin^2 I} \tag{5.130}$$

5.9 THE E-LAYER IONIZATION AND RECOMBINATION

As seen from Fig. 5.18, O_2^+ and N_2^+ are the main ion species produced by solar radiation between 130 and 90 km. We will therefore study the steady-state balance for these ions.

The O_2^+ ions produced will recombine with an electron with a dissociative recombination coefficient α_2 (Table 5.4) as before

$$O_2^+ + e \xrightarrow{\alpha_2} O + O \tag{5.131}$$

or they can rearrange with neutral constituents such as NO and N_2

$$O_2^+ + NO \xrightarrow{k_3} NO^+ + O_2 \tag{5.132}$$

$$O_2^+ + N_2 \xrightarrow{k_4} NO^+ + NO \tag{5.133}$$

with reaction rates k_3 and k_4, respectively (Table 5.4), forming NO^+ ions which are quite abundant in the E-region (Fig. 5.22). The molecular nitrogen ions can also rearrange with O and form NO^+ ions with a reaction rate k_5 (Table 5.4):

$$N_2^+ + O \xrightarrow{k_5} NO^+ + N \tag{5.134}$$

or they recombine with an electron with a dissociative recombination coefficient α_3 (Table 5.4):

$$N_2^+ + e \xrightarrow{\alpha_3} N + N \tag{5.135}$$

The NO^+ ions are lost through dissociative recombination with an electron:

$$NO^+ + e \xrightarrow{\alpha_1} N + O \tag{5.136}$$

Sec. 5.9]   **The E-layer ionization and recombination**   229

The dissociative recombination coefficient for this reaction is α_1 (Table 5.4). For the O_2^+ ions, production must be equal to loss as follows:

$$q(O_2^+) = \alpha_2[O_2^+] \cdot n_e + k_3[O_2^+] \cdot [NO] + k_4 \cdot [O_2^+] \cdot [N_2] \qquad (5.137)$$
$$= (\alpha_2 \cdot n_e + k_3[NO] + k_4[N_2]) \cdot [O_2^+] = l(O_2^+)$$

For the N_2^+ ions we have similarly

$$q(N_2^+) = \alpha_3[N_2^+] \cdot n_e + k_5[N_2^+][O] = l(N_2^+) \qquad (5.138)$$

and finally for the NO^+ ions:

$$q(NO^+) = k_3[O_2^+] \cdot [NO] + k_4 \cdot [O_2^+] \cdot [N_2] + k_5 \cdot [N_2^+] \cdot [O] \qquad (5.139)$$
$$= (k_3[NO] + k_4[N_2]) \cdot [O_2^+] + k_5[N_2^+] \cdot [O]$$

$$l(NO^+) = \alpha_1[NO^+] \cdot n_e \qquad (5.140)$$

The total production of ions due to solar radiation must be equal to the production of electrons:

$$q(n_e) = q(O_2^+) + q(N_2^+) = \alpha_3[N_2^+] \cdot n_e + \alpha_2[O_2^+] \cdot n_e \qquad (5.141)$$
$$+ (k_3 \cdot [NO] + k_4 \cdot [N_2]) \cdot [O_2^+] + k_5[N_2^+] \cdot [O]$$

and implementing (5.139) and (5.140) and implying they are equal:

$$q(n_e) = \alpha_1[NO^+] \cdot n_e + \alpha_2[O_2^+] \cdot n_e + \alpha_3[N_2^+] \cdot n_e \qquad (5.142)$$
$$= (\alpha_1[NO^+] + \alpha_2[O_2^+] + \alpha_3[N_2^+]) \cdot n_e$$

At any time there must be charge neutrality, and therefore

$$[NO^+] + [O_2^+] + [N_2^+] = n_e \qquad (5.143)$$

or

$$\frac{[NO^+]}{n_e} + \frac{[O_2^+]}{n_e} + \frac{[N_2^+]}{n_e} = 1 \qquad (5.144)$$

Introducing r_1, r_2 and r_3 as the relative abundance ratios of the respective ions

$$r_1 = [NO^+]/n_e \quad \text{(a)}$$
$$r_2 = [O_2^+]/n_e \quad \text{(b)} \qquad (5.145)$$
$$r_3 = [N_2^+]/n_e \quad \text{(c)}$$

we find from (5.144):

$$r_1 + r_2 + r_3 = 1 \qquad (5.146)$$

Finally we derive from (5.142) by introducing (5.145):

$$q(n_e) = \alpha_1 r_1 n_e^2 + \alpha_2 r_2 n_e^2 + \alpha_3 r_3 n_e^2 \qquad (5.147)$$
$$= (\alpha_1 r_1 + \alpha_2 r_2 + \alpha_3 r_3) n_e^2$$

By now introducing the effective recombination coefficient

$$\alpha_{\text{eff}} = \alpha_1 r_1 + \alpha_2 r_2 + \alpha_3 r_3 \qquad (5.148)$$

the electron production rate is given by

$$q(n_e) = \alpha_{\text{eff}} n_e^2 \qquad (5.149)$$

and the E-region electron density profile behaves as an α-profile.

As observed by many rocket probes passing through the E-region, the O^+ and N_2^+ ions (Fig. 5.22) decrease very rapidly below 150 km and can for all practical purposes be neglected in the E-region. It is also common practice to infer a mean ion mass of 30.5 a.m.u. at these altitudes. When neglecting N_2^+ we therefore have

$$r_1 + r_2 = 1 \qquad (5.150)$$

and the mean ion mass is given by

$$r_1 \cdot M_{NO^+} + r_2 \cdot M_{O_2^+} = \overline{M} \qquad (5.151)$$

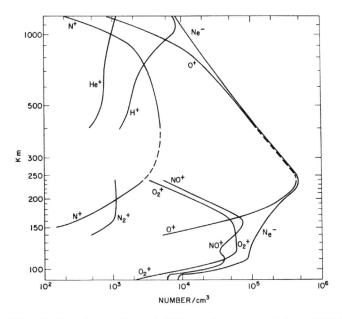

Fig. 5.22. Rocket observations of different ion species below 1000 km for daytime at solar minimum. (From Johnson, 1966.)

or
$$r_1 \cdot 30 + r_2 \cdot 32 = 30.5 \tag{5.152}$$

Solving for r_1 and r_2 give:
$$r_1 = 3/4 \quad \text{and} \quad r_2 = 1/4 \tag{5.153}$$

The effective recombination coefficient can then be expressed as follows:
$$\alpha_{\text{eff}} = 0.25 \cdot \alpha_1 + 0.75 \cdot \alpha_2 \tag{5.154}$$

where α_1 and α_2 are given in Table 5.4.

5.10 THE TIME CONSTANT OF THE RECOMBINATION PROCESS

Let us now go back to the electron density continuity equation (5.49) but still neglect any transport processes so that the equation can be simplified to read:
$$\frac{dn_e}{dt} = q_e - l_e \tag{5.155}$$

where q_e and l_e are the production and loss terms of the electrons respectively.

For a β-profile we know that the loss rate is linear and given by:
$$l_e = \beta n_e \tag{5.156}$$

and therefore
$$\frac{dn_e}{dt} = q_e - \beta n_e \tag{5.157}$$

Let us now assume that the production q_e is suddenly shut off at a time $t = 0$ ($q_e = q_0$ at $t = 0$). Then the electron density will decay by time as:
$$n_e = n_{e0} \exp(-\beta t) \tag{5.158}$$

where $n_{e0} = q_0/\beta$ is the density at $t = 0$ (Fig. 5.23a), and β is the factor determining how fast the electron density is reduced to $1/e$. On the other hand, if the production q_e is turned on at time $t = 0$ and is constant over a length of time ($q_e = q_0$), then the electron density will grow according to
$$n_e = n_\infty (1 - \exp(-\beta t)) \tag{5.159}$$

where
$$n_\infty = q_0/\beta \tag{5.160}$$

after a time $T \gg 1/\beta$ (Fig. 5.23b). For a time varying $q(t)$, however, the electron density at time t will be given by:
$$n_e(t) = \exp(-\beta t) \int_0^t q(t') \exp(\beta t') dt' \tag{5.161}$$

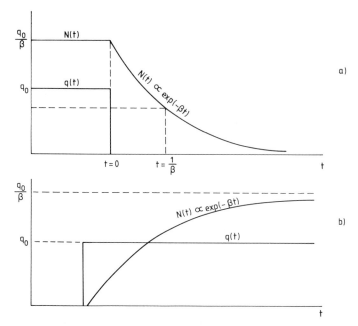

Fig. 5.23. Time behaviour of the electron density at a given height if (a) the production is turned off at $t = 0$, and (b) if it is turned on at $t = 0$.

In the E-region we have just shown that the loss rate is proportional to n_e^2 and therefore the continuity equation again for n_e, neglecting transport, is:

$$\frac{dn_e}{dt} = q_e - \alpha_{\text{eff}} n_e^2 \tag{5.162}$$

If we follow the same examples as for the β-profile and assume that q_e is turned off at $t = 0$, then the electron density will decay as:

$$n_e = \frac{n_{e0}}{n_{e0} \cdot \alpha_{\text{eff}} \cdot t + 1} \tag{5.163}$$

when $n_{e0} = \sqrt{q_0/\alpha_{\text{eff}}}$ is the electron density at $t = 0$. To introduce a time constant one can estimate the time elapsed until $n_e = \frac{1}{2} \cdot n_{e0}$ and find $t_{1/2} = 1/n_{e,0}\alpha_{\text{eff}}$ which depends on the electron density at $t = 0$. For an ionization source that is turned on at $t = 0$ and stays constant for a length of time, however, we find:

$$n_e = \sqrt{\frac{q_0}{\alpha_{\text{eff}}}} \tanh\left(\sqrt{q_0 \cdot \alpha_{\text{eff}}} \cdot t\right) = n_{e0} \tanh(n_{e0} \cdot \alpha_{\text{eff}} \cdot t) \tag{5.164}$$

where q_0 is the constant production rate. We notice for $t = 1/2 \cdot n_{e,0} \cdot \alpha_{\text{eff}}$ that:

$$n_e = 0.46 \, n_{e,0} \approx 0.5 \, n_{e,0} \tag{5.165}$$

The conventional use of the time constant concept is therefore not so obvious in the E-region.

For small fluctuations in n_e we may have:

$$n_e = n_0 + n'_e \tag{5.166}$$

where n_0 is the background electron density and n'_e is a small variation on the background. If $n'_e \ll n_0$, we get:

$$\frac{dn_e}{dt} = \frac{dn'_e}{dt} = q_e - \alpha(n_0 + n'_e)^2 = q_e - \alpha n_0^2 - 2\alpha n_0 \cdot n'_e \tag{5.167}$$

when n'^2_e is neglected compared to n_0^2.

$$\frac{dn'_e}{dt} = (q_e - \alpha n_0^2) - 2\alpha n_0 \cdot n'_e \tag{5.168}$$

and we are back to a linear equation in n'_e where the time constant will be:

$$\tau = \frac{1}{2\alpha n_0} \tag{5.169}$$

This is the same result for τ as obtained for (5.163), therefore is this value of τ often mentioned in the literature as the recombination time constant of the E-region. When dealing with these parameters we have to discriminate between the different options to avoid confusion.

5.11 THE D-REGION IONIZATION AND RECOMBINATION

The lower part of the ionosphere below 90 km and down to about 60 km is usually referred to as the D-region. The pressure is about 10^6 times as large as at typical F-region heights and collisions are dominant. The photochemistry is very complex and is not fully understood. Four important ionization sources in the D-region are identified, and the contribution from them to the ion production is illustrated in Fig. 5.24.

1. The Lyman-α line at 1215 Å penetrates to the D-region and ionizes nitric oxide (NO), a minor constituent in the neutral atmosphere.

2. The EUV radiation spectrum between 1027 and 1118 Å ionizes excited oxygen molecules $O_2(^1\Delta g)$.

3. Hard X-rays between 2 and 8 Å ionizes all constituents, thereby acting mainly on N_2 and O_2.

4. Cosmic rays similarly ionizes all constituents.

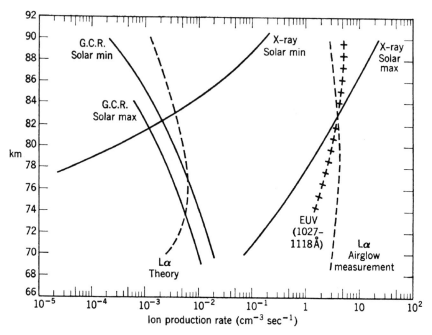

Figure 5.24. Illustrative example of ionization produced by X-ray, EUV radiation, galactic cosmic rays (G.C.R.) and hydrogen L_α radiation. (From Whitten and Poppoff, 1971.)

Relativistic electrons (100 keV – 1 MeV) may penetrate deep into the D-region on occasions and the same is true for protons in the energy range 1–100 MeV, and these particles will also ionize the atmosphere.

From the point of view of radio communication the D-region is of great importance because most of the radio wave absorption in the HF and MF regime occurs in this region.

Due to the high pressure in the D-region, conventional techniques for studying the ionosphere are not applicable. For instance, ionosonde sounding is not possible because the collision frequency is so high that it exceeds the critical frequency in general, and the absorption is heavy at all reflected frequencies. For this reason, rocket techniques have been particularly important for D-region studies. The problem with the rockets, however, is that the time spent by them in the D-region together on upleg and downleg is extremely short (10–20 sec). Modern radar techniques, however, have recently opened up new fields of observational methods which are promising for future D-region studies.

The positive ions formed directly by solar radiation in the D-region are O_2^+, N_2^+ and NO^+. The N_2^+ ion rapidly charge exchanges with O_2 to form an O_2^+ ion by the reaction rate k_6 (Table 5.4) as follows:

$$N_2^+ + O_2 \xrightarrow{k_6} N_2 + O_2^+ \tag{5.170}$$

Sec. 5.11]   **The D-region ionization and recombination**   235

The O_2^+ ion formed will quickly charge transfer to NO^+ according to rate k_3 or k_4 (see Table 5.4). Because the concentration of NO is very small compared to the density of O_2 (one in about 10^6) at 90 km, a large number of O_2^+ is expected to exist in the D-region at daytime.

From mass spectrometric analyses of ion species in the D-region (Fig. 5.25) it is seen that N_2^+ (28^+) is much less abundant than NO^+ (30^+) and O_2^+ (32^+) above 80 km. Ions with molecular mass 19 and 37 are thought to be hydrated ions, $H^+(H_2O)$ and $H_3O^+\cdot(H_2O)$; they are very abundant below 80 km but decrease rapidly above 82 km or so. They are generally found where the water vapour concentration exceeds 10^{15} m^{-3}. Heavy ions with molecular masses above 45 are observed below 75 km. These are believed to be metallic ions, probably debris from meteor showers. Above about 85 km it should, however, be safe to assume that NO^+ and O_2^+ are the dominant positive ions.

An important feature of the D-region is the formation of negative ions such as O_2^-. These are produced by a three-body reaction with a reaction rates β_1 and β_2 (Table 5.4) as follows:

$$O_2 + M + e \xrightarrow{\beta_1,\beta_2} O_2^- + M \qquad (5.171)$$

where M is any one molecule capable of increasing the chance of reaction by removing excess kinetic energy from the reactants.

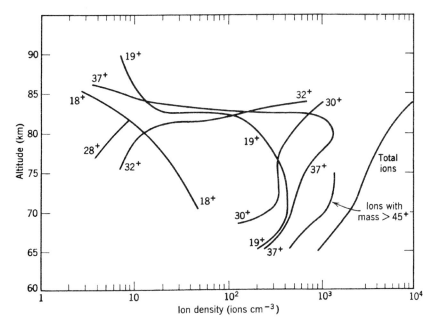

Fig. 5.25. Concentration of positive ions detected by mass spectrometer in the D-region. The ions are identified by their molecular mass number. (From Narcisi and Bailey, 1965.)

The electron affinity of O_2 is small, only 0.45 eV, the electron might therefore be rapidly detached again by a quantum of low energy such as infrared or visible light

$$O_2^- + h\nu \xrightarrow{\rho} O_2 + e \qquad (5.172)$$

The rate coefficient for this process is assigned by ρ (Table 5.4). Other processes that can destroy the O_2^- ions can be detachment by collisions with neutrals

$$O_2^- + M \xrightarrow{\gamma_1,\gamma_2} O_2 + M + e \qquad (5.173)$$

$$O_2^- + O \xrightarrow{\gamma_3} O_3 + e \qquad (5.174)$$

The reaction rates are defined by γ_1, γ_2 and γ_3 (Table 5.4), where γ_1 applies if $M = [N_2]$ and γ_2 applies if $M = [O_2]$, and γ_3 applies for the last reaction, respectively. Other possible loss processes for the negative ions in the D-region are ion–ion recombination which can be expressed as

$$O_2^+ + O_2^- \xrightarrow{\alpha_{i1}} O_2 + O_2 \qquad (5.175)$$

$$O_2^+ + O_2^- + M \xrightarrow{\alpha_{i2}} O_2 + O_2 + M \qquad (5.176)$$

where the reaction rates are α_{i1} and α_{i2}, respectively (Table 5.4). By now we should have a relatively complete picture of the major chemical processes taking place in the D-region. On the basis of several studies of the charged constituents of the D-region it is found (Fig. 5.26) that the amount of negative ions increases sharply by decreasing height. While there is less than one negative ion per 100 electrons at 90 km at daytime, there may be more than 100 negative ions per electron at 50 km, an increase by 10^4 over 40 km of altitude. At night-time there may be even larger concentrations of negative ions.

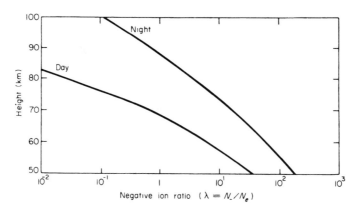

Fig. 5.26. Negative ion–electron concentration ratio λ below 90 km for daytime and night-time conditions. The model is based on molecular oxygen ions only. (From Rishbeth and Garriott, 1969.)

Sec. 5.11]   The D-region ionization and recombination

To give a full treatment of all ion species will be very complicated and time-consuming, and probably not so illustrative. Let us therefore identify the electron, positive and negative ion densities in the D-region by n_e, n^+ and n^- respectively, without differentiating between the different ion species. Let us furthermore adapt an effective recombination loss rate between positive ions and electrons by α_D, between positive and negative ions by α_i, an effective electron attachment rate by β, an effective electron detachment rate by γ, and a radiative detachment rate by ρ. We will in the following neglect radiative recombination as a possible loss rate in the D-region. Also any transport term will be ignored.

The continuity equation for the electrons will now be given by:

$$\frac{dn_e}{dt} = q - \alpha_D n^+ \cdot n_e - \beta n_e \cdot n_n + \gamma \cdot n^- \cdot n_n + \rho \cdot n^- \quad (5.177)$$

where n_n is the number densities of the neutrals. For the positive ions we have a similar continuity equation:

$$\frac{dn^+}{dt} = q - \alpha_D n^+ \cdot n_e - \alpha_i n^- \cdot n^+ \quad (5.178)$$

where q is the same production rate as for the electrons due to solar radiation. The negative ions will obey the following continuity equation:

$$\frac{dn^-}{dt} = -\alpha_i n^- \cdot n^+ + \beta n_e \cdot n_n - \gamma n^- \cdot n_n - \rho n^- \quad (5.179)$$

We have to demand charge neutrality in the plasma, therefore

$$n^+ = n_e + n^- \quad (5.180)$$

By introducing the ratio between negative ions and electrons λ

$$\lambda = \frac{n^-}{n_e} \quad (5.181)$$

we have

$$n^+ = (1 + \lambda) \cdot n_e \quad (5.182)$$

From the continuity equation (5.178) for the positive ions we get by inserting (5.182):

$$(1+\lambda)\frac{dn_e}{dt} + n_e \frac{d\lambda}{dt} = q - \alpha_D n^+ \cdot n_e - \alpha_i n^- \cdot n^+$$
$$= q - (\alpha_D + \alpha_i \cdot \lambda) \cdot (1+\lambda) n_e^2 \quad (5.183)$$

and

$$\frac{dn_e}{dt} = \frac{q}{1+\lambda} - (\alpha_D + \lambda \alpha_i) n_e^2 - \frac{n_e}{1+\lambda}\frac{d\lambda}{dt} \quad (5.184)$$

Introducing this to the equation for the electrons, (5.177), we find:

$$q - \alpha_D(1+\lambda)n_e^2 - \beta n_n \cdot n_e + \gamma\lambda n_n n_e + \rho\lambda n_e \qquad (5.185)$$
$$= \frac{q}{1+\lambda} - (\alpha_D + \lambda\alpha_i)n_e^2 - \frac{n_e}{1+\lambda}\frac{d\lambda}{dt}$$

and rearranging

$$\frac{1}{1+\lambda}\frac{d\lambda}{dt} = \beta \cdot n_n - \lambda\left[\rho + \gamma n_n + (\alpha_i - \alpha_D)n_e + \frac{q}{(1+\lambda)n_e}\right] \qquad (5.186)$$

These two equations for n_e and λ implicitly assume that the ionosphere can be described by one ionized species, one attaching species, one recombining species etc. Another way to comprehend the results derived is to assume that all reaction rates and concentrations are properly weighted composites of a multi-species atmosphere.

The attachment and detachment processes often occur so rapidly that they nearly balance each other. Under most conditions λ is fairly constant, one value at night and one in the day, as shown in Fig. 5.26 for different heights between 50 and 90 km. We therefore in the situation when λ is constant by time neglect the effect of dissociative recombination and ion production and find as our first approximation

$$\beta n_n - \lambda(\rho + \gamma n_n) = 0 \qquad (5.187)$$

and

$$\lambda = \frac{\beta n_n}{\rho + \gamma n_n} \qquad (5.188)$$

At night the detachment rate due to chemical reactions dominates above solar radiation so that $\gamma n_n > \rho$ and

$$\lambda = \frac{\beta}{\gamma} \qquad (5.189)$$

In the daytime, however, the opposite is true and

$$\lambda = \frac{\beta}{\rho}n_n \qquad (5.190)$$

Now we know from Table 5.4 that β is proportional to n_n ($[O_2]$ or $[N_2]$) and λ will therefore increase by decreasing height in agreement with Fig. 5.26. This increase, however, will be stronger in the daytime than at night-time.

We will, however, no longer assume balance between attachment and detachment and continue our analyses of λ and n_e.

We will still assume that λ is constant except during sunset and sunrise. Then the electron density will be given by:

$$\frac{dn_e}{dt} = \frac{q}{1+\lambda} - (\alpha_D + \lambda\alpha_i)n_e^2 \qquad (5.191)$$

Sec. 5.11]   **The D-region ionization and recombination**   239

Since λ is constant, this equation for n_e has the same form as (5.162), the continuity equation, for the electron density in the E-region, and n_e behaves in many respects as an α-profile.

The effective recombination coefficient can now be identified as

$$\alpha'_{\text{eff}} = \alpha_D + \lambda\alpha_i \tag{5.192}$$

and the production rate

$$q' = \frac{q}{1+\lambda} \tag{5.193}$$

is reduced by the factor $(1+\lambda)$ with respect to q in (5.162). This factor can indeed be very small in the lower part of the D-region where $\lambda \approx (100-1000)$. In a steady state for the electron density $(dn_e/dt = 0)$ we also see that

$$q \approx (1+\lambda)(\alpha_D + \lambda\alpha_i)n_e^2 = \alpha''_{\text{eff}} n_e^2 \tag{5.194}$$

where

$$\alpha''_{\text{eff}} = (1+\lambda)\alpha'_{\text{eff}} \tag{5.195}$$

α''_{eff} and α'_{eff} are often interchanged in the literature, but they are only equal when $\lambda \approx 0$. It is important to be aware of this in order to avoid confusion as α''_{eff} and α'_{eff} can differ by a factor of 1000.

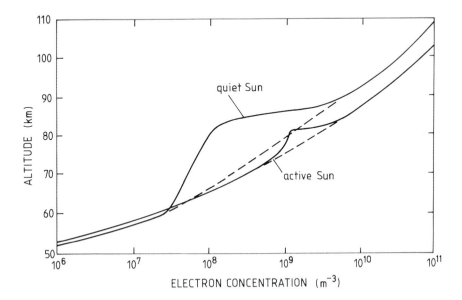

Fig. 5.27. Schematic electron concentration profiles in the D-region for quiet and active solar conditions.

It should in the preceding paragraph have been well demonstrated that the D-region is very complicated as many different chemical reactions are involved with a variety of neutral and ionized species. Fig. 5.27 illustrates schematically some typical electron density profiles observed in the D-region for active and quiet solar conditions. The electron density decreases usually from 10^{10} m^{-3} at 90 km to less than 10^7 m^{-3} below 60 km. It is of special interest to notice the ledge in the profile often observed between 80 and 90 km where the density can decrease by one magnitude at least within a couple of kilometres. This ledge is believed to occur due to the presence of hydrated ions in this height region. Since the hydrated ions are very effective in recombining with the electrons, they will rapidly destroy the electrons as soon as they are created.

From radio wave propagation studies it has been observed that the electron density in the D-region is fairly constant at day and night but changes rather rapidly at sunset and sunrise. This is illustrated rather schematically in Fig. 5.28.

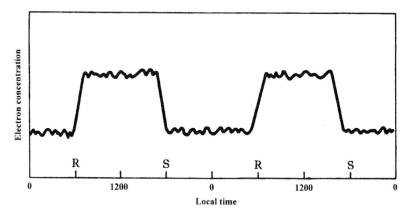

Fig. 5.28. A schematic diagram showing the behaviour of electron concentration in the lower part of the ionospheric D-region with time. R and S represent time for sunrise and sunset respectively. (From Ratcliffe, 1972.)

At first it may appear that the sharp increase at sunrise and the similar sharp decrease at sunset may just be due to the Earth's shadow passing through the D-region. This is, however, not so simple as the rapid variations are more related to the time when the shadow passes through the ozone layer (Fig. 5.29) at at height of about 50 km above the ground. The sudden change in the electron density is then ascribed to an increase in the concentration of atomic oxygen since the ultraviolet radiation is no longer absorbed in the ozone layer and can therefore dissociate the O_2 molecules in the D-region. The oxygen atoms can then detach with the O_2^- ions and release free electrons and form ozone.

$$O_2^- + O \xrightarrow{\gamma_3} O_3 + e \qquad (5.196)$$

where γ_3 is the reaction rate given in Table 5.4. Thereby the amount of electrons is increased on the expense of negative ions at sunrise. At sunset the solar EUV radiation will be absorbed by the ozone layer, and the dissociation of the oxygen molecules comes to a halt. The detachment process cannot continue to release free electrons and the electron density drops.

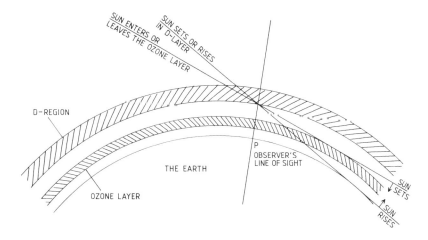

Fig. 5.29. A diagram showing the geometry of sun rays close to sunset and sunrise to illustrate the importance of the ozone layer for the electron density behaviour in the lower D-region.

5.12 THE PLASMASPHERE

In the lower ionosphere the motion of the plasma is strongly dominated by the interaction with the neutral atmosphere through collisions. Moving upward in the ionosphere, however, the influence from electric field of external origin and diffusion is starting to prevail.

In the lower ionosphere the molecular ions such as N_2^+, NO^+ and O_2^+ are dominating. In the upper ionosphere, however, the lighter ions such as O^+, H^+ and He^+ become more abundant. The lower ionosphere is formed, as we know, by ion production due to solar irradiation below 200 nm and loss due to different recombination and chemical processes. At higher altitudes, however, the transport processes are becoming more and more dominant.

During daytime, plasma is produced or lost by different processes in the E- and F-region creating the horizontally stratified slowly varying plasma in the ionosphere. Above the F-region peak we expect the plasma to be in a state dominated by diffusive equilibrium, and because of this the electron gas, due to its light mass, diffuses faster than the ions creating the ambipolar electric field which enhances the ion diffusion, and thereby results in an equilibrium diffusion state of the plasma.

Since this equilibrium is in general established between the electrons and the O^+ ions, the lighter ions such as H^+ and He^+ may experience an outward acceleration due to this ambipolar electric field, as we have seen.

Since hydrogen atoms and molecules can escape the gravitational field of the Earth, the Earth is surrounded with a hydrogen gas called the geocorona which interacts with the oxygen atoms via charge exchange processes such as:

$$H + O^+ \longrightarrow H^+ + O \tag{5.197}$$

$$H^+ + O \longrightarrow H + O^+ \tag{5.198}$$

This leads to a transition between oxygen-dominated and hydrogen-dominated plasma somewhere between 500 km and a few thousand kilometers above ground. Since the hydrogen ions actually also can escape the Earth's gravity field, the simple steady-state diffusive equilibrium situation cannot be maintained, and a net upward flow of plasma is possible from the topside ionosphere during the day. The closed magnetic field lines at lower latitudes will act as an enclosure for the upstreaming plasma and help to establish a reservoir of the plasma produced by the photoionization during daytime out at great distances from the Earth (Fig. 5.30). At night this plasma can flow back down the magnetic field lines to refill the ionosphere as the plasma there is lost by recombination. The result is a complex interaction between the ionosphere and a region of hydrogen plasma trapped by the dipole magnetic field, the plasmasphere.

The situation is complicated by the fact that neutral winds in the upper ionosphere as well as electric fields both in the ionosphere and in the magnetosphere may strongly modify the plasma flow.

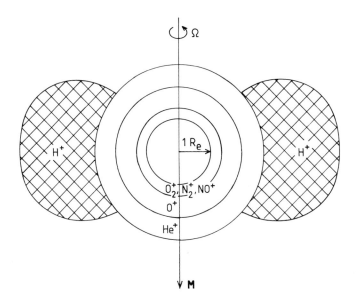

Fig. 5.30. A schematic illustration of the predominant ion species in different regions in the ionosphere and plasmasphere. The dimensions are not to scale, the plasmasphere is supposed to reach to about 5 R_e in the equatorial plane, however.

5.13 FERRARO'S THEOREM

The first problem we have to face in this context is the fact that the plasma on the closed field lines are corotating with the Earth. That this indeed is so, can be understood from the proof of the so-called Ferraro's theorem:

> *In an equilibrium situation the angular velocity of the highly conducting plasma enclosed by the magnetic dipole field of a body rotating around the magnetic axis will be constant along the magnetic field lines, and the plasma is corotating with the body.*

To prove Ferraro's theorem for isorotation we assume that the magnetic field of the Earth can be described by a dipole magnetic field being coaxial to the Earth's rotation axis.

We apply Maxwell's equations describing the magnetic field as given in Chapter 2:

$$\frac{\partial \mathbf{B}}{\partial t} = -\nabla \times \mathbf{E} \tag{5.199}$$

$$\nabla \cdot \mathbf{B} = 0 \tag{5.200}$$

$$\nabla \times \mathbf{B} = \mu_0 \mathbf{j} \tag{5.201}$$

and the generalized Ohm's law to describe the current

$$\mathbf{j} = \sigma \cdot (\mathbf{E} + \mathbf{v} \times \mathbf{B}) \tag{5.202}$$

where it is assumed that σ is uniform. For an equilibrium situation ($\partial/\partial t = 0$)

$$0 = -\nabla \times \mathbf{E} = -\nabla \times \left(\frac{\mathbf{j}}{\sigma} - \mathbf{v} \times \mathbf{B} \right) = -\frac{1}{\sigma} \nabla \times \mathbf{j} + \nabla \times (\mathbf{v} \times \mathbf{B}) \tag{5.203}$$

For a very high conductivity $\sigma \to \infty$ we find:

$$\nabla \times (\mathbf{v} \times \mathbf{B}) = 0 \tag{5.204}$$

If we now choose a cylindrical coordinate system $(\hat{\mathbf{r}}, \hat{\varphi}, \hat{\mathbf{z}})$ so that $\hat{\mathbf{z}}$ is parallel to the rotation axis, then the velocity at a distance r from the rotation axis

$$\mathbf{v} = r\Omega \hat{\varphi} \tag{5.205}$$

where Ω is the angular velocity. The magnetic field can be described by

$$\mathbf{B} = B_r \hat{\mathbf{r}} + B_\varphi \hat{\varphi} + B_z \hat{\mathbf{z}} \tag{5.206}$$

where B_r, B_φ and B_z are the components along $\hat{\mathbf{r}}$, $\hat{\varphi}$, and $\hat{\mathbf{z}}$ respectively.

In this coodinate system we now have

$$\mathbf{v} \times \mathbf{B} = \Omega r (B_z \hat{\mathbf{r}} - B_r \hat{\mathbf{z}}) \tag{5.207}$$

Now applying the curl operator in cylindrical coordinates:

$$\nabla \times (\mathbf{v} \times \mathbf{B}) = \nabla \times (\Omega r B_z \hat{\mathbf{r}} - \Omega r B_r \hat{\mathbf{z}}) \qquad (5.208)$$

$$= \frac{1}{r} \left\{ -\frac{\partial}{\partial \varphi}(\Omega r B_r) \hat{\mathbf{r}} \right.$$

$$+ \left[\frac{\partial}{\partial r}(\Omega r B_r) + \frac{\partial}{\partial z}(\Omega r B_z) \right] \hat{\varphi} - \left. \frac{\partial}{\partial \varphi}(\Omega r B_z) \hat{\mathbf{z}} \right\}$$

We will now take advantage of the axial symmetry of the dipole magnetic field, so that

$$\frac{\partial B_r}{\partial \varphi} = \frac{\partial B_\varphi}{\partial \varphi} = \frac{\partial B_z}{\partial \varphi} = 0 \qquad (5.209)$$

and likewise is $\partial \Omega / \partial \varphi = 0$ for symmetric reasons, and since $\partial r / \partial \varphi = 0$ we have:

$$\nabla \times (\mathbf{v} \times \mathbf{B}) = \left[\frac{1}{r} \frac{\partial}{\partial r}(\Omega r B_r) + \frac{1}{r} \frac{\partial}{\partial z}(\Omega r B_z) \right] \hat{\varphi} \qquad (5.210)$$

$$= \left[\left(\Omega \frac{1}{r} \frac{\partial}{\partial r}(r B_r) + \Omega \frac{\partial B_z}{\partial z} \right) \right.$$

$$+ \left. \left(B_r \frac{\partial \Omega}{\partial r} + B_z \frac{\partial \Omega}{\partial z} \right) \right] \hat{\varphi} = 0$$

Since $\partial \Omega / \partial \varphi = \partial B_\varphi / \partial \varphi = 0$ we can form

$$\nabla \times (\mathbf{v} \times \mathbf{B}) = \left[\Omega \left(\frac{1}{r} \frac{\partial}{\partial r}(r B_r) + \frac{1}{r} \frac{\partial B_\varphi}{\partial \varphi} + \frac{\partial B_z}{\partial z} \right) \right. \qquad (5.211)$$

$$+ \left. \left(B_r \frac{\partial \Omega}{\partial r} + B_\varphi \frac{\partial \Omega}{\partial \varphi} + B_z \frac{\partial \Omega}{\partial z} \right) \right] \hat{\varphi}$$

$$= (\Omega \nabla \cdot \mathbf{B} + \mathbf{B} \cdot \nabla \Omega) \hat{\varphi} = \mathbf{B} \cdot \nabla \Omega \hat{\varphi} = 0$$

since $\nabla \cdot \mathbf{B} = 0$. Therefore $\mathbf{B} \cdot \nabla \Omega = 0$, and the gradient in Ω along $\mathbf{B}$ is zero. The angular velocity is therefore constant along the magnetic field line. Ferraro's theorem is proven.

5.14 THE MAGNETOSPHERIC CONVECTION CLOSE TO THE EARTH

We now see that, due to this corotation of the plasma on closed field lines, the motion of the plasma in the inner magnetosphere will partly be dominated by this corotation and partly by the magnetospheric convection (Fig. 5.31). While the corotation represents a constant rotation speed at a given distance, the magnetospheric convection is not constant, but rather strongly dependent on the interactions between the solar wind and the magnetosphere. The point on the evening side where the corotation velocity $\mathbf{v}_R$ is exactly opposite to the convection velocity $\mathbf{v}_m$ will therefore move inwards and outwards from the Earth in concordance with

Sec. 5.14] **The magnetospheric convection close to the Earth** 245

the variation in the potential V_T across the tail. The situation is very similar to the laminar fluid flow around a rotating cylinder. In the stagnation region in the evening sector the convection velocity will be very low, and plasma can be brought to a halt. Therefore the plasmasphere often has a bulge in this sector. Actually the plasma may be forced to move in closed loops in the evening side depending on the history of the cross-tail potential and its instantaneous magnitude.

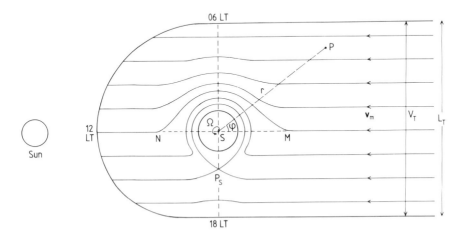

Fig. 5.31. A diagram illustrating the analogy between the convection pattern in the magnetospheric equatorial plane including the corotation of the plasmasphere, and the streamlines in a laminar flow around a cylinder. The plasma flow has a uniform flow velocity $\mathbf{v}_m$ directed towards the Sun far down the tail. The Earth rotates with an angular velocity Ω. The stagnation point P_S as well as a general point P in the flow is indicated. The width of the tail is L_T.

We can by the analogy to the streamline function in hydrodynamical flow describe the convection flow lines in a more mathematical sense.

From Maxwell's equation we know that the electric field can be derived from a potential. Let this be ϕ, then:

$$\mathbf{E} = -\nabla \phi = -\frac{\partial \phi}{\partial r}\hat{\mathbf{r}} - \frac{1}{r}\frac{\partial \phi}{\partial \varphi}\hat{\boldsymbol{\varphi}} - \frac{\partial \phi}{\partial z}\hat{\mathbf{z}} \qquad (5.212)$$

The magnetic field in the equatorial plane is directed perpendicular to the plane along the z-direction and therefore

$$\mathbf{B} = B\hat{\mathbf{z}} \qquad (5.213)$$

The velocity due to the $\mathbf{E} \times \mathbf{B}$ drift in the equatorial plane is now

$$\mathbf{v} = \mathbf{E} \times \mathbf{B}/B^2 = \frac{1}{B}\mathbf{E} \times \hat{\mathbf{z}} = -\frac{1}{B}\left[-\frac{\partial \phi}{\partial r}\hat{\boldsymbol{\varphi}} + \frac{1}{r}\frac{\partial \phi}{\partial \varphi}\hat{\mathbf{r}}\right] \qquad (5.214)$$

and the velocity is totally in the equatorial plane.

From the hydrodynamical analogy considering the streamlines in laminar flow around a rotating cylinder we have for the velocity field potential or the stream function:

$$\phi_v = -\frac{C_1}{r} + C_2 r \sin\varphi \tag{5.215}$$

where C_1 and C_2 are constants determined by the rotation of the cylinder and the flow velocity, and φ and r are as indicated in Fig. 5.31. We can introduce this because we are treating the flow only in the equatorial plane, such that the cross-section in Fig. 5.31 is equivalent to the cross-section of the flow around an infinitely long cylinder. By then introducing this stream function in the expression for $\mathbf{v}$, we get:

$$\begin{aligned}\mathbf{v} &= -\frac{1}{B}\left[-\frac{\partial}{\partial r}\left(-\frac{C_1}{r}+C_2 r\sin\varphi\right)\hat{\varphi}\right.\\ &\quad + \left.\frac{1}{r}\frac{\partial}{\partial\varphi}\left(-\frac{C_1}{r}+C_2 r\sin\varphi\right)\hat{\mathbf{r}}\right]\\ &= \left(\frac{C_1}{r^2}+C_2\sin\varphi\right)\frac{1}{B}\hat{\varphi} - \frac{C_2}{B}\cos\varphi\,\hat{\mathbf{r}}\end{aligned} \tag{5.216}$$

We notice that for very large distances ($r \to \infty$)

$$\mathbf{v} = \frac{C_2}{B}(\sin\varphi\,\hat{\varphi} - \cos\varphi\,\hat{\mathbf{r}}) \tag{5.217}$$

which implies that $\mathbf{v}$ is parallel to the MSN line in Fig. 5.31 and therefore

$$C_2 = E \quad \text{and} \quad \mathbf{v}(r=\infty) = \mathbf{v}_m \tag{5.218}$$

where $\mathbf{v}_m$ is the uniform convection velocity far down the tail away from Earth. For $r = R_e$ we have

$$\mathbf{v} = \left(\frac{C_1}{R_e^2}+E\sin\varphi\right)\frac{1}{B}\hat{\varphi} - \frac{E}{B}\cos\varphi\,\hat{\mathbf{r}} = \frac{C_1}{BR_e^2}\hat{\varphi} + \mathbf{v}_m \tag{5.219}$$

Since the velocity due to rotation of the cylinder at the surface of the cylinder is given by

$$\mathbf{v}_R = \Omega R_e \hat{\varphi} \tag{5.220}$$

we must have

$$C_1 = B\Omega R_e^3 \tag{5.221}$$

We finally have

$$\mathbf{v} = \left(\frac{\Omega R_e^3}{r^2}+\frac{E}{B}\sin\varphi\right)\hat{\varphi} - \frac{E}{B}\cos\varphi\,\hat{\mathbf{r}} \tag{5.222}$$

Sec. 5.14]   The magnetospheric convection close to the Earth   247

We notice that for $\varphi = \pm \pi/2$ the radial velocity vanishes, which means that **v** is parallel to $\mathbf{v}_m$ at the dawn meridian, while at the dusk meridian ($\varphi = -\pi/2$) the azimuthal component depends on the relation between $\Omega R_e^3/r$ and E/B. It can actually be zero when

$$E = \frac{B\Omega R_e^3}{r^2} \tag{5.223}$$

If this happens at the surface of the cylinder

$$E = B\Omega R_e = E_R \tag{5.224}$$

which is the corotating electric field at the Earth's equator. At a distance $r = L \cdot R_e$ we notice:

$$E = \frac{B\Omega R_e}{L^2} = \frac{1}{L^2} E_R \tag{5.225}$$

If the stagnation point is occurring at $L = 2$, then the E-field is $1/4$ of the corotating field at equator. Therefore a large convection electric field which occurs

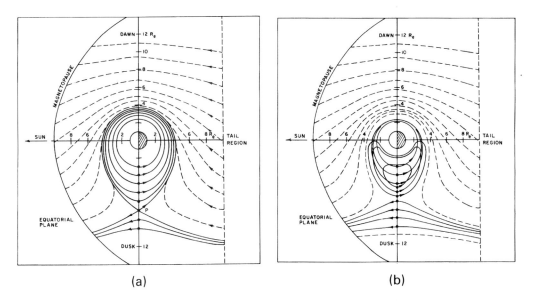

(a)   (b)

Fig. 5.32. The broken lines represent equipotential lines for a convection type electric field as outlined in section 5.14 with $C_2 = 1.8$ keV R_e^{-1}. These curves represent the drift paths of the thermal electrons in the convection region. The drift direction is indicated by arrows. (a) The solid line represents the drift paths for electrons injected with 1 keV energy at the different points, marked with dots at the dusk meridian. The drift paths separate into two distinguishable groups, corotating and conventional convection. (b) Drift paths of protons injected with 1 keV energy along the dusk meridian. Three types of drift paths are recognized: corotational, "vortices" not enclosing the Earth, and common convectional. (From Röderer, 1970.)

at disturbed conditions in the magnetosphere, will push the stagnation point P_S towards the Earth and erode the plasmasphere.

We also notice that the potential ϕ at large distances along the dawn–dusk meridian will be positive on the dawn side and negative on the dusk side. The magnitude of the potential depends on the width L_T of the tail. If the stagnation point P_S in Fig. 5.31 occurs at 4 R_e and the width of the tail is 50 R_e, the dawn–dusk potential becomes close to 47 kV. Moving the stagnation point to 3 R_e would almost double the potential ($\sim$ 83 kV) for the same dimension of L_T.

From what we have seen for the gradient and curvature drift of charged particles we notice that it is strongly dependent on their energy. Low-energy plasma drifts are basically due to an electric field perpendicular to **B** and follow equipotential surfaces. High-energetic particles can experience quite complicated drift patterns due to gradient and curvature drift. Fig. 5.32a shows the drift pattern of low-energy electrons, dashed lines, according to a model potential as discussed above. The full lines trace 1 keV electron along their drift paths as they are injected at different points along the dusk meridian (marked by dots in the figure).

In Fig. 5.32b similar drift paths for 1 keV protons injected at the dusk meridian. 3 different types of drifts prevail: corotational drift paths, "vortices" not enclosing the Earth, and convectional drift paths, as for low-energetic particles.

5.15 EQUATORIAL FOUNTAIN EFFECT

The Earth is a rotating magnet surrounded by an atmosphere where the upper part consists of a partly ionized plasma. The motion of this plasma will be strongly influenced by the magnetic field in such a way that the plasma moves more easily along the field lines than perpendicular to them. The field lines therefore can act as conductors, and it is then of interest to investigate whether the rotation of the magnetic field actually can lead to currents along the field lines between the hemispheres.

Let us now derive the magnitude of the electric field associated with the rotation in vacuum of a magnetized sphere of radius R and conductivity σ. In the closed current loop drawn in Fig. 5.33 which consists of the conducting sphere where $\mathbf{E}' = \mathbf{E} + \mathbf{v} \times \mathbf{B} = 0$ and the conducting wire where there will be a resulting electromotive force due to the rotating B-field and $\mathbf{E} \neq 0$. This must then give rise to a potential difference between A and B in the conducting wire and on the sphere as observed outside the sphere.

We will assume that the magnetic field is dipolar and aligned along the rotation axis, then at the surface of the sphere

$$B_r = -2H_0 \sin \lambda_B \tag{5.226}$$

$$B_\lambda = H_0 \cos \lambda_B \tag{5.227}$$

where H_0 is the magnetic field strength at the equator of the sphere and λ_B is the latitude of point B.

Sec. 5.15] Equatorial fountain effect 249

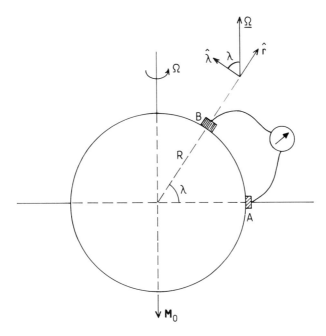

Fig. 5.33. The principle for a unipolar inductor including a rotating magnetized conducting sphere with radius R and a conductor connected to the surface of the sphere by sliding contacts. A potential difference between the points A and B is established.

The potential difference between the two contacts at A and B, where A is at equator, will be

$$V = -\int_A^B \mathbf{E} \cdot d\mathbf{s} = -\int_0^{\lambda_B} (\mathbf{E} \cdot \hat{\boldsymbol{\lambda}}) R_e \, d\lambda = -R_e \int_0^{\lambda_B} E_\lambda d\lambda \qquad (5.228)$$

From the stationary system (the conductor) we have

$$\mathbf{E} = -(\boldsymbol{\Omega} \times \mathbf{R}_e) \times \mathbf{B} \qquad (5.229)$$

where Ω is the rotation angle of the sphere. In polar coordinates:

$$\boldsymbol{\Omega} = \Omega \sin \lambda \, \hat{\mathbf{r}} + \Omega \cos \lambda \, \hat{\boldsymbol{\lambda}} \qquad (5.230)$$

and

$$\mathbf{E} = \Omega R_e B_r \cos \lambda \, \hat{\boldsymbol{\lambda}} - \Omega R_e B_\lambda \cos \lambda \, \hat{\mathbf{r}} \qquad (5.231)$$

Therefore

$$V = -R_e \int_0^{\lambda_B} \Omega R_e B_r \cos \lambda \, d\lambda \qquad (5.232)$$

$$= +H_0 R_e^2 \Omega \int_0^{\lambda_B} 2 \sin \lambda \cos \lambda \, d\lambda = +H_0 R_e^2 \Omega \sin^2 \lambda_B$$

With proper values for the Earth we find the potential difference between the north pole and the equator to be

$$V \sim 9 \times 10^4 \text{ V} = 90 \text{ kV} \tag{5.233}$$

when $R_e = 6.38 \times 10^6$ m, $\Omega = 7.27 \times 10^{-5}$ s^{-1} and $H_0 = 3.0 \times 10^{-5}$ tesla. We also notice that the total electric field is in the meridional plane and given by

$$\begin{aligned} E &= \Omega R_e H_0 \cos\lambda (1 + 3\cos^2\lambda)^{1/2} \\ &= 14 \cos\lambda (1 + 3\cos^2\lambda)^{1/2} \text{ mv/m} \end{aligned}$$

At the equator there is another effect that comes into play when discussing the behaviour of the upper ionosphere, and that is the so-called fountain effect.

At equator an eastward electric field is established in the F-region which creates an upward moving ($\mathbf{E} \times \mathbf{B}$) plasma flow (Fig. 5.34). At the higher altitudes the recombination for the upwelling plasma is slowed down compared to the situation at the lower altitudes just after sunset. The equatorial plasma will rise in the F-region until the pressure forces are high enough that, assisted by gravity, the plasma starts to slide down the magnetic field lines toward the tropical ionosphere, where enhanced plasma densities are observed at both sides of the equator (Figs. 5.4 and 5.5).

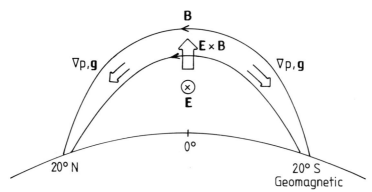

Fig. 5.34. A simplified sketch to illustrate the fountain effect at geomagnetic equator. An eastward field at geomagnetic equator drives the plasma vertically upward. Pressure gradients and gravity drives it down again along the magnetic field lines toward higher latitudes. (After Kelley, 1989.)

We realize that the plasma within the plasmasphere and the high altitude F-region is distributed under the influence of a large variety of forces and processes. Due to the large variation in the magnetospheric convection, the distance to the separation between corotating plasma and plasma convected sunward in the magnetosphere beyond the Earth can change by a large amount, and thereby change the volume of the plasmasphere. The corotating electric field can be compared to the ionospheric field of magnetospheric origin (Fig. 5.35) when assuming that the magnetic field lines are equipotentials and applying the mapping factor derived in Chapter 4. We notice that the magnetospheric field corresponding to 1 mv/m at the equator will, due to the mapping factor, be comparable to the corotating field near an L value close to 3. The weaker electric field, however, corresponding to 0.4 mv/m becomes equal to the corotating field at an L value close to 6.

We therefore realize that the corotation field can dominate out to 6–7 R_e, while the magnetospheric field dominates outside this region.

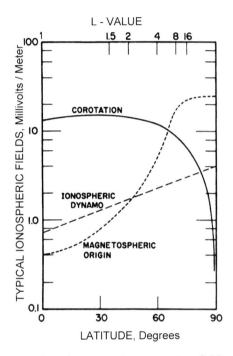

Fig. 5.35. A comparison between the corotation field and electric fields of magnetospheric origin for different geomagnetic latitudes and L values. The magnetospheric fields are mapped down to the ionosphere assuming equipotential field lines. The ionospheric dynamo electric field is also shown for reference. (From Mozer, 1973.)

Since the extraterrestrial source of the magnetospheric electric field, in contrast to the planetary rotation source is so strongly variable and irregular, the situation inside the plasmasphere will also vary. This is demonstrated in Fig. 5.36 where the plasma density as function of radial distance from the Earth's centre in the equatorial plane is shown for different degrees of disturbance in the magnetosphere. The plasma density is found to drop by two orders of magnitude or more between L equal to 3 and 6 (the plasmapause). The drop is found closer to the Earth the stronger the disturbance (the higher the Kp index). Inside the plasmapause the plasma is corotating while outside of this it is controlled by the magnetospheric convection.

Due to the different effects just mentioned the plasmapause does not stay at a constant distance from the Earth's centre throughout a day, and this is illustrated in Fig. 5.37, where it is seen that on the average the plasmasphere reaches out to beyond 5 R_e during the evening (the bulge) and contracts to inside 3.5 R_e in the morning hours. On disturbed days the plasmapause in the daytime can be eroded inside 3 R_e, and for very quiet days it can reach almost out to 6 R_e in the evening.

The position of the plasmapause also marks the transition in the abundance of the different ions as shown in Fig. 5.38, where the O^+ ions, dominant inside 2 R_e, fall off by almost 6 orders of magnitude within half an Earth radius and the H^+ ions become dominant outside 2 R_e.

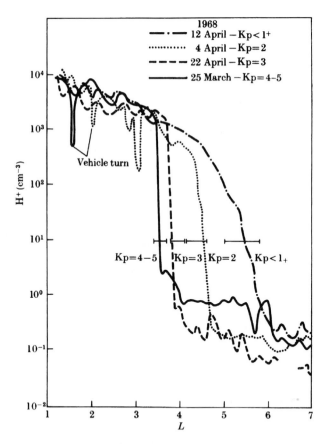

Fig. 5.36. Observed H^+ densities for different distances from the Earth surface in the equatorial plane showing a sharp decrease at distances between 3.5 and 6 R_e, for different levels of magnetospheric disturbance. The sharp discontinuity marks the position of the plasmapause. (From Chappel et al., 1970a.)

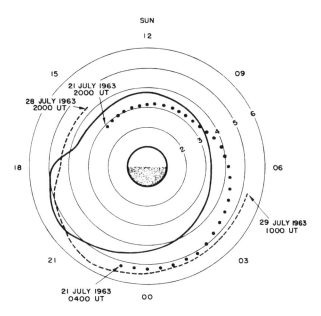

Fig. 5.37. Equatorial cross-section of the plasmasphere showing the position of the plasmapause for different time of day during average conditions. (From Carpenter, 1966.)

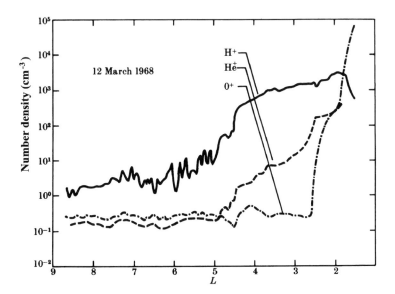

Fig. 5.38. A typical plasmasphere crossing at the noon sector showing the variation in H^+, He^+ and O^+ densities as function of the equatorial distance. (From Chappell et al., 1970b.)

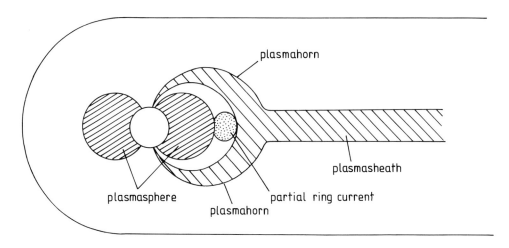

Fig. 5.39. A simplified sketch to illustrate the position of the plasmasphere in relation to the plasmasheath and the radiation belt.

Fig. 5.39 shows a cross-section in the midday–midnight meridian plane of the magnetosphere in order to illustrate the position of the plasmasphere in relation to other characteristic plasma population regions in the near-Earth environment. At the centre of the magnetospheric tail there is the plasmasheath in which the cross-tail current is flowing, and this sheath is extending in two horn-like regions toward the higher latitudes. Between the plasmasphere and the inner edge of the plasmasheath there is an area of large abundance of energetic protons, the radiation belts related to the ring current.

5.16 REFERENCES

Baron, M., Watt, T., Rino, C. and Petriceks, J. (1971) *DNA Project 617 Radar, First Auroral-Zone Results,* DNA 2826F, Stanford Research Institute, Menlo Park.
Carpenter, D. L. (1966) *J. Geophys. Res.,* **71**, 693–709.
Chappel, C. R., Harris, K. K. and Sharp, G. W. (1970a) *J. Geophys. Res.,* **75**, 50–56.
Chappel, C. R., Harris, K. K. and Sharp, G. W. (1970b) *J. Geophys. Res.,* **75**, 3848–3861.
Croom, S. A., Robbins, A. R. and Thomas, J. D. (1959) *Nature,* **184**, 2003–2004.
Eccles, D. and King, J. W. (1969) *Proc. Inst. elect. electronics Engr.,* **57**, 1012.
Evans, J. V. (1967) *Planet. Space Sci.,* **15**, 1387–1405.
Giraud, A. and Petit, M. (1978) *Ionospheric Techniques and Phenomena,* D. Reidel, Dordrecht, The Netherlands.
Hall, C. (1989) Private communication.

Hargreaves, J. K. (1979) *The Upper Atmosphere and Solar–Terrestrial Relations*, Van Nostrand Reinhold, London.

Hinteregger, H. E., Hall, L. A. and Schmidtke, G. (1965) *Space Research*, Vol. 5, pp. 1175–1190, North-Holland Publ., Amsterdam.

Johnson, C. Y. (1966) *J. Geophys. Res.*, **71**, 330.

Kelley, M. C. (1989) *The Earth's Ionosphere, Plasma Physics and Electrodynamics*, Academic Press, New York.

Mozer, F. S. (1973) *Rev. Geophys. Space Phys.*, **11**, 755–765.

Narcisi, R. S. and Bailey, A. D. (1965) *J. Geophys. Res.*, **70**, 3687–3700.

Ratcliffe, J. A. (1972) *An Introduction to the Ionosphere and the Magnetosphere*, Cambridge at the University Press.

Richmond, A. (1987) in *The Solar Wind and the Earth*, Akasofu, S.-I. and Kamide, Y. (Eds.), pp. 123–140, Terra Sci. Publ. Comp., Japan.

Rishbeth, H. and Garriott, O. K. (1969) *Introduction to Ionospheric Physics*, Academic Press, London.

Roble, R. G. and Emery, B. A. (1983) *Planet. Space Sci.*, **31**, 597–614.

Röderer, J. G. (1970) *Dynamics of Geomgnetically Trapped Radiation*, Springer-Verlag, New York.

Van Zandt, T. E. and Knecht, R. W. (1964) in *Space Physics*, Le Galley and Rosen (Eds.), John Wiley, New York.

Whitten, R. C. and Poppoff, I. G. (1971) *Fundamentals of Aeronomy*, John Wiley, New York.

Wright, J. W. (1962) *Nature*, **194**, 462.

5.17 EXERCISES

1. Prove the relations given by (5.38).

2. Prove the relations given by (5.57) and (5.62).

3. Assume that we have an electron density profile given by:

$$n_e(x) = n_{e0} \exp(1 - x - \sec\chi \cdot \exp(-x))$$

   (a) Find the total electron content in a unit column from ground to infinity of this profile.

   Assume instead this profile is given by

   $$n_e(x) = n_{e0} \exp\left[\frac{1}{2}(1 - x - \sec\chi \exp(-x))\right]$$

   (b) Find the total electron content of this profile in a unit column from ground to infinity.

4. Prove the relations given by (5.101) and (5.102).

5. We have found the following two continuity equations to apply for n_e and n^+ in the D-region:

$$\frac{dn_e}{dt} = q - \alpha_D n^+ n_e - \beta n_e \cdot n_n + \gamma n^- n_n + \rho n^-$$

$$\frac{dn^+}{dt} = q - \alpha_D n^+ n_e - \alpha_i n^+ n^-$$

We assume charge neutrality.

(a) Set up the continuity equation for the negative ions.
(b) Determine the steady-state solution of $\lambda = n^-/n_e$.
(c) Let $\lambda = \lambda_D =$ constant in time and show that the electron density continuity equation can be expressed as:

$$\frac{dn_e}{dt} = -\frac{\lambda_D + 1}{n_0} A n_e^2 - \left(B - \frac{C}{\lambda_0}\right) n_e$$

where $n_0 = n_e$ when $t = 0$. Determine A, B and C.

(d) Solve this continuity equation for n_e.

6. Prove the relation given by (5.164).

6
Dynamics of the neutral atmosphere

6.1 THE GENERAL EQUATIONS

Since the ionized part of the atmosphere represents only a very small fraction of the total atmospheric density, at least in the height region between 80 and 300 km, it is clear that the motion of the neutral atmosphere will have a profound influence on the motion of the ionospheric plasma due to their coupling by collisions. We will therefore first discuss some of the principles determining the dynamics of the neutral atmosphere.

We will treat the atmosphere as an ideal gas and apply the continuity equation, the equation of motion and the equation of state for an ideal gas as follows:

$$\frac{d\rho}{dt} + \rho \nabla \cdot \mathbf{u} = 0 \tag{6.1}$$

$$\rho \frac{d\mathbf{u}}{dt} + \nabla p = \mathbf{F} \tag{6.2}$$

$$p = \rho RT \tag{6.3}$$

where all quantities have their usual meaning. $R = R_0/M'$ is the individual gas constant per unit mass as noticed in section 3.5. For the atmosphere below say 90 km where $M' = 28.8$ kg/kmol

$$R_{\text{air}} = 289 \text{ J/K·kg} \tag{6.4}$$

An energy conservation law must be included for which we turn to the second law of thermodynamics

$$\frac{dS}{dt} = \frac{1}{T}\frac{dQ}{dt} = \frac{\dot{Q}}{T} \tag{6.5}$$

which states that the change of the entropy by time is equal to the heating rate per unit temperature of the system. The heating rate can be expressed by means of different terms such as q_T, the primary production rate by external or internal sources, the thermal loss by radiation L_T, and the heat flux ϕ_T associated with advection and conduction. We then can write:

$$\dot{Q} = q_T - L_T - \nabla \phi_T \tag{6.6}$$

From the first law of thermodynamics we have

$$dQ = d\mathcal{E} + p\,dV \tag{6.7}$$

where V is the volume and $\mathcal{E}$ represents the internal energy of the system, respectively. $\mathcal{E}$ is given by

$$d\mathcal{E} = C_v dT \tag{6.8}$$

where C_v is the heat capacity at constant volume. Since $\rho = m/V$ we have that

$$dV = -m\,\frac{d\rho}{\rho^2} \tag{6.9}$$

For an ideal gas which obeys the ideal gas law we derive from (6.7)

$$dQ = C_v dT - pm\,\frac{d\rho}{\rho^2} = C_v dT - mR_T\,\frac{d\rho}{\rho} \tag{6.10}$$

The entropy now becomes:

$$dS = \frac{dQ}{T} = C_v\,\frac{dT}{T} - mR\,\frac{d\rho}{\rho} \tag{6.11}$$

Dividing (6.11) by m we obtain

$$ds = c_v\,\frac{dT}{T} - R\,\frac{d\rho}{\rho} \tag{6.12}$$

where s and c_v are the entropy and heat capacity per unit mass at constant volume, respectively.

Now introducing

$$c_p - c_v = c_v(\gamma - 1)R \tag{6.13}$$

where c_p is the heat capacity per unit mass at constant pressure and γ is the adiabatic constant, we obtain from (6.12)

$$ds = c_v\left[\frac{dT}{T} + (1-\gamma)\,\frac{d\rho}{\rho}\right] \tag{6.14}$$

Integrating (6.13) gives:

$$s = c_v \cdot \ln T \cdot \rho^{1-\gamma} + \text{const.} \tag{6.15}$$

With the help of the ideal gas law again we replace ρ with the pressure

$$\begin{aligned} s &= c_v \cdot \ln T^\gamma \cdot p^{-(\gamma-1)} + \text{const.} \\ &= c_p \ln T \cdot p^{-(\gamma-1)/\gamma} + \text{const.} \end{aligned} \tag{6.16}$$

By now introducing the potential temperature which is a widely used concept in meteorology:

$$\theta = T \cdot \left(\frac{p}{p_s}\right)^{-\frac{\gamma-1}{\gamma}} \tag{6.17}$$

Here p_s is a constant reference pressure equal to 1000 mb. The potential temperature of an air parcel is the temperature the parcel would attain if it was brought adiabatically to the pressure level p_s from an arbitrary pressure level p with temperature T. Then we obtain:

$$s = c_p \cdot \ln \theta + \text{const.} \tag{6.18}$$

The potential temperature is therefore a representation of the entropy. If the entropy is constant, so is the potential temperature and vice versa. The heat capacities per unit mass for the air is found to be

$$c_v = 712 \text{ J/K kg}$$
$$c_p = 996 \text{ J/K kg}$$

when the adiabatic constant $\gamma = 1.4$ and the molecular mass $M' = 28.8$ are used.

6.2 THE VERTICAL MOTION

It is of special interest to study the vertical winds although these are usually very small compared to the horizontal winds and also difficult to measure. They are, however, important as they play a significant role in the vertical transport of kinetic energy and also of gravity potential energy. Furthermore, the vertical winds control to a large extent the vertical distribution of the different constituents as we have seen in relation to the diffusion. Let us assume that there are no horizontal gradients in the density but that the atmosphere is vertically stratified ($\partial/\partial x = \partial/\partial y = 0$).

For a quasi-steady state ($\partial/\partial t = 0$) we derive from the continuity equation (6.1)

$$(\mathbf{u} \cdot \nabla)\rho + \rho \nabla \cdot \mathbf{u} = 0 \tag{6.19}$$

and

$$w \frac{\partial \rho}{\partial z} + \rho \nabla \cdot \mathbf{u} = 0 \tag{6.20}$$

where we have used $\mathbf{u} = (u, v, w)$. By multiplying by p and dividing by ρ we get

$$p \nabla \cdot \mathbf{u} = -\frac{p}{\rho} \frac{\partial \rho}{\partial z} w \tag{6.21}$$

From the equation of state (6.3) we have

$$\frac{\partial p}{\partial z} = R \left(\rho \frac{\partial T}{\partial z} + T \frac{\partial \rho}{\partial z} \right) \tag{6.22}$$

and solving for $\partial \rho / \partial z$ gives:

$$\frac{\partial \rho}{\partial z} = \frac{1}{T}\left(\frac{1}{R}\frac{\partial p}{\partial z} - \rho \frac{\partial T}{\partial z}\right) \quad (6.23)$$

By inserting this into (6.21) we have

$$p\nabla \cdot \mathbf{u} = -\frac{p}{\rho}\frac{1}{T}\left(\frac{1}{R}\frac{\partial p}{\partial z} - \rho \frac{\partial T}{\partial z}\right)w \quad (6.24)$$

$$= -\left(\frac{\partial p}{\partial z} - \frac{p}{T}\frac{\partial T}{\partial z}\right)w$$

The vertical velocity is now given by

$$w = -\frac{p\nabla \cdot \mathbf{u}}{\frac{\partial p}{\partial z} - \frac{p}{T}\frac{\partial T}{\partial z}} \quad (6.25)$$

By introducing the partial derivative $\partial T/\partial p$ and the barometric equation (3.39) we have:

$$w = \frac{p\nabla \cdot \mathbf{u}}{g\rho\left(1 - \frac{p}{T}\frac{\partial T}{\partial p}\right)} = \frac{H\nabla \cdot \mathbf{u}}{\left(1 - \frac{p}{T}\frac{\partial T}{\partial p}\right)} \quad (6.26)$$

The vertical velocity is now expressed by the divergence of $\mathbf{u}$ and the variation in temperature by pressure. We have also introduced the scale height from (3.40) into (6.26). For an isothermal atmosphere where $\partial T = 0$ we find

$$w = H\frac{\partial w}{\partial z} \quad (6.27)$$

and

$$w = w_0 \exp\left(\frac{z - z_0}{H}\right) \quad (6.28)$$

when $w = w_0$ at a reference height z_0. The vertical velocity therefore increases by height for an expansion in an isothermal atmosphere. This comes about since the density decreases by height, and in order to keep the energy flux constant the velocity must increase.

For an adiabatic expansion where

$$\frac{p}{T}\frac{\partial T}{\partial p} = \frac{\gamma - 1}{\gamma} \quad (6.29)$$

we find

$$w = \gamma H \nabla \cdot \mathbf{u} \quad (6.30)$$

and

$$w = w_0^* \exp\left(\frac{z - z_0}{\gamma H}\right) \quad (6.31)$$

Since $\gamma > 1$, the vertical velocity will increase more slowly by height in the adiabatic situation than for an expansion in an isothermal atmosphere.

The term $p\nabla \cdot \mathbf{u}$ represents a rate of mechanical work against the pressure force due to expansion or compression. We now want to relate this to the change of energy in the system.

From the entropy relation given by (6.16) we can derive the rate of entropy as:

$$\frac{ds}{dt} = c_v \left(\frac{1}{T} \frac{dT}{dt} + (1-\gamma) \frac{1}{\rho} \frac{d\rho}{dt} \right) = \frac{1}{T} \frac{dq}{dt} \tag{6.32}$$

where q is the heating rate per unit mass. We can rewrite this equation:

$$c_v \frac{dT}{dt} - \frac{RT}{\rho} \frac{d\rho}{dt} = \frac{dq}{dt} = \dot{q} \tag{6.33}$$

where we have used the relation $c_v(1-\gamma) = -R$. By inserting (6.1) into (6.33) above we find:

$$RT \, \nabla \cdot \mathbf{u} = \dot{q} - c_v \frac{dT}{dt} = \dot{q} - \dot{\varepsilon} \tag{6.34}$$

where ε is the internal energy per unit mass of the gas.

Finally

$$p\nabla \cdot \mathbf{u} = \frac{p}{RT}(\dot{q} - \dot{\varepsilon}) = \rho(\dot{q} - \dot{\varepsilon}) \tag{6.35}$$

The left-hand side represents the rate at which work is done by the system if the right side is positive, and on the system if the right side is negative. The vertical velocity can now be expressed by the rate of this work:

$$w = \frac{\dot{q} - \dot{\varepsilon}}{g\left(1 - \frac{p}{T} \frac{\partial T}{\partial p}\right)} \tag{6.36}$$

For an adiabatic process where $\dot{q} = 0$ we now have:

$$w = -\frac{\gamma \dot{\varepsilon}}{g} = -\frac{\gamma c_v}{g} \frac{dT}{dt} \tag{6.37}$$

If the internal temperature increases by time, then work must be done on the system by the surroundings and the air must contract downward. In the opposite sense, if the temperature decreases, the work is done by the system and the atmosphere expands upward. Since we have only allowed for vertical motion, this is equivalent to considering the motion of a column of air in a cylinder with a fixed base.

For external heat allowed to exchange with the system the situation can be far more dramatic, especially if $\dot{q}$ is large. This occurs quite frequently within the auroral region of the upper atmosphere and then strong vertical motion can be expected.

6.3 THE EQUATION OF MOTION OF THE NEUTRAL GAS

Since we are going to discuss the motion of the atmosphere of a rotating planet, it is convenient to express the kinetic equations in a reference frame rotating with the planet at an angular velocity $\mathbf{\Omega}$. We start out with the general equation of motion for a neutral gas

$$\frac{d\mathbf{u}}{dt} = -\frac{1}{\rho}\nabla p + \mathbf{g} + \mathbf{f} \tag{6.38}$$

Here we have split up the external force into the force of gravity $\mathbf{g}$ and any other accelerating force $\mathbf{f}$ per unit mass.

We now denote by $\mathbf{u}_f$ the velocity observed in an inertial frame of reference, for example one fixed to the centre of the Earth. The velocity $\mathbf{u}_r$, observed in a reference frame rotating with an angular velocity $\mathbf{\Omega}$ around an axis through the centre of the Earth and at a distance $\mathbf{r}$ from this centre, is according to the laws of mechanics, related to $\mathbf{u}_f$ by (Fig. 6.1):

$$\left(\frac{d\mathbf{r}}{dt}\right)_f = \left(\frac{d\mathbf{r}}{dt}\right)_r + \mathbf{\Omega} \times \mathbf{r} \tag{6.39}$$

and

$$\mathbf{u}_f = \mathbf{u}_r + \mathbf{\Omega} \times \mathbf{r} \tag{6.40}$$

The acceleration of this air parcel in the fixed frame of reference is now, since $\mathbf{\Omega}$ is constant, given by:

$$\begin{aligned}
\mathbf{a}_f &= \left(\frac{d\mathbf{u}_f}{dt}\right)_f = \left(\frac{d}{dt}(\mathbf{u}_r + \mathbf{\Omega} \times \mathbf{r})\right)_f \\
&= \left(\frac{d}{dt}\mathbf{u}_r\right)_f + \left(\frac{d}{dt}(\mathbf{\Omega} \times \mathbf{r})\right)_f \\
&= \left(\frac{d\mathbf{u}_r}{dt}\right)_r + \mathbf{\Omega} \times \mathbf{u}_r + \left(\frac{d}{dt}\mathbf{\Omega}\right)_f \times \mathbf{r} + \mathbf{\Omega} \times \left(\frac{d\mathbf{r}}{dt}\right)_f \\
&= \mathbf{a}_r + \mathbf{\Omega} \times \mathbf{u}_r + \mathbf{\Omega} \times \mathbf{u}_r + \mathbf{\Omega} \times (\mathbf{\Omega} \times \mathbf{u}_r) \\
&= \mathbf{a}_r + 2\mathbf{\Omega} \times \mathbf{u}_r + \mathbf{\Omega} \times (\mathbf{\Omega} \times \mathbf{r})
\end{aligned} \tag{6.41}$$

where $\mathbf{a}_r$ is the acceleration in the rotating reference frame. This acceleration can be expressed as

$$\mathbf{a}_r = \mathbf{a}_f - 2\mathbf{\Omega} \times \mathbf{u}_r - \mathbf{\Omega} \times (\mathbf{\Omega} \times \mathbf{r}) \tag{6.42}$$

Since the equation of motion in the fixed frame of reference is given by

$$\mathbf{a}_f = \left(\frac{d\mathbf{u}}{dt}\right)_f = -\frac{1}{\rho}\nabla p + \mathbf{g} + \mathbf{f} \tag{6.43}$$

Sec. 6.3] The equation of motion of the neutral gas 263

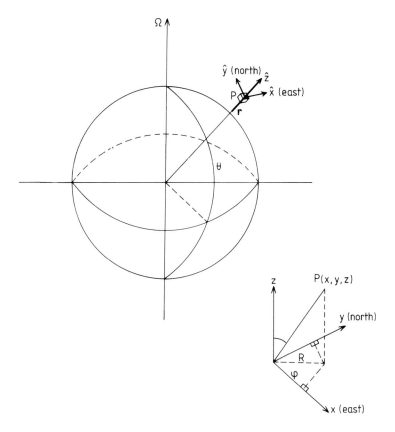

Fig. 6.1. The coordinate system used to describe the motion of an air parcel at a distance r from the Earth's centre and rotating with the Earth at an angular velocity Ω.

the equation of motion in a reference system following the Earth's rotation becomes

$$\mathbf{a} = \frac{d\mathbf{u}}{dt} = -\frac{1}{\rho}\nabla p + \mathbf{g} + \mathbf{f} - 2\mathbf{\Omega} \times \mathbf{u} - \mathbf{\Omega} \times (\mathbf{\Omega} \times \mathbf{r}) \qquad (6.44)$$

when deleting the index r. Expanding the final vector product we have

$$\frac{d\mathbf{u}}{dt} = -\frac{1}{\rho}\nabla p + \mathbf{g} + \mathbf{f} - 2\mathbf{\Omega} \times \mathbf{u} - (\mathbf{\Omega} \cdot \mathbf{r})\mathbf{\Omega} + \Omega^2 \mathbf{r} \qquad (6.45)$$

It is then customary to introduce the potential function

$$\chi = -\mathbf{g} \cdot \mathbf{r} - \frac{1}{2}\Omega^2 r^2 \cos^2 \theta \qquad (6.46)$$

where θ is the latitude angle and r the radial distance to the point P as seen from the Earth's centre (Fig. 6.1).

Finally the equation of motion takes the form:

$$\frac{d\mathbf{u}}{dt} + 2\mathbf{\Omega} \times \mathbf{u} + \frac{1}{\rho}\nabla p + \nabla\chi = \mathbf{f} \tag{6.47}$$

Here $2(\mathbf{\Omega} \times \mathbf{u})$ is called the Coriolis acceleration while $\mathbf{\Omega} \times (\mathbf{\Omega} \times \mathbf{r})$ is the centrifugal acceleration.

For an observer fixed on the Earth we have the rotation frequency $\Omega = 7.27 \times 10^{-5}$ s^{-1} and since the Earth radius is $R_e = 6.37 \times 10^6$ m, we find $\Omega^2 R_e = 3.37 \times 10^{-2}$ m/s^2 which represents the maximum centrifugal acceleration (at equator). This can be neglected compared to $g = 9.81$ m/s^2.

We notice that in the northern hemisphere where $\mathbf{\Omega}$ points out of the ground, the Coriolis force will tend to deflect the motion to the right of the direction of the velocity vector. The Coriolis force is quite weak since for a velocity of 1000 m/s, which is very high for neutral air velocity, the maximum acceleration corresponds to 1.49×10^{-2} g. For motions persisting over long periods, however, the deflection of the velocity is quite important and creates in meteorological terms the cyclones in the northern hemisphere. These are important elements in the weather system. For practical reasons we will now choose a Cartesian coordinate system (x, y, z) fixed on the Earth's surface as indicated in Fig. 6.1 and denote the velocity components (u, v, w) along the different axes. u, v and w are counted positive eastward, northward and upward, respectively. Since we have neglected the centrifugal force, the acceleration of gravity is given by

$$\mathbf{g} = -g\hat{\mathbf{z}} \tag{6.48}$$

The angular velocity vector is given by:

$$\mathbf{\Omega} = \Omega \sin\theta\, \hat{\mathbf{z}} + \Omega \cos\theta\, \hat{\mathbf{y}} \tag{6.49}$$

and we find for each component of the acceleration:

$$\frac{du}{dt} - 2\Omega v \sin\theta + 2\Omega w \cos\theta + \frac{1}{\rho}\frac{\partial p}{\partial x} = f_x \quad \text{(a)} \tag{6.50}$$

$$\frac{dv}{dt} + 2\Omega u \sin\theta + \frac{1}{\rho}\frac{\partial p}{\partial y} = f_y \quad \text{(b)}$$

$$\frac{dw}{dt} - 2\Omega u \cos\theta + \frac{1}{\rho}\frac{\partial p}{\partial z} + g = f_z \quad \text{(c)}$$

6.4 THE GEOSTROPHIC AND THERMAL WINDS

We will now study some of the steady-state solutions ($d/dt = 0$) by assuming first of all that there is no vertical motion ($w = 0$), then that all other external forces except gravity are negligible and finally that the hydrostatic equation (3.39) applies, i.e.:

$$\frac{\partial p}{\partial z} = -\rho g$$

For the horizontal motion we obtain the so-called geostrophic wind solution:

$$u = -\frac{1}{2\rho\Omega \sin\theta}\frac{\partial p}{\partial y} \quad \text{(a)} \tag{6.51}$$

$$v = \frac{1}{2\rho\Omega \sin\theta}\frac{\partial p}{\partial x} \quad \text{(b)}$$

We notice that since $\sin\theta = 0$ at equator, there can be no geostrophic wind there. u and v change sign, however, across equator. The geostrophic equations describe the motion of an air parcel that is initially moving from a high to low pressure area. This motion is slow enough for the Coriolis force to deflect the air motion so that it finally becomes parallel to the isobars. We notice this because

$$\mathbf{v}\cdot\nabla p = u\frac{\partial p}{\partial x} + v\frac{\partial p}{\partial y} = 0 \tag{6.52}$$

and $\mathbf{v}$ is perpendicular to the pressure gradients which by definition are perpendicular to the isobars.

By introducing the ideal gas law in the geostrophic wind equations we find the so-called thermal wind equation:

$$\frac{u}{T} = -\frac{R}{2\Omega \sin\theta}\frac{1}{\rho RT}\frac{\partial p}{\partial y} = -\frac{R}{2\Omega \sin\theta}\frac{\partial}{\partial y}(\ln p) \quad \text{(a)} \tag{6.53}$$

$$\frac{v}{T} = \frac{R}{2\Omega \sin\theta}\frac{\partial}{\partial x}(\ln p) \quad \text{(b)}$$

By now differentiating with respect to height we obtain

$$\frac{\partial}{\partial z}\left(\frac{u}{T}\right) = -\frac{R}{2\Omega \sin\theta}\frac{\partial}{\partial z}\left(\frac{\partial}{\partial y}\ln p\right) = -\frac{g}{2\Omega T^2 \sin\theta}\frac{\partial T}{\partial y} \quad \text{(a)} \tag{6.54}$$

$$\frac{\partial}{\partial z}\left(\frac{v}{T}\right) = +\frac{R}{2\Omega \sin\theta}\frac{\partial}{\partial z}\left(\frac{\partial}{\partial x}\ln p\right) = \frac{g}{2\Omega T^2 \sin\theta}\frac{\partial T}{\partial x} \quad \text{(b)}$$

These equations relate the vertical wind shears in the geostrophic wind to the horizontal gradients in the temperature.

Consider the northern hemisphere close to ground where it is hot at the equator and cold at the poles. Then $\partial T/\partial y$ will be less than zero there so that $(\partial/\partial z)(u/T)$ becomes larger than zero. Up toward the tropopause therefore the eastward wind will increase in strength. Even if the wind is westward at the ground, it may become eastward at some height in the troposphere. In the southern hemisphere, because both $\partial T/\partial y$ as well as $\sin\theta$ change sign there, the wind will also increase towards the east as one ascends in the troposphere.

In the stratosphere the situation can be different. The summer pole is continually heated at a higher temperature than at equator which in turn is hotter than the totally unilluminated winter pole in the stratosphere. Thus, the thermal winds in the stratosphere blow in the opposite direction to the tropospheric wind.

The wind pattern and temperature distribution up to the mesosphere is fairly well-known, and except for the lowest 1000 metres of the atmosphere the zonal winds appear to be adequately described by the thermal wind equation (Mourgatroyd, 1957). There is a problem, however, related to the warm winter mesosphere as already mentioned in Chapter 3. It is at about 65 km that the summer and winter temperatures are roughly equal. Above this height in the mesosphere the winter temperatures are higher.

There are also meridional prevailing winds leading to a net latitudinal flow of air which must be balanced by return flows at other heights. This will set up large meridional wind cells. We will return to these wind systems below.

6.5 THE EQUATION OF MOTION IN POLAR COORDINATES

Let us first, however, go back to the equation of motion and solve it in cylindrical coordinates (R, φ, z) as shown in Fig. 6.1. Here z is vertical, R is the radial distance in the horizontal plane and φ is the azimuthal angle measured positive from east toward north. The components of the velocity vector are (v_r, v_φ, v_z).

In this coordinate system we have the following unit vectors:

$$\hat{\mathbf{R}} = \cos\varphi\,\hat{\mathbf{x}} + \sin\varphi\,\hat{\mathbf{y}} \quad\text{(a)}$$

$$\hat{\boldsymbol{\varphi}} = -\sin\varphi\,\hat{\mathbf{x}} + \cos\varphi\,\hat{\mathbf{y}} \quad\text{(b)} \qquad (6.55)$$

$$\hat{\mathbf{z}} = \hat{\mathbf{z}} \quad\text{(c)}$$

The angular velocity vector is given by

$$\boldsymbol{\Omega} = \Omega(\cos\theta\sin\varphi\,\hat{\mathbf{R}} + \cos\theta\cos\varphi\,\hat{\boldsymbol{\varphi}} + \sin\theta\,\hat{\mathbf{z}}) \qquad (6.56)$$

where θ is the local latitude. We also find that

$$\frac{d\hat{\mathbf{R}}}{dt} = (-\sin\varphi\,\hat{\mathbf{x}} + \cos\varphi\,\hat{\mathbf{y}})\frac{d\varphi}{dt} = \frac{1}{R}v_\varphi\,\hat{\boldsymbol{\varphi}} \qquad (6.57)$$

where v_φ is the azimuthal velocity. Similarly we have

$$\frac{d\hat{\boldsymbol{\varphi}}}{dt} = -\frac{1}{R}v_\varphi\,\hat{\mathbf{R}} \qquad (6.58)$$

The equation of motion on component form is then:

$$\frac{dv_R}{dt} + 2\Omega(\cos\theta\cos\varphi\,v_z - \sin\theta\,v_R) - \frac{v_\varphi^2}{R} = f_R \quad\text{(a)} \qquad (6.59)$$

$$\frac{dv_\varphi}{dt} + 2\Omega(\sin\theta\,v_R - \cos\theta\sin\varphi\,v_z) + \frac{v_R v_\varphi}{R} = f_\varphi \quad\text{(b)}$$

$$\frac{dv_z}{dt} + 2\Omega(\cos\theta\cos\varphi\,v_\varphi - \cos\theta\cos\varphi\,v_R) = f_z \quad\text{(c)}$$

where f_R, f_φ and f_z are the R, φ and z components of the external forces, respectively.

Let us again consider a steady state and no vertical wind ($v_z = 0$). When the only external force in the horizontal plane is the pressure gradient ($\mathbf{f} = -(1/\rho)\nabla p$), then we have:

$$2\Omega v_R \sin\theta + \frac{v_\varphi^2}{R} = \frac{1}{\rho}\frac{\partial p}{\partial R} \quad \text{(a)} \qquad (6.60)$$

$$2\Omega v_R \sin\theta + \frac{v_R v_\varphi}{R} = -\frac{1}{\rho R}\frac{\partial p}{\partial \varphi} \quad \text{(b)}$$

If the system is axial symmetric ($\partial p/\partial \varphi = 0$), then the only motion is azimuthal. If we now suppose that the lines of constant p are concentric circles, then in the northern hemisphere v_φ will be negative, i.e. clockwise if the pressure is highest in the centre ($\partial p/\partial R < 0$). This is the situation in the so-called anticyclones. If the pressure is lowest in the centre, however, ($\partial p/\partial R > 0$), then the motion becomes anticlockwise ($v_\varphi > 0$) as is the situation in cyclonic areas. We also notice that for the vertical case

$$2\Omega \cos\theta \sin\varphi\, v_\varphi = -g - \frac{1}{\rho}\frac{\partial p}{\partial z} \qquad (6.61)$$

such that the hydrostatic equation is modified by a Coriolis term due to the azimuthal motion.

In Fig. 6.2 we have illustrated the clockwise rotation around a high pressure region (anticyclone) and an anticlockwise rotation around a low pressure region (cyclone) in the northern hemisphere, and the opposite rotation direction in the southern hemisphere.

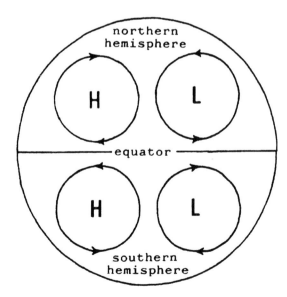

Fig. 6.2. A simplified sketch showing the difference in main rotation direction of the geostrophic wind in the two hemispheres. (From Tohmatsu, 1990.)

6.6 THE WIND SYSTEMS AT LOWER ALTITUDES

It is the inhomogeneous heat input to the Earth and its atmosphere which represents the energy source of the global wind system. The most striking feature is the large difference in solar radiation between the polar and equatorial regions as well as between the nightside and the dayside of the Earth. The large scale gradients in the global temperature this creates, represent fundamental driving forces in the atmospheric dynamics. Due to the uneven distribution of land and sea and the local variations in the albedo will there be a large variety of local effects in the total global wind system, and a detailed description of the winds can not be completed by the simplified equations we so far have presented. Since we are primarily interested in the behaviour of the neutral atmosphere at ionospheric altitudes, we will, however, only give a superficial treatment of the wind system in the weather zone and below the mesosphere, and consider the mean annual wind system.

First of all we should realize that although it has been known for generations that the solar heat input on the average is much higher for the equatorial region than at the poles, it is only recently that it has been possible to measure by means of satellites the difference in the incoming and outgoing radiation for different regions of the globe. In Fig. 6.3 the daily mean values of these radiating quantities for one

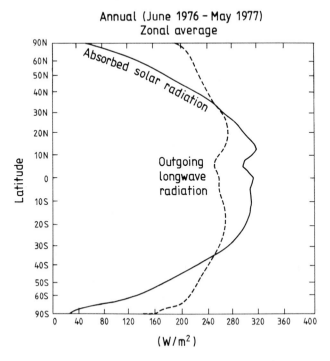

Fig. 6.3. Variation with latitude of average incoming and outgoing radiation for the year June 1, 1976 to May 31, 1977. (From NOAA–NESS, 1979.)

Sec. 6.6] The wind systems at lower altitudes 269

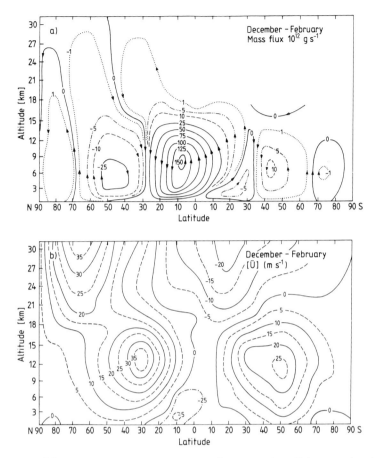

Fig. 6.4. (a) Mean meridional circulation for December–February showing three distinct cells in each latitude. The data represent latitudinal average of the meridional component. Numerals labelled to the contours are mass flux in units of 10^{12} g s^{-1}. (b) East–west component of winds averaged around the Earth for December–February. Positive numbers mean eastward winds. Numerals labelled to the contours are wind speed in units of m/s. Positive is eastward. (After Neiburger et al., 1982.)

year are illustrated. These observations show that while there is a net radiation input of about 300 W/m^2 to the Earth per day between the latitudes 30° N and 30° S, there is a net radiation loss above 35° of latitudes in both hemispheres. It is clear that the global wind system is set up to average out some of this large heat imbalance.

As early as 1735 Hadley proposed that the large difference in global heat input at equator and at the poles was redistributed by two gigantic meridional wind cells created by ascending air at the equator and descending air at the poles with equatorward winds at ground level and poleward winds in the upper atmosphere.

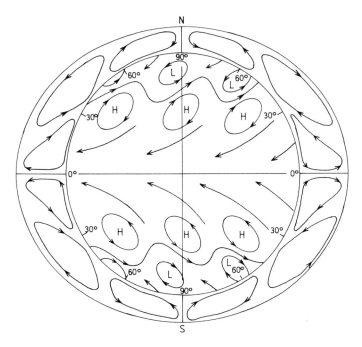

Fig. 6.5. Schematic diagram showing the main features of the global tropospheric wind system. The horizontal pattern is indicated inside the circle representing the globe, while the meridional cells are indicated outside the circle. The scale is arbitrary in the vertical direction but the outer circle is representing the tropopause. (After Neiburger et al., 1982.)

Such large wind cells cannot exist due to the effect of the Coriolis force and the law of conservation of angular momentum and the friction forces between the atmosphere and the ground. The winds must have strong zonal components as well and the Hadley cells must be broken up. From modern observations of the mean meridional winds in the lower atmosphere it is clear that several cells exist in both hemispheres (Fig. 6.4a).

The Hadley cell is there but it is limited below about 30° of latitude. In the winter hemisphere, however, the Hadley cell has a tendency to extend into the summer hemisphere in the lower part of the troposphere, and it stretches poleward at higher altitude along the so-called polar front. Between 35° and 65° of latitude there is a cell in each hemisphere called the Ferrel cell. Polewards of 70° again there is a new cell with a similar direction of flow as the original Hadley cell. The Ferrel cells therefore do in effect break up the Hadley cell.

Similar observations of the zonal winds as illustrated in Fig. 6.4b show that these are much stronger than the meridional winds, and two regions, one at about 30° of latitude in the winter hemisphere and one at about 45° of latitude in the summer hemisphere, have enhanced eastward velocities ($\sim$ 30 m/s) around the

Sec. 6.7]   **The wind systems of the upper atmosphere**   271

tropopause (∼ 12 km) called the jet streams. These are located at the poleward border of the Hadley cell and are to a first degree a consequence of the thermal wind equation.

Finally, for the wind system in the lower troposphere a schematic illustration is given in Fig. 6.5 showing the main features of the general circulation at different latitudes. In the equatorial region up to about 30° of latitude the Hadley cell and the westward trade winds dominate. At middle latitudes where the Ferrel cells are situated, there are dominating eastward winds in both hemispheres. These winds are modulated by the subtropical high pressure and the polar low pressure regions. Across the polar regions the winds are predominantly westward in both hemispheres.

6.7 THE WIND SYSTEMS OF THE UPPER ATMOSPHERE

We will now leave the lower atmosphere and discuss the wind system in the mesosphere and thermosphere. In Fig. 6.6 the average zonal winds for January are shown below 130 km. We recognize the strong jet streams at the tropopause.

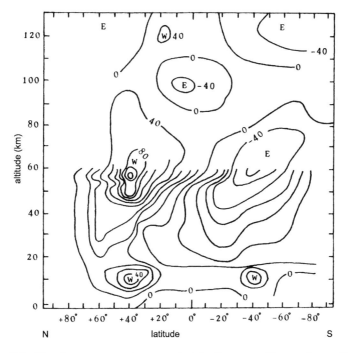

Fig. 6.6. Zonal average east–west wind in January. Numerals on the contours are wind speed in units of m s^{-1} below 130 km altitude. Positive numbers represent eastward winds. Notice the meteorological convention of W representing westerly, i.e. winds coming from the west. (From Tohmatsu, 1990.)

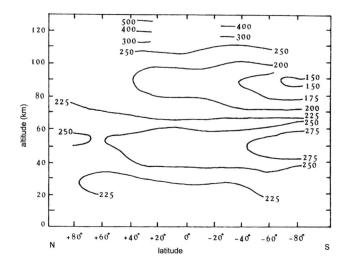

Fig. 6.7. Meridional distribution of the mean atmospheric temperature in January. Numerals labelled on the contours are in units of K. Notice that the mesosphere is warmer in winter time than in summertime. (From Tohmatsu, 1990.)

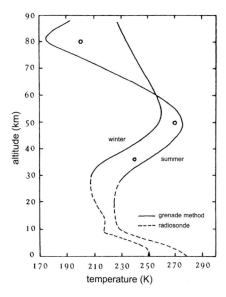

Fig. 6.8. Vertical distributions of the mean atmospheric temperature in summer and winter measured at Fort Churchill (59° N). The solid and dashed curves are those measured by the grenade method and radiosondes, respectively. Open circles represent the average values at White Sands (23° N). (From Stroud et al., 1959.)

Sec. 6.7]   The wind systems of the upper atmosphere   273

Note that by meteorological convection E and W stands for easterly and westerly, which actually means winds coming from the east and west, respectively. This is the opposite sense to what is the convention in ionospheric physics where E and W means eastward and westward blowing winds, respectively.

In the mesosphere with its centre around 60 km, however, there is a strong eastward blowing wind in the winter hemisphere and a similar but weaker westward blowing wind in the summer hemisphere. The maximum wind speed is observed close to 40° of latitude in both hemispheres. This wind pattern again appears to be well explained by the thermal wind equation. As can be seen from the temperature distribution presented in Fig. 6.7, the temperatures are high in the stratosphere

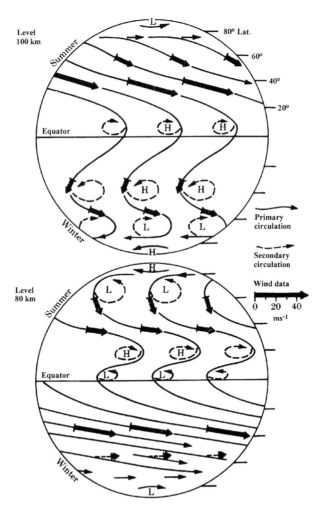

Fig. 6.9. Schematic models of the planetary neutral air circulation at 80 and 100 km. (From Kochanski, 1963.)

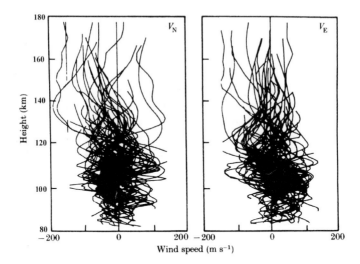

Fig. 6.10. The superposition of sixty observations of the north–south and east–west components of the wind as a function of height between 80 and 180 km at mid-latitudes. (From Rosenberg, 1968.)

in the summer hemisphere, while they are low in the winter hemisphere. The difference is about 50 K at about 40 km. In the mesosphere, however, the situation is the opposite where the temperature in the polar hemisphere at about 80 km is about 230 K, while it is less than 180 K in the summer hemisphere as illustrated in Fig. 6.8 where the vertical temperature distribution as observed from Fort Churchill (59° N) in summer and winter are compared. In contrast to the average zonal wind in the troposphere, which are eastward in both hemispheres, the average zonal wind in the mesosphere have opposite directions.

The meridional winds in the mesosphere appear to have an average net flow from the summer hemisphere to the winter hemisphere. Large variability in this height region seems to be present as can be observed by comparing the schematic wind diagrams for 80 and 100 km, respectively, in Fig. 6.9. This variability in the mesospheric wind is even more clearly borne out in Fig. 6.10 where sixty mid-latitude observations in the height region 80 to 180 km are superposed on each other giving a strong impression of variable wind zonally as well as meridionally.

6.8 OBSERVATIONS OF THE NEUTRAL WIND

While the winds at the ground can be sensed by the human body and relatively easily measured, the wind at higher altitudes is a rather elusive parameter to observe. In the stratosphere and lower mesosphere our knowledge of the neutral wind has been derived from balloon and meteorological rocket measurements. One method quite often used has been observations of anomalous propagation of sound waves from small grenade detonations released by rockets. Higher up in the mesosphere and lower thermosphere meteor trail tracking by radars have produced considerable insight in the neutral wind system at these height regions. Observations of the motion of noctilucent clouds have also been used to deduce neutral winds in the mesopause region.

At higher altitudes in the lower thermosphere information about the neutral wind has, in addition to the meteor trail observations, been derived from chemiluminescent cloud releases from rockets at twilight. The solar radiation which is resonance-scattered by the alkali metal atoms can easily be observed from ground in clear weather. Extended height profiles of the neutral wind in the lower thermosphere can also be obtained by chemical trimethyl aluminium (TMA) releases from rockets. As these chemical releases have to be made at twilight, limited time coverage is available for these methods.

For the F-region and partly also the E-region neutral winds at high-latitude optical observations of the Doppler shift in auroral emission lines from oxygen is a widely used method. But again the ability of the method to give extended time coverage of the winds is limited to the period of dark and clear polar sky. Optical Doppler measurements of airglow emissions have also been used to study F-region neutral winds at middle and lower latitudes.

The incoherent scatter method for deriving F-region and E-region neutral winds is the only method that can give continuous coverage. These methods, as we will see, are rather indirect and are still related to large uncertainties in the choice of model atmospheres especially at high latitudes.

In the upper thermosphere observations of atmospheric drag on satellites have been used to derive the temperature distribution in these higher regions, and from these again neutral wind patterns have been derived by applying these temperature observations as an input to the pressure term in the mobility equation of the neutral gas.

For the E-region neutral winds, probably the most important basis for our understanding has been the analysis of geomagnetic variations on the ground. These have a very long history which has resulted in a larger amount of work attempting to resolve the different factors contributing to the magnetic variations – one of the most important of these being the E-region neutral wind set-up by tidal forces.

6.9 COLLISIONS BETWEEN PARTICLES

Let us assume a gas where we have two kinds of gas particles, m and n, which are colliding with each other. Each collision must be related to a mutual force. We will now, for simplicity, assume that this force per unit mass acting on the m-type particles due to the collision with the n-type particles is proportional to the velocity difference $\mathbf{u}_m - \mathbf{u}_n$ between the two kinds such that

$$\mathbf{f}_{m,n} = -\nu_{mn}(\mathbf{u}_m - \mathbf{u}_n) \tag{6.62}$$

is the force reducing $\mathbf{u}_m$ when $\mathbf{u}_m > \mathbf{u}_n$ and vice versa. ν_{mn} is called the collision frequency for momentum transfer. The acceleration of the m-type particle due to collisions with the n-type particles is now:

$$\frac{d\mathbf{u}_m}{dt} = \mathbf{f}_{mn} = -\nu_{m,n}(\mathbf{u}_m - \mathbf{u}_n) \tag{6.63}$$

Since the momentum must be conserved during a collision, we have

$$\rho_m \mathbf{f}_{m,n} + \rho_n \mathbf{f}_{n,m} = 0 \tag{6.64}$$

since $\mathbf{f}_{m,n}$ and $\mathbf{f}_{n,m}$ are forces per unit mass, and $\mathbf{f}_{n,m}$ is the force on particle of type n due to the collision with particle of type m. Inserting for $\mathbf{f}_{m,n}$ and $\mathbf{f}_{n,m}$ gives:

$$\rho_m \nu_{m,n}(\mathbf{u}_m - \mathbf{u}_n) + \rho_n \nu_{n,m}(\mathbf{u}_n - \mathbf{u}_m) = 0 \tag{6.65}$$

$$(\rho_m \nu_{m,n} - \rho_n \nu_{n,m})(\mathbf{u}_m - \mathbf{u}_n) = 0 \tag{6.66}$$

For any $\mathbf{u}_m$ and $\mathbf{u}_n$ therefore

$$\rho_m \nu_{m,n} = \rho_n \nu_{n,m} \tag{6.67}$$

Assume now for simplicity that we only consider one-dimensional velocities and that there is only one kind of particle type n, then the equation of motion for the type m particle can be written:

$$\frac{du_m}{u_m - u_n} = -\nu_{m,n} \, dt \tag{6.68}$$

and the solution for u_m is:

$$u_m = u_m^0 \exp(-\nu_{m,n} t) + u_n(1 - \exp(-\nu_{m,n} t)) \tag{6.69}$$

if $u_m = u_m^0$ at $t = 0$. The equilibrium solution ($t \to \infty$) for u_m is:

$$u_m = u_n \tag{6.70}$$

The characteristic relaxation time is defined as:

$$\tau_{m,n} = \frac{1}{\nu_{m,n}} \tag{6.71}$$

In the opposite case the velocity of the type n particle is given by:

$$u_n = u_n^0 \exp(-\nu_{n,m} t) + u_m(1 - \exp(-\nu_{n,m} t)) \tag{6.72}$$

and the relaxation time is:

$$\tau_{n,m} = \frac{1}{\nu_{n,m}} = \frac{\rho_n}{\rho_m \nu_{m,n}} \tag{6.73}$$

We therefore notice that if $\rho_m \gg \rho_n$, the n-type particle will reach the equilibrium velocity $\mathbf{u}_m$ much faster than the m-type particle will reach its equilibrium velocity $\mathbf{u}_n$. In such a situation the n-type particles will feel the presence of the m-type

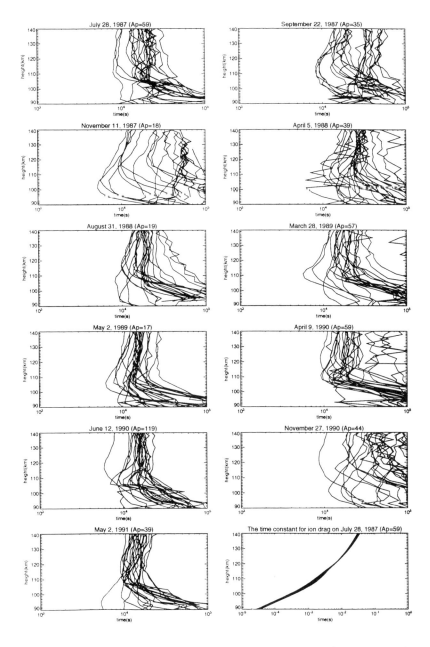

Fig. 6.11. A comparison of the time constant τ_{in}, for the ions to approach the neutral velocity (the lower right panel) and τ_{ni} for the neutrals to approach the ion velocity for different days in the E-region above Tromsø (the 11 other panels). (From Nozawa and Brekke, 1995.)

particles much more strongly than the latter type feels the former. This is in simple terms the effect of drag from one kind of gas particles to another.

The mean velocity is defined as:

$$\mathbf{u} = \frac{\rho_m \mathbf{u}_m + \rho_n \mathbf{u}_n}{\rho} \tag{6.74}$$

where

$$\rho = \rho_m + \rho_n \tag{6.75}$$

Now, since $\dot{\rho} = 0$ we obtain

$$\rho \dot{\mathbf{u}} = -(\rho_m \nu_{m,n}(\mathbf{u}_m - \mathbf{u}_n) + \rho_n \nu_{n,m}(\mathbf{u}_n - \mathbf{u}_m)) = 0 \tag{6.76}$$

Therefore the mean velocity $\mathbf{u}$ is constant and

$$\mathbf{u}(\infty) = \mathbf{u}(t) = \mathbf{u}(0) \tag{6.77}$$

The collision term being proportional to the relative velocities between each type of particle contributes to a force which tends to drive each individual gas velocity toward a total mean velocity which is constant, i.e. an equalization in velocity takes place.

For the collisional interaction between ions and neutrals in the ionosphere it is of interest to compare the time constants with which the ions approach the neutral motion (τ_{in}) and vice versa (τ_{ni}). Such a comparison is presented for the E-region in Fig. 6.11 for different days at an auroral zone station (Tromsø). We notice that while τ_{in} is always less than 10^{-4} s, τ_{ni} is hardly less than 10^4 s. A constant ion velocity acting for several hours is therefore needed in order to bring the neutrals into a motion along with the ions. On the other hand, the ions will almost immediately adjust their velocities to the neutral motion. Therefore the ion motion can be used as a good tracer of the neutral motion.

6.10 COLLISIONS IN GASES WITH DIFFERENT TEMPERATURES

It is customary to treat the gas of the upper atmosphere as a plasma or fluid, assuming the collisions between individual particles are frequent enough for an MHD description to be valid.

To discuss the collisional interactions between individual kinds of particles m and n in the ionospheric plasma is not a straightforward problem, and it is not the purpose of this book to outline this in detail here.

By performing a statistical analysis where the macroscopic behaviour of the plasma is considered to be due to a mean behaviour of individual particles, one introduces the distribution function $f_m(\mathbf{r}, \mathbf{u}, t)$. This describes the distribution of particles of kind m having a velocity $\mathbf{u}$ at position $\mathbf{r}$ at time t. The Boltzmann equation describing the variation of this function in space and time can be written as:

$$\frac{\partial f_m}{\partial t} + v_i \frac{\partial f_m}{\partial x_i} + a_i \frac{\partial f_m}{\partial u_i} = \left(\frac{\partial f_m}{\partial t}\right)_c \tag{6.78}$$

where the index i indicates 3 individual coordinates, $i = 1, 2, 3$, and the term to the right is the collision term which is crucial for determining the shape of f_m.

By using this equation, however, and forming statistical averages, it is possible to find conservation laws for most parameters related to the behaviour of the particles of type m.

Boltzmann introduced the following term:

$$\left(\frac{\partial f_m}{\partial t}\right)_c = \sum_n \int (f'_m f'_{n_1} - f_m f_{n_1}) \, gb \, db \, d\varepsilon \, d^3 v_1 \tag{6.79}$$

where f_m and f_{n_1} are the distribution functions of particles of kind m and n before the collision, respectively, and f'_m and f'_{n_1} are the corresponding parameters after the collision. g is the relative velocity between the two kinds of particles, b is the impact parameter, ε is an angle to account for all directions of impact, and $d^3 v_1$ is the velocity space for the particles of kind n.

For so-called Maxwellian gas particles, where the reaction force between the particles can be expressed as

$$F_R = \frac{\kappa}{r^5} \tag{6.80}$$

where κ is an arbitrary constant, it can be shown that for such particles in thermodynamic equilibrium the collision term in the energy equation results in a heat transfer rate

$$\dot{Q}_m = \frac{\rho_m \nu_{mn}}{m_m + m_n} \left[3\kappa(T_n - T_m) + m_n(\mathbf{u}_n - \mathbf{u}_m)^2\right] \tag{6.81}$$

where T_n and T_m are the temperatures of the particles of kind n and m, respectively, and m_n and m_m are their respective masses. Equation (6.81) shows that as long as the temperatures or the velocities of the two kinds of particles are different, a heat transfer between them will occur.

In the ionosphere where the two kinds of particles are neutrals and ions, the last term in (6.81) is often referred to as the Joule heating term due to ohmic losses by ionospheric currents. Assuming one type of ions (i) and one type of neutrals (n) only, we obtain for the heating rate:

$$\dot{Q}_i = \frac{\rho_i \nu_{in}}{m_i + m_n} \left[3\kappa(T_n - T_i) + m_n(\mathbf{u}_n - \mathbf{v}_i)^2\right] \tag{6.82}$$

where we have introduced $\mathbf{v}_i$ and $\mathbf{u}_n$ as the ion and neutral velocities, respectively.

6.11 DRAG EFFECTS

Let us consider the ionosphere where we have one kind of ion and one kind of neutral with velocities $\mathbf{v}_i$ and $\mathbf{u}_n$, respectively. Assume an initial situation where $\mathbf{u}_n = 0$ and assume that an electric field perpendicular to $\mathbf{B}$ induces an ion velocity (equation (4.96)):

$$\mathbf{v}_{i0} = \mathbf{E}_\perp \times \mathbf{B}/B^2 \tag{6.83}$$

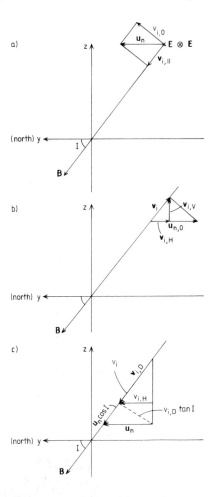

Fig. 6.12. Diagrams illustrating the different drag effects taking place between the neutral and the ionized gas species in the upper atmosphere. (a) The neutral gas is initially at rest. An electric field perpendicular to **B** is associated with an ion drift $\mathbf{v}_{i,0}$. Since the neutrals are restricted to moving mainly horizontally, there will be a horizontal neutral velocity $\mathbf{u}_n$. This neutral velocity will drag the ions along the **B**-field, and the ions will obtain a velocity component $\mathbf{v}_{i,\parallel}$ parallel to **B**. A steady-state situation occurs when the ions and the neutrals are moving horizontally with the same speed. (b) The ions are initially at rest and the neutral gas moves horizontally with velocity $\mathbf{u}_{n,0}$. The component of the neutral velocity along **B** drags the ions along until they reach a steady-state velocity $\mathbf{v}_i$ equal to this component. (c) The neutral gas is initially at rest while the plasma diffuses downward along **B** with a velocity $\mathbf{v}_{i,D}$. The horizontal component of this velocity will drag the neutrals along until they reach a velocity $\mathbf{u}_n$. This horizontal neutral motion will again drag the ions down along the **B**-field, and a steady state occurs when the horizontal neutral and ion velocities are equal.

Drag effects

If $\mathbf{E}_\perp$ is directed eastward (Fig. 6.12a), then v_{i0} will be upward and northward in the magnetic meridional plane. There will then be a horizontal ion velocity in this plane. Since the neutral gas is considered to be in hydrostatic equilibrium, we can neglect any motion of the neutrals in the vertical direction. The horizontal ion velocity component will then drag the neutrals along after some time. Since the neutrals effectively drag the ions with them, the ions will also obtain a component $\mathbf{v}_{i,\|}$, parallel to the magnetic field. A stationary state will be obtained when the ions and the neutrals are moving horizontally with the same speed. This occurs when

$$u_n = \frac{v_{i,0}}{\sin I} = \frac{1}{\sin I} \frac{E}{B} \tag{6.84}$$

where I is the inclination angle of $\mathbf{B}$. The time constant for this to happen in the ionosphere is, however, very long due to the small ρ_i/ρ_n ratio. Assume instead the ions are initially at rest and that a neutral wind $\mathbf{u}_{n0}$ is blowing horizontally in the negative y-direction (south) in the northern hemisphere (Fig. 6.12b). The component of the wind parallel to the magnetic field will drag the ions until they reach a steady-state velocity $\mathbf{v}_i$ along the field which is given by

$$v_i = u_{n,0} \cos I \tag{6.85}$$

The ions therefore attain a vertical velocity given by

$$v_{i,V} = u_{n,0} \cos I \sin I \tag{6.86}$$

and a horizontal velocity

$$v_{i,H} = u_{n,0} \cos^2 I \tag{6.87}$$

Let us again assume that $\mathbf{u}_n = 0$ initially and that the plasma is diffusing downward with a velocity $v_{i,D}$ (Fig. 6.12c) due to gravity and pressure gradients. Since the magnetic field is present, the ions are forced to follow the field lines. The ion motion will therefore attain a horizontal component given by:

$$v_{i,H} = v_{i,D} \cos I \tag{6.88}$$

This will drag the neutrals along until they eventually obtain a horizontal velocity $\mathbf{u}_n$. Then the neutrals will also drag the ions along the B-field by the component $u_n \cos I$. Equilibrium is obtained when the ions and neutrals are moving horizontally with the same speed. This happens when

$$(v_{i,D} + u_n \cos I) \cos I = u_n \tag{6.89}$$

$$u_n = \frac{v_{i,D} \cos I}{\sin^2 I} \tag{6.90}$$

and then the plasma velocity along the field becomes:

$$\mathbf{v}_i = \mathbf{v}_{i,D}/\sin^2 I \tag{6.91}$$

6.12 THERMOSPHERIC NEUTRAL WINDS

As already mentioned atmospheric drag on satellites has been extensively used to deduce the temperature distribution in the upper thermosphere, especially at the thermopause. Fig. 6.13 is a representation of the inferred isotherms at the thermopause on the basis of a large number of satellite passes. The temperature distribution is rather symmetric around equator but is shifted about one hour to the east of the subsolar point, probably due to the sluggishness in the neutral gas. For this special diagram a maximum temperature of about 1300 K is observed at the equator at about 1300 local time. This maximum temperature, however, is sensitive to the solar cycle variations and to magnetospheric disturbances. In fact, it does appear that the temperature at the thermopause is more sensitive to magnetic disturbances than to variations in the solar EUV flux.

The daily global temperature distribution as shown in Fig. 6.13 leads to a daytime expansion of the atmosphere which is called the "diurnal bulge". The horizontal pressure gradients around this "diurnal bulge" provide the driving force for the thermospheric winds. From this type of temperature distribution it is then possible to derive the large-scale pressure gradients and from those again to calculate the resulting neutral wind when applying an appropriate equation of motion for the neutrals. Fig. 6.14 gives an early example of such calculations where we notice that the velocities are directed from the hot dayside across the pole to the nightside, i.e. the wind is very close to being perpendicular to the isotherms. The wind speeds are of the order of 40 m/s.

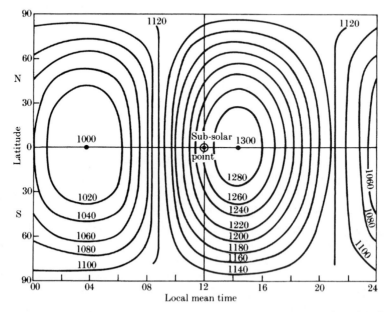

Fig. 6.13. The inferred isotherms at the thermopause. The numbers labelled to the contours is in units of K. (From Jacchia, 1965.)

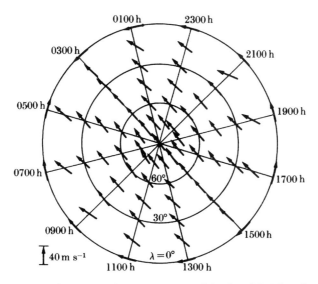

Fig. 6.14. The inferred wind system at an altitude of 300 km based on a temperature distribution of the thermopause similar to the one shown in Fig. 6.13. (From Kohl and King, 1967.)

Let us for simplicity neglect all other terms than the pressure gradient force and **f** in the equation of motion for the neutral gas (equation (6.38)). Then for a steady-state solution of the horizontal winds as Fig. 6.14 represents:

$$\mathbf{f} = \frac{1}{\rho}\nabla p \tag{6.92}$$

If now **f** is due only to collisions $(-\nu_{ni}(\mathbf{u}_n - \mathbf{v}_i))$ between the neutral gas and the ions in the thermosphere and assuming that the ions are stationary, then:

$$\mathbf{u}_n = -\frac{1}{\rho \cdot \nu_{ni}}\nabla p = -\frac{\kappa}{m\nu_{ni}}\nabla T \tag{6.93}$$

where we have neglected horizontal variations in the neutral density, and ν_{ni} is the neutral–ion collision frequency. In this simplified situation we therefore find that the neutral wind should blow down the temperature gradients, i.e. perpendicularly to the isotherms as indicated in Fig. 6.13. We also notice, however, that only for very high collision frequencies can it be legitimate to neglect all other terms in the equation of motion except the pressure force and the collision term. For smaller collision frequencies the situation will be more complicated.

We immediately then realize that a more realistic treatment makes the equation of motion for the neutrals far more difficult to solve, especially when we notice that the ions are not at all stationary at times at high latitudes when they are acted on by electric fields propagating from the magnetosphere. Furthermore, neutral air motion at one height can carry the ions along which will in turn set up polarization fields that can propagate to other heights and other latitudes along magnetic field lines where they can act on the ion motion. The collisions between the ions and the neutrals will always be present forcing the different species to drag each other.

Returning again to the equation of motion (6.38) and dividing the force term **f** up into a potential term and a viscosity and a collision term, we can write:

$$\frac{\partial \mathbf{u}_n}{\partial t} = -(\mathbf{u} \cdot \nabla)\mathbf{u}_n - 2\mathbf{\Omega} \times \mathbf{u}_n + \mathbf{g} \qquad (6.94)$$
$$- \frac{1}{\rho}\nabla p - \nabla \psi + \frac{\mu}{\rho}\nabla^2 \mathbf{u}_n - \nu_{ni}(\mathbf{u}_n - \mathbf{v}_i)$$

Here ψ is a potential due to gravity and centrifugal force, μ the coefficient of viscosity and $\mathbf{v}_i$ the ion velocity. In order to get a better understanding of the neutral wind behaviour we need to know more about the different terms. Especially since we observe that the velocity of the ion gas enters the equation of motion for the neutrals, the equation of motion for the ions should also be solved simultaneously to give a self-consistent picture. Let us therefore introduce the equation of motion for the ions:

$$m_i \frac{\partial \mathbf{v}_i}{\partial t} = -m_i(\mathbf{v}_i \cdot \nabla)\mathbf{v}_i + q\mathbf{E} + q\mathbf{v}_i \times \mathbf{B} - m_i\nu_{in}(\mathbf{v}_i - \mathbf{u}_n) \qquad (6.95)$$

where m_i is the ion mass, q the ion charge and $\mathbf{E}$ the electric field. We have neglected the Coriolis force, gravity, potential and pressure forces together with viscosity since these are considered to be small in the thermosphere compared to the electric field, Lorentz force and collision terms. We now observe that since the electric field and Lorentz force enter into the equation of motion for the ions, these may also effect the motion of the neutrals through the collision term between ions and neutrals. In a situation where **E** is very large, the ion velocity may dominate and the neutrals act as a drag on the ions. In the opposite case when the **E**-field is negligible, the ions may act as a drag on the neutrals. Since the **E**-field is so variable, especially at high latitudes, it is realized that the neutral gas motion can be a strongly modified and dynamic feature of the thermosphere.

Since there are so many dynamic parameters affecting the neutral gas motion, it is an almost insurmountable problem to obtain a comprehensive view of the situation. We have to make some assumptions and hope that the results we derive have a validity general enough that we can use them as a rule of thumb.

In Fig. 6.15 calculations obtained for a high-latitude station (Chatanika, Alaska, 62.5° N, −145° W) pertinent to the different terms in the equation of motion for the neutrals are illustrated. The potential term $\nabla \psi$, however, is neglected.

We notice from Fig. 6.15 which applies for an altitude close to 300 km, that the ion drag and the pressure terms are dominant during daytime for both the meridional and the zonal components. At night, however, all terms are almost of equal importance in the meridional component, although the pressure and advection terms dominate the total force. For the zonal component in night-time, however, the pressure term together with the ion drag and the Coriolis terms tend to dominate.

Sec. 6.12] Thermospheric neutral winds

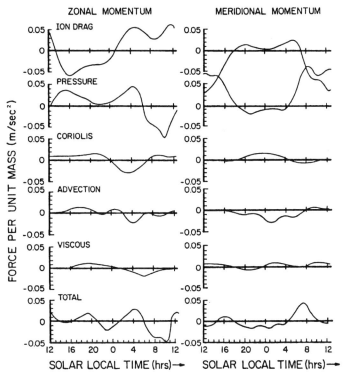

Fig. 6.15. Analysis of forces for the neutral gas over the Alaska station (62.5° N, −145° E) calculated for conditions corresponding to December 4, 1981 for the F-region height at about 300 km. (From Killeen and Roble, 1986.)

In the auroral zone and high latitudes where the ion velocity may become much larger than the neutral velocity, it is not legitimate to neglect the ion velocity in the collision term. In a steady state to a first approximation the neutral wind will therefore be:

$$\mathbf{u}_n = -\frac{1}{\rho \nu_{ni}} \nabla p + \mathbf{v}_i \tag{6.96}$$

This implies that the ion velocity can have a strong effect on the neutral velocity. This is illustrated in Fig. 6.16 where a large equatorward surge in the neutral velocity occurs at night-time having a maximum of close to 200 m/s. This is apparently related to an enhancing effect of the pressure term and the ion drag term in the mobility equation. They both happen to force the neutrals in the same direction at this time over Alaska. A southward motion would be expected from the pressure gradients alone but the ion motion, which for this particular day,

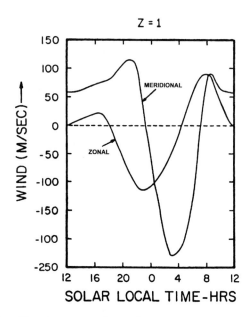

Fig. 6.16. F-region neutral wind predictions for a station in Alaska corresponding to conditions as of December 4, 1981 at F-region heights (300 km). Of particular interest is the strong equatorward surge in the meridional component after midnight. A common feature observed at this station is apparently a consequence of the pressure and ion drag terms acting together on the neutral gas. (From Killeen and Roble, 1986.)

Dec. 4, 1981, was forced by a cross-polar cap electric field corresponding to about 60 kV, enhanced the equatorward motion in the morning hours.

At higher latitudes, within the polar cap where it is possible to observe the Doppler shifted OI (6300 Å) line throughout a 24-hour period in midwinter, the neutral wind is found to be nearly cross-polar from day to night (Fig. 6.17). In this diagram the circle marks the latitude of observation (Longyearbyen, Svalbard, 78° N), and the centre of the circle is the magnetic north pole. The diagram does not represent a snapshot of the wind vectors since the observations are collected over a 24-hour period. Under the assumption, however, that the wind pattern does not change during this period of observations, it can be interpreted as an instantaneous pattern.

In Fig. 6.18a more recent model calculations of the global F-region neutral wind circulation field are demonstrated as a function of longitude and latitude toward a background of isothermal contours. This can be compared with the one shown in Fig. 6.14. In this model (Fig. 6.18a) there is no potential drop across the polar cap such that the ions act as a drag on the neutrals. Again we notice that the main wind field is perpendicular to the isotherms from the hot subsolar point at about 1600 LT toward the cooler nightside. The cross-polar cap neutral wind

Sec. 6.12] Thermospheric neutral winds 287

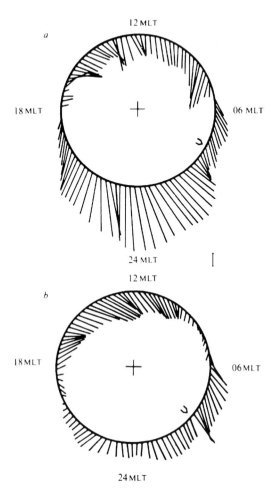

Fig. 6.17. Neutral wind patterns in the polar thermosphere obtained for two days, January 21 and 29, 1979. The circle and the centre represent the latitude of Longyearbyen, Svalbard (78° N) and the geographic north pole, respectively. An arrow pointing toward the centre is a northward wind and tangentially toward the right from north is eastward. The scale bar corresponds to a wind velocity of 100 m/s. (From Smith and Sweeney, 1980.)

is rather outstanding, reaching velocities of the order of 100 m/s. When, however, a cross-polar cap potential of about 60 kV is applied (Fig. 6.18b), the wind field as well as the temperature distribution changes character; the latter is related to local heating due to enhanced collisions between the neutrals and the ions (Joule heating). The velocities across the polar region are strongly enhanced (notice the change in scales by a factor of more than 3 on the velocity arrows). At

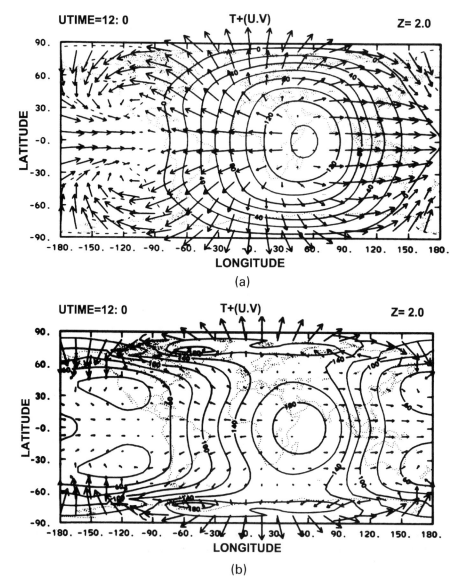

Fig. 6.18. Calculated global circulation and temperatures along a constant pressure surface corresponding to about 300 km for the case of solar heating as the only driving force mechanism. (a) The contours describe the temperature perturbation in K and the arrows are giving wind directions. The length of the arrows gives the wind speed with a maximum of 100 m/s. (b) The same as in (a) except that a magnetospheric convection source with cross-tail potential corresponding to 60 kV is included. The wind speed represented by the length of arrows now corresponds to 336 m/s. (From Dickinson et al., 1984.)

Sec. 6.12] Thermospheric neutral winds 289

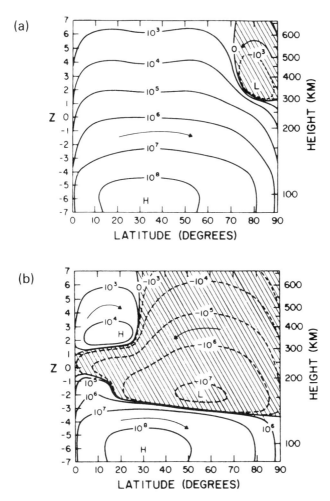

Fig. 6.19. The calculated contours of mass flow stream function. The numbers on the contours are in g/s. The difference between the contour values gives the global flux mass between the two contours. (a) For solar heating as the only driving source. (b) For solar heating, Joule heating and momentum source also included. (From Dickinson et al., 1975.)

night-time the high-latitude velocities are more strongly equatorward related to the surge, as noticed already in relation to Fig. 6.16. In the daytime, however, at middle latitudes the velocities are strongly reduced and the zonal component is reversed from eastward to westward. These effects are due to enhanced Joule heating at high latitudes caused by strong currents in the auroral zone.

That the electromagnetic disturbances at high latitudes can have a significant effect at thermospheric heights is illustrated more clearly in Fig. 6.19. In panel (a) the meridional circulation for quiet conditions is mainly forced by solar heating which produces a large Hadley cell with warm air rising at the equator, flowing towards the poles and sinking down again to lower heights. Above the polar regions (> 70° latitude), however, a small cell with the opposite flow direction is set up above 300 km altitude due to heat influx from the magnetosphere outside the plasmapause. In panel (b), however, where a large magnetic storm takes place, the contrary cell expands in altitude (down to 120 km) as well as in latitude (down to equator around 200 km of altitude). Air rich in molecular constituents is rising in the thermosphere above the high-latitude regions and is being brought at high altitude toward the equator where the air descends to about 150 km. Ions with high velocities are observed flowing outward from the thermosphere along the magnetic field lines during such stormy events.

6.13 THE E-REGION WINDS

By now moving down in altitude to study the E-region winds, it is again useful to compare the different terms in the equation of motion (6.38). In Fig. 6.20 the outcome of such a comparison for the same high-latitude station as shown in Fig. 6.15 is presented.

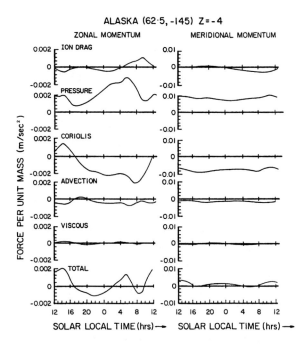

Fig. 6.20. Similar to the data shown in Fig. 6.15 but applied at 120 km approximately. (From Killeen and Roble, 1986.)

Sec. 6.13] The E-region winds 291

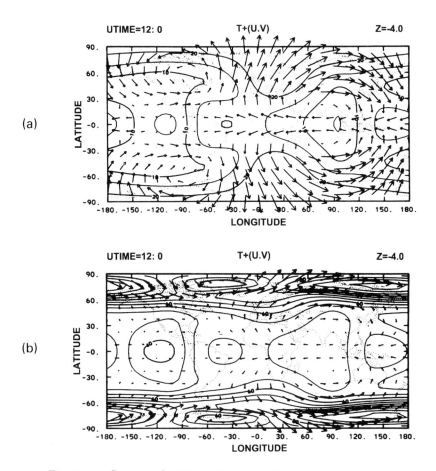

Fig. 6.21. Same as for Fig. 6.18 except that the constant pressure level corresponds to close to 120 km. (a) The maximum wind speed corresponds to 32 m/s. (b) The maximum wind speed corresponds to 79 m/s. (From Dickinson et al., 1984.)

Contrary to the F-region situation where the pressure and ion drag term appeared to be dominant forcing mechanisms to the neutral wind, it is the pressure and Coriolis term that rule over the E-region neutral air motion. A global model calculation of the E-region neutral wind at about 120 km of altitude is illustrated in Fig. 6.21a against isothermal contours. It is noticed that in contrast to the F-region winds which are mainly directed perpendicularly to the isotherms, the E-region neutral air motion more closely follows these contour lines. A geostrophic component is therefore present in the E-region neutral winds. The E-region wind speeds are also only a third of the F-region speeds on the average. A 60 kV cross-polar cap potential is enforced on the ion motion for the wind data shown in Fig. 6.21b. The high-latitude winds are now predominantly eastward and the cross-polar cap motion is reduced. At low latitudes, however, there do not appear to be marked changes in the E-region wind field.

6.14 OBSERVATIONS OF E-REGION NEUTRAL WINDS

Due to the strong coupling between the neutrals and the ions in the E-region, observations of the ion motion can sometimes be used as a tracer for the neutral motion. Considering the equation of motion for the ions (equation (6.95)). At a steady state we can solve for the neutral velocity in a simple manner:

$$\mathbf{u}_n = \mathbf{v}_i - \frac{q}{m_i \nu_{in}} \mathbf{v}_i \times \mathbf{B} - \frac{q}{m_i \nu_{in}} \mathbf{E} \tag{6.97}$$

where advection has been neglected. At high latitude where the magnetic field is almost vertical, this equation applies to the horizontal component of $\mathbf{u}_n$ while the vertical component is assumed equal to the ion motion along the magnetic field line. We notice then that when ν_{in} is very large, the neutral velocity is equal to the ion velocity. Observing the ion motion by some radio or radar technique therefore allows for deriving the neutral motion in a rather direct manner. For smaller values of ν_{in} the complete coupling between ions and neutrals as given by (6.97) has to be accounted for in order to derive $\mathbf{u}_n$ from measurements of $\mathbf{v}_i$. By means of an incoherent scatter radar such as EISCAT close to Tromsø, Norway (70° N) one can obtain almost simultaneous observations of the E- and F-region velocities. In the F-region (300 km) where the collision term will be negligible to the ion motion compared to the E-field and Lorentz force term, the electric field can be derived for a steady-state solution:

$$\mathbf{E} = -\mathbf{v}_i^F \times \mathbf{B} \tag{6.98}$$

where $\mathbf{v}_i^F$ is the F-region ion velocity observed simultaneously with the E-region ion velocity. The neutral wind in the E-region can then be derived with an appropriate choice of a collision frequency model.

Fig. 6.22 illustrates the average neutral wind (x, y, z) components in the geographical frame of reference as function of time at 6 E-region heights for quiet days ($Ap \leq 16$) in the four seasons. From such time series the mean velocity components as well as the average 24-, 12-, 8- and 6-hour tidal components can be derived for each of the six heights. This has been carried out for all four seasons and the results are presented in Fig. 6.23. The most outstanding feature of these components is the strong mean eastward wind between 90 and 120 km having a maximum at about 110 km of 60 m/s in summer and 20 m/s in winter. The very low mean northward and vertical wind components in the same height region are also noticeable; these components are relatively less influenced by change of season.

The eastward component of the diurnal tide increases gradually by height and reaches an amplitude close to 100 m/s above 120 km. The northward component of this tide is fairly independent of height. For the other horizontal tidal components no systematic height dependence is found. In the vertical component, however,

Sec. 6.16] Observations of E-region neutral winds 293

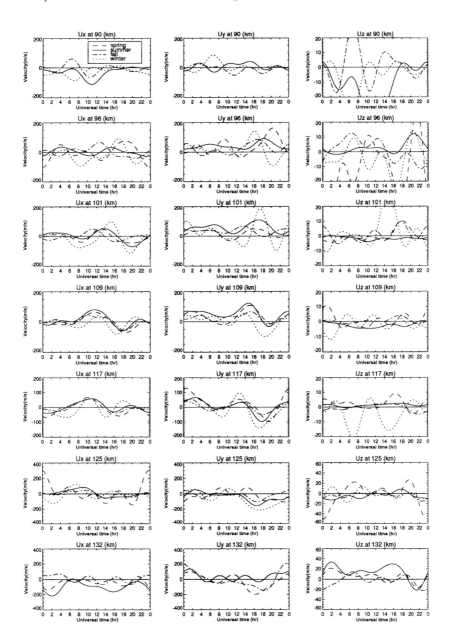

Fig. 6.22. The average neutral wind components derived by the EISCAT incoherent scatter radar at Tromsø (69.66° N, 18.94° E) at 7 E-region heights for quiet days ($Ap \leq 16$) in the four seasons. These average wind components are derived by adding the mean wind components together with the 24-, 12-, 8- and 6-hour tidal components with appropriate phases. (Courtesy Nozawa, 1994.)

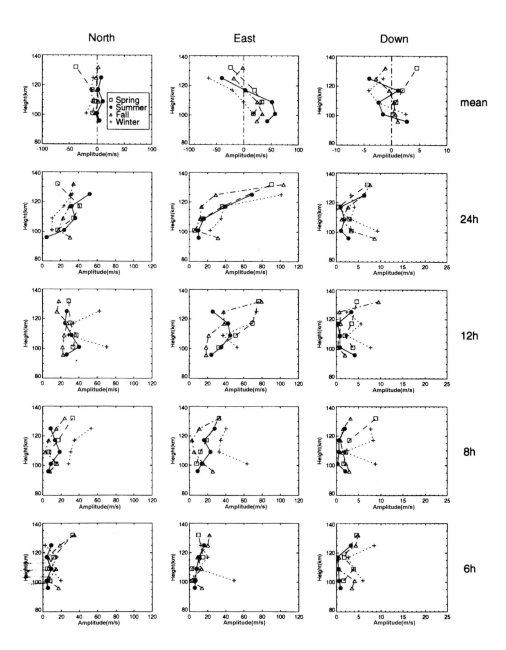

Fig. 6.23. Upper row of panels: The mean components as function of height for the four seasons. The four lower rows of panels: The 24-, 12-, 8- and 6-hour tidal components as function of height for the four seasons. (From Brekke et al., 1994.)

Sec. 6.16] Observations of E-region neutral winds 295

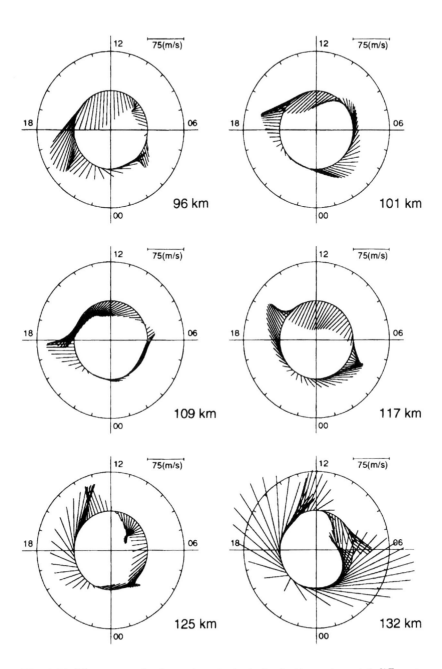

Fig. 6.24. The average horizontal neutral wind velocity vectors at 6 different E-region heights for the average autumn quiet days. These vectors are obtained by adding the appropriate north and east components in Fig. 6.22. No data exist at 90 km for the autumn days. (From Brekke et al., 1994.)

296   Dynamics of the neutral atmosphere   [Ch. 6

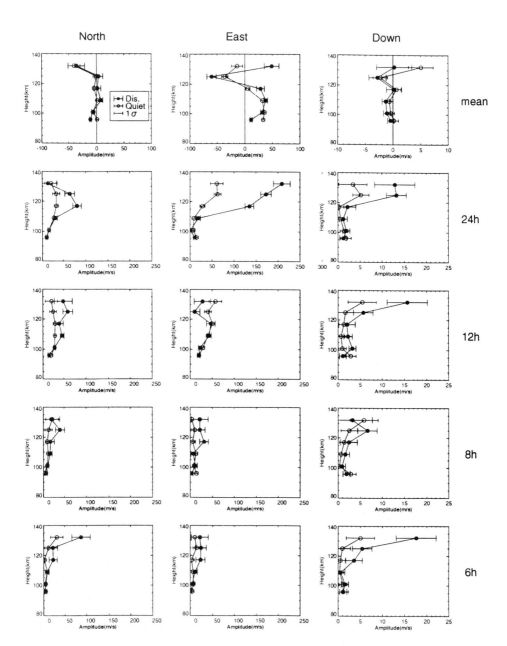

Fig. 6.25. The variation by height of the northward, eastward and downward components of the mean wind and the amplitudes of the 24-, 12-, 8- and 6-hour tidal components. The different disturbance levels are indicated by quiet ($Ap \leq 16$) and disturbed ($Ap > 16$). The standard deviations are shown by vertical bars. (From Nozawa and Brekke, 1995.)

Sec. 6.15]   **Tidal oscillations in the neutral atmosphere**   297

all tidal components except the 6-hour tide have a minimum amplitude at the height region of the strong mean eastward wind.

In Fig. 6.24 the horizontal neutral wind components presented as function of time for the 6 E-region heights for the quiet autumn days in Fig. 6.22 are shown as horizontal clock dial plots. We notice the outstanding daytime poleward wind at the 3 lower heights 101–117 km. The semidiurnal tide is also present at the same three heights and its amplitude increases by increasing height. At 125 km shorter tidal periods are also present while the semidiurnal tide tends to dominate again at 132 km. Clearly there are large variations by height in the neutral wind in the auroral E-region.

It is generally believed, in spite of the long time constants involved, as shown in Fig. 6.11, that the E-region neutral velocities will be modified during strong auroral disturbances at the higher latitudes.

A similar averaging process of the average neutral wind at different E-region heights has been performed on data obtained by EISCAT during disturbed conditions ($Ap > 16$). The mean velocity components as well as the tidal components are shown as function of height in Fig. 6.25. Data obtained from 11 disturbed days are compared with corresponding data from 24 quiet days ($Ap \leq 16$). We notice that the strong eastward mean velocity is rather unaffected by the condition of disturbance, and is more dependent on season, as shown in Fig. 6.23. The most outstanding difference between quiet and disturbed days appears to be the enhancement of the eastward diurnal tide above 110 km altitude.

6.15 TIDAL OSCILLATIONS IN THE NEUTRAL ATMOSPHERE

As we have noticed in the measurements of the neutral wind derived from radar observations in the auroral zone ionospheric E-region, there appears to be a weft of diurnal and semidiurnal oscillations or tidal motions present in the velocity components.

These tides at ionospheric heights are believed to originate due to the diurnal variation of atmospheric heating and cooling especially in the middle altitude. It is preliminary due to the presence of the ozone layer but also the water vapour. Models of the atmospheric heating rates as function of altitude and latitude are presented in Fig. 6.26. The heating rate for both constituents maximizes at the equatorial region and decreases to zero at the poles. While the water vapour contributes strongest to the heating at the ground and decreases exponentially with height, the ozone layer gives rise to a maximum heating rate at about 55 km. Negligible heating due to the ozone content is present below 20 km and above 80 km according to these models.

In order to describe the tidal motion in the upper atmosphere, however, we will return to the set of equations describing the neutral dynamics and investigate more in details what the periodicities of these oscillations are. We will then soon realize that in the atmosphere motion there is a large variety of tidal and wave components.

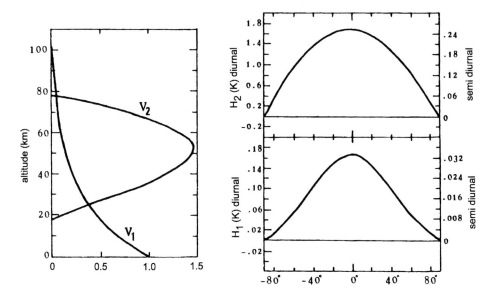

Fig. 6.26. Altitude and latitude distribution of thermal excitation due to water vapour (V_1, H_1) and ozone (V_2, H_2). The abscissa to the left is a dimensionless quantity composed of the molecular weight M, the heating function $\hat{Q}(J)$ and the tidal frequency ω. (From Lindzen, 1968.)

Let the wind velocity be separated into two components $\mathbf{u}_0$ and $\mathbf{u}_1$ representing the steady-state solution and a perturbation to the steady state, respectively. Then we have:

$$\mathbf{u}_n = \mathbf{u}_0 + \mathbf{u}_1 \qquad (6.99)$$

A perturbation in $\mathbf{u}_n$ can also create perturbations in the other parameters and vice versa, which we then split up in a similar manner:

$$\rho = \rho_0 + \rho_1 \qquad (6.100)$$

$$p = p_0 + p_1 \qquad (6.101)$$

$$T = T_0 + T_1 \qquad (6.102)$$

$$s = s_0 + s_1 \qquad (6.103)$$

for the density, pressure, temperature and entropy per unit mass, respectively.

Introducing these parameters to the respective equations, we can separate those describing the steady state from those describing the perturbed situation when all second-order terms in the perturbations are neglected.

Sec. 6.15]   Tidal oscillations in the neutral atmosphere

Then for the steady state when assuming $\mathbf{u}_0 = 0$, that $\mathbf{f} = \mathbf{f}_1$ is the perturbing force only and $\mathbf{g} = -g\hat{\mathbf{z}}$, we have from the equation of motion in the vertical direction (equation (6.38)):

$$\frac{1}{\rho}\frac{\partial p_0}{\partial z} = -g = \mathbf{g} \cdot \hat{\mathbf{z}} \tag{6.104}$$

where $\hat{\mathbf{z}}$ is the vertical unit vector. This equation is recognized as the barometric equation (3.39). From the ideal gas law, the equation of state (6.3), we have

$$p_0 = \rho_0 R T_0 \tag{6.105}$$

and from the second law of thermodynamics (equation (6.16)):

$$s_0 = c_v \ln T_0^\gamma \cdot p_0^{-(\gamma-1)} \tag{6.106}$$

For the first-order perturbation equations we obtain from the equation of motion (6.45):

$$\frac{\partial \mathbf{u}_1}{\partial t} + 2\mathbf{\Omega} \times \mathbf{u}_1 + \frac{1}{\rho_0}\nabla p_1 + \frac{g}{\rho_0}\rho_1 \hat{\mathbf{z}} = \mathbf{f}_1 \tag{6.107}$$

From the second law of thermodynamics (equation (6.4)) assuming no horizontal gradients in s

$$\frac{\partial s_1}{\partial t} + w_1 \frac{\partial s_1}{\partial z} = \frac{\dot{q}}{T_0} \tag{6.108}$$

From the continuity equation (6.19):

$$\frac{\partial \rho_1}{\partial t} + w_1 \frac{\partial \rho_0}{\partial z} + \rho_0 \nabla \cdot \mathbf{u}_1 = \frac{\partial \rho_1}{\partial t} + \rho_0(\nabla \cdot \mathbf{u}_1 + \mathbf{g} \cdot \mathbf{u}_1) = 0 \tag{6.109}$$

From the equation of state for an ideal gas (equation (3.26)):

$$\frac{p_1}{p_0} - \frac{\rho_1}{\rho_0} - \frac{T_1}{T_0} = 0 \tag{6.110}$$

From (6.16) we find when applying (6.13):

$$\frac{T_1}{T_0} = \frac{s_1}{c_p} + \frac{R}{c_p}\frac{p_1}{p_0} \tag{6.111}$$

Introducing the temperature ratio from (6.110) we find:

$$\frac{p_1}{p_0} - \frac{\rho_1}{\rho_0} = \frac{R}{c_p}\frac{p_1}{p_0} + \frac{s_1}{c_p} \tag{6.112}$$

$$\frac{\rho_1}{\rho_0} = \frac{p_1}{p_0}\left(1 - \frac{R}{c_p}\right) - \frac{s_1}{c_p} \tag{6.113}$$

By applying (6.13):

$$\frac{\rho_1}{\rho_0} = \frac{1}{\gamma}\frac{p_1}{p_0} - \frac{s_1}{c_p} \tag{6.114}$$

When introducing (6.114) to (6.107), we obtain:

$$\frac{\partial \mathbf{u}_1}{\partial t} + 2\mathbf{\Omega} \times \mathbf{u}_1 + \frac{1}{\rho_0}\left(\nabla p_1 - \frac{1}{\gamma}\frac{p_0}{p_0}p_1 \mathbf{g}\right) + \frac{s_1}{c_p}\mathbf{g} = \mathbf{f}_1 \tag{6.115}$$

when remembering that $\mathbf{g} = -g\hat{\mathbf{z}}$. Dividing (6.109) by ρ_0:

$$\frac{1}{\rho_0}\frac{\partial \rho_1}{\partial t} + \frac{w_1}{\rho_0}\frac{\partial \rho_0}{\partial z} + \nabla \cdot \mathbf{u}_1 = 0 \tag{6.116}$$

and inserting (6.114) gives:

$$\frac{1}{\gamma p_0}\frac{\partial p_1}{\partial t} - \frac{1}{c_p}\frac{\partial s_1}{\partial t} + \frac{w_1}{\rho_0}\frac{\partial \rho_0}{\partial z} + \nabla \cdot \mathbf{u}_1 = 0 \tag{6.117}$$

For the steady-state situation the adiabatic law applies:

$$p_0 \rho_0^{-\gamma} = \text{const.} \tag{6.118}$$

and

$$\frac{\partial p_0}{\partial z}\rho_0^{-\gamma} - \gamma p_0 \rho_0^{-\gamma-1}\frac{\partial \rho_0}{\partial z} = 0 \tag{6.119}$$

which can be rearranged by help of (6.104):

$$\frac{\partial \rho_0}{\partial z} = \frac{\rho_0}{\gamma p_0}\frac{\partial p_0}{\partial z} = -\frac{\rho_0^2}{\gamma p_0}g \tag{6.120}$$

Then by introducing this last result into (6.117) and multiplying by γp_0, we find:

$$\frac{\partial p_1}{\partial t} - \frac{p_0}{c_v}\frac{\partial s_1}{\partial t} - w_1 \rho_0 g + \gamma p_0 \nabla \cdot \mathbf{u}_1 = 0 \tag{6.121}$$

$$\frac{\partial p_1}{\partial t} + \gamma p_0 \nabla \cdot \mathbf{u}_1 + \rho_0 \mathbf{g} \cdot \mathbf{u}_1 = \frac{p_0}{c_v}\frac{\partial s_1}{\partial t} \tag{6.122}$$

When now substituting for s_1 by (6.4):

$$\frac{\partial s_1}{\partial t} = \frac{\dot{q}}{T_0} \tag{6.123}$$

we finally have:

$$\frac{\partial p_1}{\partial t} + \gamma p_0 \nabla \cdot \mathbf{u}_1 + \rho_0 \mathbf{g} \cdot \mathbf{u}_1 = \frac{p_0}{c_v}\frac{\dot{q}}{T_0} = \frac{R}{\rho_0 c_v}\dot{q} = (\gamma - 1)\frac{\dot{q}}{\rho_0} \tag{6.124}$$

A complete discussion of the tidal motion requires use of spherical coordinates. Although we do not have the intention to treat this problem in its completeness,

Tidal oscillations in the neutral atmosphere

we recast the equations just derived into spherical coordinates (r, θ, φ), where the velocity perturbation vector will be given by its coordinates $(v_r, v_\theta, v_\varphi)$. Then we get the following set of equations when neglecting the potential that could cause tidal motions

$$\frac{\partial p_1}{\partial r} = -g\rho_1 - \rho_0 \frac{\partial \psi}{\partial t} \quad \text{(a)} \tag{6.125}$$

$$\frac{\partial v_\theta}{\partial t} - 2\Omega \cos\theta \cdot v_\varphi = -\frac{1}{R_e} \frac{\partial}{\partial \theta} \left(\frac{p_1}{\rho_0} + \psi\right) \quad \text{(b)}$$

$$\frac{\partial v_\varphi}{\partial t} + 2\Omega \cos\theta \cdot v_\theta = -\frac{1}{R_e} \frac{\partial}{\partial \varphi} \left(\frac{p_1}{\rho_0} + \psi\right) \quad \text{(c)}$$

$$\frac{\partial s_1}{\partial t} + v_r \frac{\partial s_0}{\partial r} = \frac{\dot{q}}{T_0} \tag{6.126}$$

$$\frac{p_1}{p_0} - \frac{\rho_1}{\rho_0} - \frac{T_1}{T_0} = 0 \tag{6.127}$$

$$\frac{\partial \rho_1}{\partial t} + v_r \frac{\partial \rho_0}{\partial r} + \rho_0 \nabla \cdot \mathbf{u}_1 = 0 \tag{6.128}$$

$$\frac{T_1}{T_0} = \frac{s_1}{c_p} + \frac{R}{c_p} \frac{\rho_1}{\rho_0} \tag{6.129}$$

We find by substituting for T_1/T_0 from (6.127) into (6.129) that:

$$s_1 = c_v \frac{p_1}{p_0} - c_p \frac{\rho_1}{\rho_0} \tag{6.130}$$

and we have from (6.13) when applying the ideal gas law (equation (3.26)):

$$s_0 = c_v \ln p_0 \cdot \rho_0^{-\gamma} \tag{6.131}$$

When using these relations in the energy equation (6.126), we obtain:

$$\frac{\partial s_1}{\partial t} + v_r \frac{\partial s_1}{\partial r} = \frac{c_v}{p_0}\frac{\partial p_1}{\partial t} - \frac{c_p}{\rho_0}\frac{\partial \rho_1}{\partial t} + v_r \frac{c_v}{p_0}\frac{\partial p_0}{\partial r} - v_r \frac{\gamma c_v}{\rho_0}\frac{\partial \rho_0}{\partial r} = \frac{\dot{q}}{T_0} \tag{6.132}$$

and further by rearranging:

$$\frac{c_v}{p_0}\left(\frac{\partial p_1}{\partial t} + v_r \frac{\partial p_0}{\partial r}\right) - \frac{c_p}{\rho_0}\left(\frac{\partial \rho_1}{\partial t} + v_r \frac{\partial \rho_0}{\partial r}\right) = \frac{\dot{q}}{T_0} \tag{6.133}$$

Dividing by c_v/p_0 and taking advantage of (6.105) gives:

$$\frac{d}{dt}p = \frac{d}{dt}(p_1 + p_0) = \gamma R T_0 \frac{d}{dt}(\rho + \rho_0) + \frac{R}{c_v}\rho_0 \dot{q} \tag{6.134}$$

Finally

$$\frac{d}{dt}p = \gamma g H \frac{d}{dt}\rho + (\gamma - 1)\rho_0 \dot{q} \tag{6.135}$$

where $H = RT_0/g$ is the scale height. This is now a linear differential equation where $\dot{q}$ is the thermotidal heating per unit mass and unit time and it represents the periodic driving force.

Since all the equations are linear, they can be solved by letting all the variables be represented by harmonic series with respect to temporal and longitudinal variations. It is then customary to start out with the continuity equation where the divergence of $\mathbf{u}_1$ is denoted by

$$\nabla \cdot \mathbf{u}_1 = -\chi \tag{6.136}$$

and then from (6.1):

$$-\frac{1}{\rho_0}\frac{d\rho}{dt} = \chi \tag{6.137}$$

By taking the time derivative of (6.131) we find:

$$\frac{1}{\rho_0}\frac{d\rho}{dt} = \frac{1}{\gamma}\frac{1}{p_0}\frac{dp}{dt} \tag{6.138}$$

the divergence of $\mathbf{u}_1$ is then also given by

$$-\frac{1}{\gamma p_0}\frac{dp}{dt} = \chi \tag{6.139}$$

By now introducing the harmonic components so that we have

$$\chi = -\frac{1}{\gamma p_0}\frac{dp}{dt} = G(\theta, z)\exp[i(\omega t + m\varphi)] \tag{6.140}$$

where $2\pi/\omega$ represents the solar or lunar day and $m = 0, \pm 1, \pm 2, \ldots$ and we have interchanged r with z. $G(\theta, z)$ is now a set of coefficients which depend on latitude as well as altitude. It is then the purpose of the analysis to solve for these coefficients for any position in the atmosphere. It turns out that the final equation can be separated in the variables z and θ, and the coefficients $G(\theta, z)$ can be expanded according to a set of functions $\Theta_n^m(\theta)$ called the Hough function. Then

$$G^m(\theta, z) = \sum_{n=1}^{\infty} L_n^m(z)\Theta_n^m(\theta) \quad \text{(a)} \tag{6.141}$$

$$\dot{Q}^m(\theta, z) = \sum_{n=1}^{\infty} J_n^m(z)\Theta_n^m(\theta) \quad \text{(b)}$$

where the heat source is also expanded by means of the Hough function.

It can be shown that $\Theta_n^m(\theta)$ constitutes a regular orthogonal system of functions so that it actually can be expanded by the Legendre's associated functions. They are therefore tabulated for different conditions.

There are two important boundary conditions imposed upon $\Theta_n^m(\theta)$. One is that $\Theta_n^m(0) = 0$, i.e. it must be zero at the pole in order to ensure finite velocities there. The second condition

$$\frac{d}{d\theta}\Theta_n^m(\theta)\bigg|_{\theta=\pi/2} = 0 \tag{6.142}$$

Sec. 6.16] The Brunt–Väisälä frequency 303

The derivative of $\Theta_n^m(\theta)$ at the equator must be zero in order to give the velocities single and finite values all over the globe. The requirement of single valuedness also lies behind the specification of m being an integer. The two possibilities for Θ at the equator give rise to two different classes of modes, the so-called symmetric mode and the antisymmetric mode.

Finally the different parameters can be expressed by a sum of harmonic functions so that

$$\begin{pmatrix} p_1 \\ \rho_1 \\ T_1 \\ \mathbf{u}_1 \end{pmatrix} = \sum_{n=1}^{\infty} \begin{pmatrix} p_1(z) \\ \rho_1(z) \\ T_1(z) \\ \mathbf{u}_1(z) \end{pmatrix} \Theta_n^m(\theta) \exp(i(\omega t + m\varphi)) \tag{6.143}$$

6.16 THE BRUNT–VÄISÄLÄ FREQUENCY

Since we already have seen that the atmosphere can be so stable that a small air bubble, which is moved out of equilibrium, actually can be forced back to equilibrium, it is likely that oscillations in the atmosphere can occur. We will now search for the natural oscillation frequency for such oscillations.

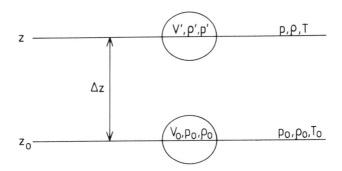

Fig. 6.27. Schematic diagram showing an air parcel in thermodynamic equilibrium at height z being lifted adiabatically an altitude Δz.

Assume that we at the altitude z_0 in the atmosphere have an air bubble with volume V_0, density ρ_0 and pressure p_0 (Fig. 6.27). This bubble has existed so long at the height z_0 that it is in thermodynamic equilibrium with the surroundings, i.e. the temperature in the bubble is T_0 as in the surrounding atmosphere at height z_0. Let now the bubble move a height distance Δz without heat exchanging with the rest of the atmosphere. At the height $z = z_0 + \Delta z$ the bubble will have a volume V^1, a density ρ^1 and a pressure p^1 while at the same height the surroundings have density, pressure and temperature equal to ρ, p and T, respectively. The mass m of the bubble is of course conserved. Since the density in the bubble is different from the surroundings, there must be a buoyancy force acting on the bubble given by:

$$O = V^1 \cdot (\rho - \rho^1) \cdot g = m \cdot g \frac{\rho - \rho^1}{\rho^1} \qquad (6.144)$$

since

$$V^1 = \frac{m}{\rho^1} \qquad (6.145)$$

According to Newton's law

$$m \Delta \ddot{z} = O = mg \frac{\rho - \rho^1}{\rho^1} \qquad (6.146)$$

where $\Delta \ddot{z}$ is the acceleration in the vertical direction.

We will now assume that since Δz is very small

$$\rho^1 = \rho_0 \qquad (6.147)$$

The density of the bubble will not change very much during the motion, and

$$\Delta \ddot{z} = g \cdot \frac{\rho - \rho^1}{\rho^1} \approx g \frac{\rho - \rho^1}{\rho_0} = g \left(\frac{\rho}{\rho_0} - \frac{\rho^1}{\rho_0} \right) \qquad (6.148)$$

The density in the surroundings at height z can be expressed as:

$$\rho = \rho_0 + \left(\frac{\partial \rho}{\partial z} \right)_{z_0} \Delta z \qquad (6.149)$$

or

$$\frac{\rho}{\rho_0} = 1 + \frac{1}{\rho_0} \left(\frac{\partial \rho}{\partial z} \right)_{z_0} \Delta z \qquad (6.150)$$

If we now assume that the atmosphere is isothermal and obeying the hydrostatic law, then;

$$\frac{1}{\rho_0} \left(\frac{\partial \rho}{\partial z} \right)_{z_0} = -\frac{1}{H_0} \qquad (6.151)$$

where H_0 is the density scale height, and when inserting (6.151) into (6.150):

$$\frac{\rho}{\rho_0} = 1 - \frac{\Delta z}{H_0}$$

The bubble expands adiabatically and therefore:

$$p_0 \rho_0^{-\gamma} = p^1 \cdot (\rho^1)^{-\gamma} = \text{const.} \qquad (6.152)$$

and

$$(\rho^1)^{-\gamma} \cdot dp^1 - \gamma p^1 (\rho^1)^{-\gamma-1} d\rho^1 = 0 \qquad (6.153)$$

Sec. 6.16] The Brunt–Väisälä frequency

$$\frac{d\rho^1}{dp^1} = \frac{1}{\gamma}\frac{\rho^1}{p^1} = \frac{1}{\gamma}\frac{m}{kT_0} = \frac{1}{u^2} \tag{6.154}$$

where u is the speed of sound. We also have to a first approximation that:

$$\rho^1 = \rho_0 + \left(\frac{\partial \rho^1}{\partial p^1}\right)_{z_0} \Delta p = \rho_0 + \frac{m}{\gamma kT_0}\Delta p \tag{6.155}$$

and

$$\frac{\rho^1}{\rho_0} = 1 + \frac{m}{\gamma kT_0}\frac{\Delta p}{\rho_0} = 1 + \frac{1}{\gamma g H_0}\frac{\Delta p}{\rho_0} \tag{6.156}$$

Finally

$$\Delta \ddot{z} = g\left(\frac{\rho}{\rho_0} - \frac{\rho^1}{\rho_0}\right) = g\left[1 - \frac{\Delta z}{H_0} - 1 - \frac{1}{\gamma g H_0}\frac{\Delta p}{\rho_0}\right]$$

$$= -\frac{g}{H_0}\left[1 + \frac{1}{\gamma g \rho_0}\frac{\Delta p}{\Delta z}\right]\Delta z \tag{6.157}$$

For small enough Δz:

$$\frac{1}{\rho_0}\frac{\Delta p}{dz} = -g \tag{6.158}$$

and

$$\Delta \ddot{z} = -\frac{g}{H_0}\left[1 - \frac{1}{\gamma}\right]\Delta z = -\frac{g(\gamma - 1)}{\gamma H_0}\Delta z \tag{6.159}$$

This is the equation for a harmonic oscillator where the oscillating frequency is defined as

$$\omega_B^2 = \frac{g(\gamma - 1)}{\gamma H_0} = \frac{mg^2}{\gamma \kappa T_0}(\gamma - 1) \tag{6.160}$$

This frequency is called the Brunt–Väisälä frequency, and it is a characteristic frequency in the atmosphere. Since the speed of sound is given by

$$c^2 = \gamma \frac{\kappa T_0}{m} \tag{6.161}$$

the Brunt–Väisälä frequency is also given by:

$$\omega_B^2 = \frac{g^2}{c^2}(\gamma - 1) \tag{6.162}$$

In Fig. 6.28 a typical height profile for the Brunt–Väisälä frequency in the atmosphere is given. In the homosphere it is of the order of 0.02–0.025 Hz or corresponding to oscillation periods of the order of 4 to 5 minutes.

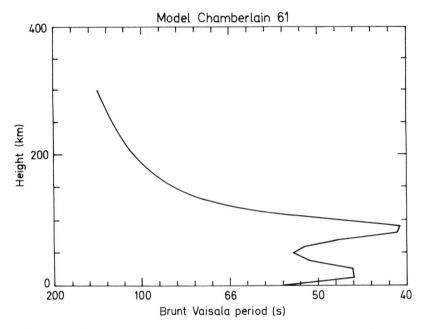

Fig. 6.28. The Brunt–Väisälä frequency as function of altitude. (Courtesy Sigernes, 1992.)

6.17 INTERNAL WAVES IN THE ATMOSPHERE

In the mesosphere and lower thermosphere there is another type of atmospheric waves different from the atmospheric tides and planetary waves. These have smaller spatial scale lengths, not greater than a few hundred kilometres, and shorter periods, less than a few hours; they are called internal gravity waves, or acoustic waves depending on the period and wavelength.

Because the scale length of these waves is so relatively short compared to the Earth's radius, we can neglect the curvature of the planet and assume a flat Earth. We will therefore choose a Cartesian coordinate system fixed on the Earth, when describing these waves. Furthermore, since the periods are so short compared to the revolution time of the Earth, we can also neglect the effects of the Coriolis force.

We will apply the basic equations for the neutral atmosphere which we already are familiar with. We will, however, simplify them for the present purpose to study internal waves and neglect any external forces except gravity, i.e. constant entropy. The basic equations then read as follows:

The continuity equation:

$$\frac{\partial \rho}{\partial t} + \nabla(\rho \mathbf{u}_n) = 0 \tag{6.163}$$

The equation of motion:

$$\frac{\partial \mathbf{u}_n}{\partial t} + (\mathbf{u}_n \cdot \nabla)\mathbf{u}_n = -\frac{1}{\rho}\nabla p + \mathbf{g} \tag{6.164}$$

The equation of state:

$$p = \rho RT = \rho \frac{R_0}{M'} T \tag{6.165}$$

The equation of energy conservation:

$$\frac{dS}{dt} = 0 \tag{6.166}$$

From the last equation we have that

$$\frac{dp}{dt} = \gamma \frac{p}{\rho}\frac{d\rho}{dt} = c^2 \frac{d\rho}{dt} \tag{6.167}$$

where $c^2 = \gamma RT$ $(= \gamma(\kappa T/m))$ is the speed of sound. We will in the following study the relationship between the propagation of these waves in the vertical and horizontal plane. In order to reduce the number of coordinates we can rotate our coordinate system in such a way that the wave vector $\mathbf{k}$ is always in the vertical plane (x, z). Our problem is reduced to two dimensions without losing any generality.

Again we will make use of the method of perturbations and let the variable parameters be superposed on a steady-state component as follows:

$$p = p_0 + p_1 \tag{6.168}$$

$$\rho = \rho_0 + \rho_1 \tag{6.169}$$

$$\mathbf{u}_n = \mathbf{u}_1 \tag{6.170}$$

where we also neglect any background neutral wind since $\mathbf{u}_0 = 0$. The steady-state equations then become:

$$\frac{\partial \rho_0}{\partial t} = 0 \tag{6.171}$$

There are no pressure gradients in the horizontal direction in the background ionosphere:

$$-\frac{1}{\rho_0}\frac{\partial p_0}{\partial x} = 0 \tag{6.172}$$

The hydrostatic law applies in the background atmosphere:

$$-\frac{1}{\rho_0}\frac{\partial p_0}{\partial z} - g = 0 \tag{6.173}$$

In the background medium we have from (6.167):

$$\frac{dp_0}{dt} = c^2 \frac{d\rho_0}{dt} \tag{6.174}$$

or since $\mathbf{u}_0 = 0$ ($d/dt = \partial/\partial t$)

$$\frac{\partial p_0}{\partial t} = c^2 \frac{\partial \rho_0}{\partial t} \tag{6.175}$$

For the first-order perturbation equations we now derive by inserting into (6.163):

$$\frac{\partial \rho_1}{\partial t} + w_1 \frac{\partial \rho_0}{\partial z} + \rho_0 \left(\frac{\partial u_1}{\partial x} + \frac{\partial w_1}{\partial z} \right) = 0 \tag{6.176}$$

From the equation of motion (6.164) we have

$$\rho_0 \frac{\partial u_1}{\partial t} = -\frac{\partial p_1}{\partial x} \tag{6.177}$$

and

$$\rho_0 \frac{\partial w_1}{\partial t} = -\frac{\partial p_1}{\partial z} - g\rho_1 \tag{6.178}$$

From the energy equation (6.167) we derive:

$$\frac{\partial p_1}{\partial t} + w_1 \frac{\partial p_0}{\partial z} = c^2 \left(\frac{\partial \rho_1}{\partial t} + w_1 \frac{\partial \rho_0}{\partial z} \right) \tag{6.179}$$

again where we have used the fact that $\mathbf{u}_0 = 0$. We will now assume that all the perturbations can be expressed as harmonically variable parameters where all are proportional to the wave function such that

$$u_1, w_1, \rho_1, p_1 \propto \exp[i(\omega t - k_x x - k_z z)] \tag{6.180}$$

Inserting this into the equations above, gives the following set of equations:

$$i\omega u_1 \rho_0 = i k_x p_1 \tag{6.181}$$

$$i\omega w_1 \rho_0 = i k_z p_1 - g\rho_1 \tag{6.182}$$

$$i\omega \rho_1 + w_1 \frac{\partial \rho_0}{\partial z} + \rho_0 (-i k_x u_1 - i k_z w_1) = 0 \tag{6.183}$$

$$i\omega p_1 + w_1 \frac{\partial p_0}{\partial z} = c^2 \left(i\omega \rho_1 + w_1 \frac{\partial \rho_0}{\partial z} \right) \tag{6.184}$$

When applying the ideal gas law on the background and for the perturbation, then we obtain:

$$\frac{p_1}{\rho_0} = \frac{RT}{p_0} p_1 = gH \frac{p_1}{p_0} \tag{6.185}$$

where H is the scale height. Furthermore we have from (6.173):

$$\frac{1}{\rho_0}\frac{\partial \rho_0}{\partial z} = \frac{1}{\rho_0 RT}\frac{\partial p_0}{\partial z} = -\frac{g}{RT} = -\frac{1}{H} \tag{6.186}$$

where we have assumed that $\partial T/\partial z = \partial T/\partial x = 0$ or that the atmosphere is isothermal. By using these relations we now derive:

$$i\omega u_1 - ik_x gH \cdot \frac{p_1}{p_0} = 0 \tag{6.187}$$

$$i\omega w_1 + g\frac{\rho_1}{\rho_0} - ik_z gH\frac{p_1}{p_0} = 0 \tag{6.188}$$

$$-ik_x u_1 - \left(\frac{1}{H} + ik_z\right)w_1 + i\omega\frac{\rho_1}{\rho_0} = 0 \tag{6.189}$$

$$\frac{\gamma-1}{H}w_1 - i\omega\gamma\frac{\rho_1}{\rho_0} + i\omega\frac{p_1}{p_0} = 0 \tag{6.190}$$

We now can express this with the help of a matrix:

$$\begin{pmatrix} i\omega & 0 & 0 & -ik_x gH \\ 0 & i\omega & g & -ik_z gH \\ -ik_x & -\left(\frac{1}{H}+ik_x\right) & i\omega & 0 \\ 0 & \frac{\gamma-1}{H} & -i\gamma\omega & i\omega \end{pmatrix} \begin{pmatrix} u_1 \\ w_1 \\ \frac{\rho_1}{\rho_0} \\ \frac{p_1}{p_0} \end{pmatrix} = 0 \tag{6.191}$$

The non-trivial solution is given when the determinant is zero; then we obtain

$$\omega^4 - (k_x^2 + k_z^2)\gamma gH\omega^2 + k_x^2 g^2(\gamma-1) - \frac{g}{H}\omega^2 = 0 \tag{6.192}$$

Now introducing $\omega_a^2 = \gamma g^2/c^2$ and the Brunt–Väisälä frequency $\omega_B^2 = [(\gamma-1)g^2]/c^2$, then we obtain

$$\omega^4 - [(k_x^2 + k_z^2)c^2 + \omega_a^2]\omega^2 + k_x^2 c^2 \omega_B^2 = 0 \tag{6.193}$$

which solved for ω^2 gives

$$\omega^2 = \frac{k^2 c^2 + \omega_a^2 \pm \sqrt{(k^2 c^2 + \omega_a^2)^2 - 4k_x^2 c^2 \omega_B^2}}{2} \tag{6.194}$$

when we have used that $k^2 = k_x^2 + k_z^2$. There are two possible solutions for ω^2 which express the permissible relationship between the frequency of the wave and its horizontal and vertical wave numbers. For the positive sign in front of the square root we obtain the solutions with the highest frequencies which are the acoustic waves. For the negative sign we get the solution for the lower frequencies which are the gravity waves. If ω^2 becomes negative, the frequency is imaginary and the wave is evanescent. There are therefore two branches where propagation modes are possible.

The angle of propagation direction is given by

$$\tan\theta = k_z/k_x \tag{6.195}$$

or

$$\cos\theta = k_x/(k_x^2 + k_z^2)^{1/2} = \frac{k_x}{k} \tag{6.196}$$

The phase velocity of the wave is expressed by

$$v_{ph} = \omega/k \tag{6.197}$$

which can be expressed by the equation for ω^2.

$$(v_{ph})^2 = \frac{1}{2}\left[c^2 + \left(\frac{\omega_a}{k}\right)^2 \pm \sqrt{\left(c^2 + \left(\frac{\omega_a}{k}\right)^2\right)^2 - 4\cos^2\theta\, c^2\omega_B^2}\right] \tag{6.198}$$

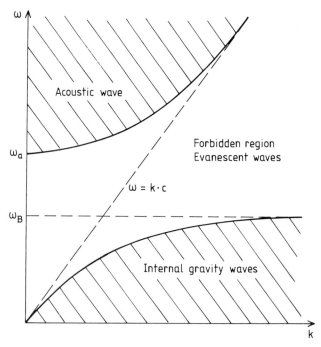

Fig. 6.29. Possible real solutions of the dispersion relation in a ω versus k diagram. $\omega_a^2 = \gamma g^2/c^2$ and $\omega_B^2 = [(\gamma-1)g^2]/c^2$.

For very large wavenumbers ($k = 2\pi/\lambda$) or small wavelengths, $v_{ph} \to c$ for the upper sign which corresponds to sonic waves. We also notice that these waves have a lower cutoff frequency when $k \to 0$ at $\omega = \omega_a$. For the lower sign, however, we get the gravity waves which have an upper frequency limit which approaches $\omega = \omega_B$ when k is very large. In Fig. 6.29 are the different regions of possible wave modes shown in a ω, k diagram. The acoustic waves and the gravity waves are separated by a so-called forbidden region of evanescent modes.

6.18 REFERENCES

Brekke, A., Nozawa, S. and Sparr, T. (1994) *J. Geophys. Res.*, **99**, 8801–8825.
Dickinson, R. E., Ridley, E. C. and Roble, R. G. (1975) *J. Atmos. Sci.*, **32**, 1737–1754.
Dickinson, R. E., Ridley, E. C. and Roble, R. G. (1984) *J. Atmos. Sci.*, **41**, 205–219.
Jacchia, L. G. (1965) in *Space Research*, King–Hele, D. G., Müller, P. and Righini, G. (Eds.), Vol. V, pp. 1152–1174, North Holland, Amsterdam.
Killeen, T. L. and Roble, R. G. (1986) *J. Geophys. Res.*, **91**, 11291–11307.
Kochanski, A. (1963) *J. Geophys. Res.*, **68**, 213–216.
Kohl, H. and King, J. W. (1967) *J. Atmos. Terr. Phys.*, **29**, 1045–1062.
Lindzen, R. S. (1968) *Proc. Roy. Soc.*, **A303**, 299.
Neiburger, M., Edinger, J. G. and Bonner, W. D. (1982) *Understanding our Atmospheric Environment*, W. H. Freeman and Company, San Francisco.
NOAA–NESS (1979) *Earth–Atmosphere Radiation Budget Analyses Derived from NOAA Satellite Data, June 1974 – February 1978*, Washington D.C.
Nozawa, S. (1994) Private communication.
Nozawa, S. and Brekke, A. (1995) *J. Geophys. Res.*, **100**, 14717–14734.
Rosenberg, N. W. (1968) *J. Atmos. terr. Phys.*, **30**, 907–917.
Sigernes, F. (1992) Private communication.
Smith, R. W. and Sweeney, P. J. (1980) *Nature*, **284**, 437–438.
Stroud, W. G., Nordberg, W., Bandeen, W. R., Bartman, F. L. and Titus, P. (1959) *J. Geophys. Res.*, **64**, 1342–1343.
Tohmatsu, T. (1990) *Compendium of Aeronomy*, Terra Sci. Publ. Comp., Tokyo.

6.19 EXERCISES

1. Choose a place in the atmosphere at latitude φ. In a Cartesian coordinate system is x positive eastward, y northward and z in the vertical direction.

    (a) Find the horizontal geostrophic wind components u and v in this coordinate system.

    Let the pressure gradient around equator look like in Fig. 6.30.

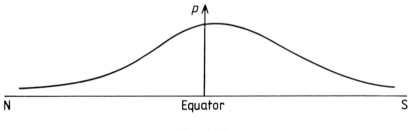

Fig. 6.30.

(b) Give the zonal wind directions in the northern and southern hemisphere.

(c) Does it make sense to talk about a geostrophic wind at the equator?

Let the pressure gradient around equator look like in Fig. 6.31.

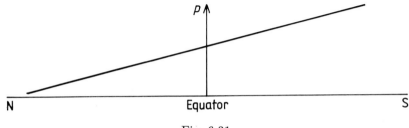

Fig. 6.31.

(d) How do the zonal wind directions look like in both hemispheres in this case?

2. The so-called potential temperature, θ, can be defined as:

$$\theta = T \cdot \left(\frac{p}{p_s}\right)^{-\frac{(\gamma-1)}{\gamma}}$$

where p_s is the pressure at some reference height.

(a) Derive this equation by using the ideal gas law and the adiabatic law.

(b) For an atmosphere in hydrostatic equilibrium, show that the following relation holds:

$$T\frac{\partial}{\partial z}\ln\theta = \frac{g}{c_p} + \frac{dT}{dz}$$

(c) What does g/c_p express?

3. Derive equation (6.50).

4. Derive equation (6.60).

5. Derive equation (6.192).

7
Currents in the ionosphere

7.1 THE STEADY-STATE APPROACH

When describing the dynamics of the ionospheric plasma, it is conventional to apply the following set of dynamic and electrodynamic equations of the plasma.

The mobility equation of the ion species j

$$\rho_j \frac{d\mathbf{v}_j}{dt} = -\nabla p_j + \rho_j \mathbf{g} + \frac{q_j \rho_j}{m_j}(\mathbf{E} + \mathbf{v}_j \times \mathbf{B}) - \sum_{\substack{k \\ j \neq k}} \rho_j \nu_{jk}(\mathbf{v}_j - \mathbf{v}_k) \quad (7.1)$$

where the summation is taken over all other species except j.

The continuity equation for the ion species j

$$\frac{\partial \rho_j}{\partial t} + \nabla(\rho_j \mathbf{v}_j) = (P_j - L_j) \quad (7.2)$$

where P_j and L_j are the production and loss term of species j.

The ideal gas law, the partial pressure of ion species j:

$$p_j = \rho_j \kappa T_j / m_j = n_j \kappa T_j \quad (7.3)$$

The Maxwell equations

$$\mathbf{E} = -\nabla \phi \quad (7.4)$$

$$\nabla \cdot \mathbf{j} = 0 \quad (7.5)$$

$$\nabla \cdot \mathbf{B} = 0 \quad (7.6)$$

$$\nabla \times \mathbf{B} = \mu_0 \mathbf{j} \quad (7.7)$$

Usually it is also assumed that the number density of positive ions are set equal to the number density of the electrons, thus

$$n = n_e \simeq \sum_{\text{ions}} n_j \tag{7.8}$$

We have, however, seen that for the D-region (7.8) may be an oversimplification. In order to give a complete description of the plasma we would have to treat each individual ion species by (7.1)–(7.3). This would, however, lead to a large set of coupled equations due to the collision terms. In order to understand the basic features of the ionospheric dynamics it is therefore more practical to consider a single ion species of mass m_i or equivalently assuming a pseudo-ion derived by proper weighting of all ions with respect to their individual mass, density, charge, velocity, and temperature.

Assuming now the ion and electron temperatures to be spatially uniform, then (7.1) for the ions and the electrons can respectively be expressed as:

$$nm_i \frac{d\mathbf{v}_i}{dt} = -\kappa T_i \nabla n + nm_i \mathbf{g} + ne(\mathbf{E} + \mathbf{v}_i \times \mathbf{B}) - nm_i \nu_{in}(\mathbf{v}_i - \mathbf{u}_n) \tag{7.9}$$

$$nm_e \frac{d\mathbf{v}_e}{dt} = -\kappa T_e \nabla n + nm_e \mathbf{g} - ne(\mathbf{E} + \mathbf{v}_e \times \mathbf{B}) - nm_e \nu_{en}(\mathbf{v}_e - \mathbf{u}_n) \tag{7.10}$$

where $n_e = n_i = n$. Here $\mathbf{E}$ is the electric field one would measure in an Earth fixed reference frame which is usually the field measured in ionospheric experiments. We have here neglected ion–ion and ion–electron collision frequencies as these are usually compared to the ion–neutral and electron–neutral collision frequencies at least in the lower part of the ionosphere (the E-region). The collision frequencies, however, play an important role in a partially ionized plasma.

An approximate formula for the ion–neutral collision frequency found in the literature is:

$$\nu_{in} = 2.6 \times 10^{-15} \, (n_n + n_i)(M'_n)^{-1/2} \tag{7.11}$$

where M'_n denotes the mean neutral molecular mass. The densities are given in m^{-3}. Sometimes an alternative expression separating the contribution from each neutral species is used especially in E-region studies:

$$\nu_{in} = k_{N_2} \cdot [N_2] + k_{O_2} \cdot [O_2] + k_O \cdot [O] \tag{7.12}$$

where k_{N_2}, k_{O_2} and k_O are the different collision rates according to the neutral and ionized species involved, and the brackets denote the density of each individual neutral species. For an average ion mass number of 30.7, Table 7.1 gives the characteristic collision rates (Kunitake and Schlegel, 1991).

Table 7.1. The collision rate coefficients in m^3/s

k_{N_2}	k_{O_2}	k_O
4.34×10^{-16}	4.28×10^{-16}	2.44×10^{-16}

An approximate formula for the electron collision frequency is:

$$\nu_e \equiv \nu_{en} + \nu_{ei} \quad (7.13)$$
$$= 5.4 \times 10^{-10} n_n \cdot T_e^{1/2} + [34 + 4.18 \cdot \ln(T_e^3/n_e)] n_e T_e^{-3/2}$$

The densities of the different species are here given in cm^{-3}, and T_e is given in kelvins.

Typical values for the ion–neutral (ν_{in}) and electron–neutral (ν_{en}) collision frequencies are shown as function of altitude between 90 and 400 km for an auroral latitude station (Tromsø) (Fig. 7.1). Also indicated for comparison are the corresponding ion (Ω_i) and electron (Ω_e) gyrofrequencies in the same height region.

When we compare the terms including velocities in the mobility equation, we find that the acceleration term on the left-hand side is of the order v_j/τ_j or v_j^2/L, where τ_j is the response time and L is a distance characteristic for velocity change. The Lorentz term on the right-hand side is of the order of $v_j \Omega_j$, where Ω_j is the gyrofrequency ($q_j B/m_j$) for ion species j, and the frictional term is of order $v_j \nu_j$ where ν_j is the collision frequency. As long as $\tau_j \gg \Omega_j^{-1}$ or $\tau_j \gg \nu_j^{-1}$ the inertia term can be neglected. The collision frequency and gyrofrequency are sufficiently high that in most problems of interest to macroscopic dynamics this is a valid simplification.

The steady-state fluid velocity for the ions may now be written as:

$$-\kappa T_i \nabla n + n m_i \mathbf{g} + n e (\mathbf{E} + \mathbf{v}_i \times \mathbf{B}) - n m_i \nu_{in}(\mathbf{v}_i - \mathbf{u}_n) = 0 \quad (7.14)$$

and for the electrons:

$$\kappa T_e \nabla n + n m_e \mathbf{g} - n e (\mathbf{E} + \mathbf{v}_i \times \mathbf{B}) - n m_e \nu_{en}(\mathbf{v}_e - \mathbf{u}_n) = 0 \quad (7.15)$$

Solving (7.14) for $\mathbf{v}_i$ we find:

$$\mathbf{v}_i = \mathbf{u}_n + \frac{1}{\nu_{in}} \mathbf{g} - \frac{\kappa T_i}{m_i \nu_{in}} \frac{\nabla n}{n} + \frac{e}{m_i \nu_{in}} (\mathbf{E} + \mathbf{v}_i \times \mathbf{B}) \quad (7.16)$$

By introducing the ion diffusion coefficient:

$$D_i = \frac{\kappa T_i}{m_i \nu_{in}} \quad (7.17)$$

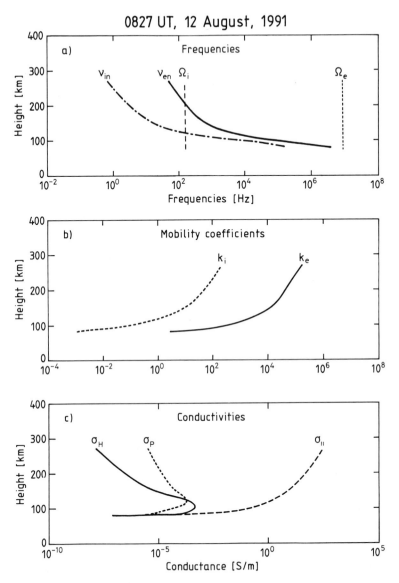

Fig. 7.1. (a) Typical altitude profiles of the ion–neutral (ν_{in}) and electron–neutral (ν_{en}) collision frequencies for an auroral zone station such as Tromsø (69°39′ N, 18°56′ E). Also indicated for comparison are the ion (Ω_i) and electron (Ω_e) gyrofrequencies. (b) Altitude profiles of the ion ($k_i = \Omega_i/\nu_{in}$) and electron ($k_e = \Omega_e/\nu_{en}$) mobility coefficients derived from the profiles given in panel (a). (c) Altitude profiles of the Pedersen (σ_P), Hall (σ_H) and parallel ($\sigma_\parallel$) conductivities as derived by EISCAT at Tromsø at 0827 UT on Aug. 12, 1991. These profiles are typical for quiet summer days. (Courtesy Blixt, 1995.)

and the ion scale height:

$$H_i = \frac{\kappa T_i}{m_i g} \tag{7.18}$$

we can write:

$$\mathbf{v}_i = \mathbf{u}_n + \mathbf{G}_i + \frac{e}{m_i \nu_{in}}(\mathbf{E} + \mathbf{v}_i \times \mathbf{B}) \tag{7.19}$$

where

$$\mathbf{G}_i = D_i \left(\frac{\hat{\mathbf{g}}}{H_i} - \frac{\nabla n}{n} \right) \tag{7.20}$$

is the velocity component due to gravity and diffusion.

We can in a similar manner solve (7.15) for $\mathbf{v}_e$ and write:

$$\mathbf{v}_e = \mathbf{u}_n + \mathbf{G}_e - \frac{e}{m_i \nu_{en}}(\mathbf{E} + \mathbf{v}_e \times \mathbf{B}) \tag{7.21}$$

where

$$\mathbf{G}_e = D_e \left(\frac{\hat{\mathbf{g}}}{H_e} - \frac{\nabla n}{n} \right) \tag{7.22}$$

is the velocity component due to gravity and diffusion. D_e $(= \kappa T_e/m_e\nu_{en})$ and H_e $(= \kappa T_e/m_e g)$ are the diffusion coefficient and scale height for the electron gas.

In the following we will neglect diffusion and gravity since we are studying the ion and electron velocities in the part of the ionosphere where the main currents are flowing and where the collision frequencies are rather large so D_e and D_i are correspondingly small.

The ion velocity given by (7.16) can then be simplified to:

$$\mathbf{v}_i = \mathbf{u}_n + \frac{e}{m_i \nu_{in}}(\mathbf{E} + \mathbf{v}_i \times \mathbf{B}) \tag{7.23}$$

By adding and subtracting a term $\mathbf{u}_n \times \mathbf{B}$ on the right-hand side and rearranging we get:

$$\mathbf{v}'_i = \frac{e}{m_i \nu_{in}} \mathbf{E}' + \frac{e}{m_i \nu_{in}} \mathbf{v}'_i \times \mathbf{B} \tag{7.24}$$

Here

$$\mathbf{E}' = \mathbf{E} + \mathbf{u}_n \times \mathbf{B} \tag{7.25}$$

and

$$\mathbf{v}'_i = \mathbf{v}_i - \mathbf{u}_n \tag{7.26}$$

are the electric field and ion velocity measured in a reference frame moving with the neutral wind, respectively.

A similar expression can be derived for the electron velocity:

$$\mathbf{v}'_e = -\frac{e}{m_i \nu_{en}} \mathbf{E}' - \frac{e}{m_e \nu_{en}} \mathbf{v}'_e \times \mathbf{B} \qquad (7.27)$$

where

$$\mathbf{v}'_e = \mathbf{v}_e - \mathbf{u}_n \qquad (7.28)$$

is the electron velocity observed from a reference frame moving with the neutral wind. Let us now assume that the neutral wind can be neglected, then the ion velocity is given by:

$$\mathbf{v}_i = \frac{k_i}{B} \mathbf{E} + \frac{k_i}{B} \mathbf{v}_i \times \mathbf{B} \qquad (7.29)$$

Here the ion mobility coefficient $k_i = \Omega_i/\nu_{in}$ has been introduced. This equation has the same form as (7.24). Therefore, to solve (7.24) we only replace $\mathbf{v}_i$ and $\mathbf{E}$ by $\mathbf{v}'_i$ and $\mathbf{E}'$, respectively, in the following. By now forming $\mathbf{v}_i \times \mathbf{B}$ from (7.29) we get

$$\mathbf{v}_i \times \mathbf{B} = \frac{k_i}{B} \mathbf{E} \times \mathbf{B} + \frac{k_i}{B}(\mathbf{v}_i \cdot \mathbf{B})\mathbf{B} - k_i B \mathbf{v}_i \qquad (7.30)$$

We also find from (7.29) that

$$\mathbf{v}_i \cdot \mathbf{B} = \frac{k_i}{B} \mathbf{E} \cdot \mathbf{B} \qquad (7.31)$$

and therefore

$$\mathbf{v}_i \times \mathbf{B} = \frac{k_i}{B} \mathbf{E} \times \mathbf{B} + \left(\frac{k_i}{B}\right)^2 (\mathbf{E} \cdot \mathbf{B})\mathbf{B} - k_i B \mathbf{v}_i \qquad (7.32)$$

By introducing (7.32) into (7.29) we find

$$\mathbf{v}_i = \frac{k_i}{B} \mathbf{E} + \left(\frac{k_i}{B}\right)^2 \mathbf{E} \times \mathbf{B} + \left(\frac{k_i}{B}\right)^3 (\mathbf{E} \cdot \mathbf{B})\mathbf{B} - k_i^2 \mathbf{v}_i \qquad (7.33)$$

Solving (7.33) for $\mathbf{v}_i$ we finally derive:

$$\mathbf{v}_i = \frac{1}{1+k_i^2}\left[\frac{k_i}{B} \mathbf{E} + \left(\frac{k_i}{B}\right)^2 \mathbf{E} \times \mathbf{B} + \left(\frac{k_i}{B}\right)^3 (\mathbf{E} \cdot \mathbf{B})\mathbf{B}\right] \qquad (7.34)$$

This equation represents a decomposition of $\mathbf{v}_i$ into three orthogonal components with respect to $\mathbf{E}$ and $\mathbf{B}$. The first term in the brackets of (7.34) represents the component along $\mathbf{E}$ while the last term represents the component along $\mathbf{B}$, and finally the middle term is along the direction perpendicular to both $\mathbf{E}$ and $\mathbf{B}$.

We also notice that by solving (7.27) while neglecting the neutral wind ($\mathbf{u}_n = 0$) we find the corresponding expression to $\mathbf{v}_e$:

$$\mathbf{v}_e = \frac{1}{1+k_e^2}\left[-\frac{k_e}{B} \mathbf{E} + \left(\frac{k_e}{B}\right)^2 \mathbf{E} \times \mathbf{B} - \left(\frac{k_e}{B}\right)^3 (\mathbf{E} \cdot \mathbf{B})\mathbf{B}\right] \qquad (7.35)$$

Here k_e ($= \Omega_e/\nu_{en}$) is the electron mobility coefficient. Fig. 7.1b presents the altitude variations of k_i and k_e between 90 and 400 km for an auroral zone station (Tromsø).

Allowing now the neutral wind $\mathbf{u}_n$ to be different from zero, we can, as mentioned above, replace $\mathbf{v}_i$, $\mathbf{v}_e$ and $\mathbf{E}$ by $\mathbf{v}'_i$, $\mathbf{v}'_e$ and $\mathbf{E}'$ in (7.34) and (7.35) and obtain

$$\mathbf{v}'_i = \mathbf{v}_i - \mathbf{u}_n = \frac{1}{1+k_i^2}\left[\frac{k_i}{B}\mathbf{E}' + \left(\frac{k_i}{B}\right)^2 \mathbf{E}' \times \mathbf{B} + \left(\frac{k_i}{B}\right)^3 (\mathbf{E}'\cdot\mathbf{B})\mathbf{B}\right] \quad (7.36)$$

$$\mathbf{v}'_e = \mathbf{v}_e - \mathbf{u}_n = \frac{1}{1+k_i^2}\left[-\frac{k_i}{B}\mathbf{E}' + \left(\frac{k_i}{B}\right)^2 \mathbf{E}' \times \mathbf{B} - \left(\frac{k_i}{B}\right)^3 (\mathbf{E}'\cdot\mathbf{B})\mathbf{B}\right] \quad (7.37)$$

From Fig. 7.1b we notice that $k_i \ll 1$ below about 100 km and therefore $\mathbf{v}_i \approx \mathbf{u}_n$. The ion velocity is then completely determined by the neutral wind, and the ions are closely coupled to the neutral gas in the bottom part of the E-region.

For a slightly larger value of k_i but still $k_i \ll 1$ we can neglect the last two terms on the right side of (7.36) such that:

$$\mathbf{v}_i - \mathbf{u}_n = k_i \frac{\mathbf{E}'}{B} \approx k_i \frac{\mathbf{E}}{B} \quad (7.38)$$

Here we have taken advantage of the fact that

$$\frac{k_i}{B}|\mathbf{u}_n \times \mathbf{B}| \ll |\mathbf{u}_n|$$

The relative ion–neutral velocity in the middle part of the E-region say between 100 and 115 km, is parallel to the electric field.

If we go to the other extreme and let $k_i \gg 1$ which corresponds to altitudes above say 180 km or the upper part of the E-region and in the F-region, then we notice that the relative ion–neutral velocity along $\mathbf{B}$ becomes

$$(\mathbf{v}_i)_\| - (\mathbf{u}_n)_\| = \frac{k_i}{B}\mathbf{E}_\| \quad (7.39)$$

When neglecting diffusion and gravity forces the relative ion–neutral velocity along $\mathbf{B}$ is proportional to the E-field component along B.

When $k_i \gg 1$ the ion velocity perpendicular direction to $\mathbf{B}$ becomes:

$$(\mathbf{v}_i)_\perp = \left(\mathbf{u}_n + \frac{\mathbf{E}'\times\mathbf{B}}{B^2}\right)_\perp = \left[\mathbf{u}_n + \frac{\mathbf{E}\times\mathbf{B}}{B^2} - \mathbf{u}_n\right]_\perp = \frac{\mathbf{E}\times\mathbf{B}}{B^2} \quad (7.40)$$

The ion velocity is along the $\mathbf{E}\times\mathbf{B}$ direction and independent of the neutral wind. We also notice from Fig. 7.1b that $k_e \gg 1$ above say 100 km and therefore

$$(\mathbf{v}_e)_\perp = \frac{\mathbf{E}\times\mathbf{B}}{B^2} \quad (7.41)$$

in most of the E-region.

We also notice that by replacing $\mathbf{u}_n$ by

$$\mathbf{u}'_i = \mathbf{u}_n + \mathbf{G}_i \tag{7.42}$$

$$\mathbf{u}'_e = \mathbf{u}_n + \mathbf{G}_e \tag{7.43}$$

in (7.19) and (7.21), respectively, we find the following expressions for (7.36) and (7.37), respectively

$$\begin{aligned}\mathbf{v}''_i &= \mathbf{v}_i - \mathbf{u}_n - \mathbf{G}_i = \mathbf{v}'_i - \mathbf{G}_i \\ &= \frac{1}{1+k_i^2}\left[\frac{k_i}{B}\boldsymbol{\mathcal{E}}'_i + \left(\frac{k_i}{B}\right)^2 \boldsymbol{\mathcal{E}}'_i \times \mathbf{B} + \left(\frac{k_i}{B}\right)^3 (\boldsymbol{\mathcal{E}}'_i \cdot \mathbf{B})\mathbf{B}\right] \end{aligned} \tag{7.44}$$

and

$$\begin{aligned}\mathbf{v}''_e &= \mathbf{v}_e - \mathbf{u}_n - \mathbf{G}_e = \mathbf{v}'_e - \mathbf{G}_e \\ &= \frac{1}{1+k_e^2}\left[-\frac{k_e}{B}\boldsymbol{\mathcal{E}}'_e + \left(\frac{k_e}{B}\right)^2 \boldsymbol{\mathcal{E}}'_e \times \mathbf{B} - \left(\frac{k_e}{B}\right)^3 (\boldsymbol{\mathcal{E}}'_e \cdot \mathbf{B})\mathbf{B}\right]\end{aligned} \tag{7.45}$$

where

$$\boldsymbol{\mathcal{E}}'_i = \mathbf{E} + (\mathbf{u}_n + \mathbf{G}_i) \times \mathbf{B} = \mathbf{E}' + \mathbf{G}_i \times \mathbf{B} \tag{7.46}$$

$$\boldsymbol{\mathcal{E}}'_e = \mathbf{E} + (\mathbf{u}_n + \mathbf{G}_e) \times \mathbf{B} = \mathbf{E}' + \mathbf{G}_e \times \mathbf{B} \tag{7.47}$$

Pressure or diffusion as well as gravity can therefore in addition to the neutral wind and the electric field also contribute to ion and electron velocities perpendicular to $\mathbf{B}$.

7.2 ROTATION OF THE ION VELOCITY BY HEIGHT IN THE IONOSPHERE

We will now neglect all other forces except the electrostatic field $\mathbf{E}$ for a while and assume that $\mathbf{E}$ is perpendicular to $\mathbf{B}$. Then (7.34) takes the simplified form:

$$\mathbf{v}_i = \frac{k_i}{1+k_i^2}\frac{\mathbf{E}}{B} + \frac{k_i^2}{1+k_i^2}\frac{\mathbf{E}\times\mathbf{B}}{B^2} \tag{7.48}$$

In a Cartesian coordinate system (x, y, z) where $\mathbf{B} = B\hat{\mathbf{z}}$ and $\mathbf{E} = E\hat{\mathbf{y}}$, $\mathbf{E} \times \mathbf{B}$ will be in the positive x-direction (Fig. 7.2). The ion velocity $\mathbf{v}_i$ will therefore make an angle θ_i with the applied electrostatic field $\mathbf{E}$, and this angle is given by

$$\tan\theta_i = k_i = \frac{\Omega_i}{\nu_{in}} \tag{7.49}$$

The velocity therefore rotates clockwise when seen along $\mathbf{B}$ as ν_{in} increases. In the ionosphere $\nu_{in} = \Omega_i$ at about 125 km altitude. At this altitude then

$$\theta_i = 45°$$

Lower down in the ionosphere where $\nu_{in} > \Omega_i$ the ion velocity will be more and more along $\mathbf{E}$ as the collision frequency increases. At 90 km where $n_n = 6 \times 10^{19}$ m^{-3} and assuming $M'_n = 28.8$ we find the ion–neutral collision frequency

Sec. 7.2] Rotation of the ion velocity by height

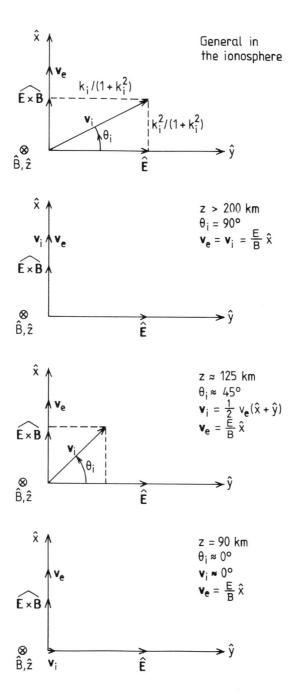

Fig. 7.2. Vector diagrams showing the variation of $\mathbf{v}_i$ and $\mathbf{v}_e$ with respect to the applied electric field for three different altitudes in the ionosphere.

to be:

$$\nu_{in} = 8 \times 10^5 \text{ s}^{-1}$$

Here we have been neglecting the density of the ions. Assuming an ion mass equal to 30.5 a.m.u. which is representative for an altitude of 90 km, and a magnetic field strength of $B = 5.7 \times 10^{-5}$ tesla which applies for auroral zone latitudes, the gyrofrequency becomes:

$$\Omega_i = \frac{eB}{m_i} = 180 \text{ s}^{-1}$$

and

$$k_i = \frac{\Omega_i}{\nu_{in}} = 2.25 \times 10^{-4}$$

The angle at 90 km is then:

$$\theta_i \approx 0.01°$$

The ion velocity in the D-region is therefore parallel to the E-field in the absence of other forces. In the upper ionosphere say at 180 km where

$$n_n = 1.3 \times 10^{16} \text{ m}^{-3}$$

and assuming $M'_n = 24$

$$\nu_{in} = 165 \text{ s}^{-1}$$

For the dominant ion species O^+ the ion gyrofrequency will be ($B = 5.7 \times 10^{-5}$ tesla)

$$\Omega_i = 340 \text{ s}^{-1}$$

and

$$k_i = 2.06$$

leaving the angle

$$\theta_i = 64.16°$$

At 250 km where $n_n = 1.3 \times 10^{15}$ m^{-3}, the angle $\mathbf{v}_i$ will make with $\mathbf{E}$ is close to 87° for $M'_n = 21$ and O^+ ions. Above 250 km therefore the ion velocity is very close to the $\mathbf{E} \times \mathbf{B}$ direction in the absence of any other forces.

We also notice that the magnitude of $\mathbf{v}_i$ is given by

$$v_i = \left\{ \left[\left(\frac{k_i}{1+k_i^2} \right)^2 + \left(\frac{k_i^2}{1+k_i^2} \right)^2 \right] \frac{E^2}{B^2} \right\}^{1/2} \qquad (7.50)$$

$$= k_i \cdot (1+k_i^2)^{-1/2} \frac{E}{B} = \sin\theta_i \frac{E}{B}$$

Rotation of the ion velocity by height

At the upper ionosphere where $\theta_i \approx 90°$

$$v_i \approx E/B$$

and in the D-region where $\theta_i \approx 0$

$$v_i \approx 0$$

The variation of $\mathbf{v}_i$ by height in the ionosphere is illustrated in Fig. 7.2.

If we do the same analysis for the electrons, we get:

$$\mathbf{v}_e = -\frac{k_e}{1+k_e^2}\frac{\mathbf{E}}{B} + \frac{k_e^2}{1+k_e^2}\frac{\mathbf{E}\times\mathbf{B}}{B^2} \tag{7.51}$$

and

$$\tan\theta_e = -k_e = -\frac{\Omega_e}{\nu_{en}} \tag{7.52}$$

The electron gyrofrequency for $B = 5.7 \times 10^{-5}$ tesla is $\Omega_e = 10^7$ s^{-1}. For a neutral density $n = 6 \times 10^{19}$ m^{-3} at 90 km and an electron temperature equal to the neutral temperature $T_e = T_n = 170$ K, we find an electron–neutral collision frequency given by:

$$\nu_{en} = 5.4 \times 10^{-16} \cdot n_n \cdot T_e^{1/2} = 4.2 \times 10^5 \text{ s}^{-1}$$

when neglecting the electron density. This gives

$$k_e = 24$$

and

$$\theta_e = 92°$$

or very close to the $\mathbf{E}\times\mathbf{B}$ direction. Since ν_{en} decreases above 90 km, θ_e will become closer and closer to the $\mathbf{E} \times \mathbf{B}$ drift. It is therefore a reasonable assumption to make that the electrons are moving in the $\mathbf{E} \times \mathbf{B}$ direction above 90 km when no other forces than the electric field is applied. Above 250 km or so the electrons as well as the ions are $\mathbf{E} \times \mathbf{B}$ drifting with the same speed when the E-field is the only applied force.

This rotation of the ion velocity with respect to the electric field and the electron velocity in the ionosphere is, as we have demonstrated above, a result of the varying collision frequency by height.

We also notice from (7.36) that the ion velocity $\mathbf{v}'_i$ observed in a frame moving with the neutral wind will rotate by height with respect to $\mathbf{E}'$ in a similar manner as $\mathbf{v}_i$ rotates with respect to $\mathbf{E}$ when the neutral velocity is zero in the Earth's fixed frame of reference.

The neutral wind, however, changes direction quite dramatically with height, and therefore it is not very practicable to refer to the neutral wind frame when discussing the plasma motion.

7.3 THE CURRENT DENSITY IN THE IONOSPHERE

The current density at a given height in the ionosphere is given by

$$\mathbf{j} = n \cdot e \cdot (\mathbf{v}_i - \mathbf{v}_e) \tag{7.53}$$

when we assume singly charged ions and charge neutrality. Note that the dimension of $\mathbf{j}$ is A/m² in standard units. Inserting (7.34) and (7.35) for $\mathbf{v}_i$ and $\mathbf{v}_e$ in (7.53) we find the following expression for the current density

$$\mathbf{j} = n \cdot e \left\{ \left(\frac{k_e}{1+k_e^2} + \frac{k_i}{1+k_i^2} \right) \frac{\mathbf{E}_\perp}{B} \right. \tag{7.54}$$
$$\left. - \left(\frac{k_e^2}{1+k_e^2} - \frac{k_i^2}{1+k_i^2} \right) \frac{\mathbf{E} \times \mathbf{B}}{B^2} + (k_e + k_i) \frac{\mathbf{E}_\parallel}{B} \right\}$$

where $\mathbf{E}_\perp$ and $\mathbf{E}_\parallel$ are the electric field components perpendicular to and parallel with $\mathbf{B}$ ($\mathbf{E} = \mathbf{E}_\perp + \mathbf{E}_\parallel$).

Now introducing the conductivities:

$$\sigma_P = \frac{ne}{B} \left(\frac{k_e}{1+k_e^2} + \frac{k_i}{1+k_i^2} \right) \quad \text{(a)}$$

$$\sigma_H = \frac{ne}{B} \left(\frac{k_e^2}{1+k_e^2} - \frac{k_i^2}{1+k_i^2} \right) \quad \text{(b)} \tag{7.55}$$

$$\sigma_\parallel = \frac{ne}{B} (k_e + k_i) \quad \text{(c)}$$

Here σ_P, σ_H and $\sigma_\parallel$ are the height-dependent Pedersen, Hall and parallel conductivities, respectively. Equation (7.54) can then be expressed as

$$\mathbf{j} = \sigma_P \mathbf{E}_\perp - \sigma_H \frac{\mathbf{E} \times \mathbf{B}}{B} + \sigma_\parallel \mathbf{E}_\parallel \tag{7.56}$$

Fig. 7.3 presents some altitude profiles of the electron density as observed at an auroral zone station (Tromsø) for a quiet summer day. Also presented are the Pedersen and Hall conductivity profiles.

In Fig. 7.4 are shown some more examples of the Pedersen and Hall conductivity profiles as observed at an auroral zone station (Tromsø). During auroral particle precipitation events the enhanced electron densities can increase the conductivities at their maxima by a factor 10 above the typical quiet time value. We notice that while the maximum in the Pedersen conductivity is approximately determined by the height where $\Omega_i = \nu_{in}$, the maximum in the Hall conductivity is more closely related to the peak in the electron density profile. For this reason the Hall conductivity is more sensitive to the high-energy auroral particle precipitation which most strongly enhances the ionization below 125 km.

Sec. 7.3] The current density in the ionosphere 325

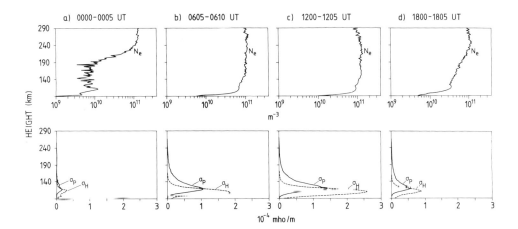

Fig. 7.3. Upper half: Typical quiet summer electron density profiles observed at night, morning, daytime and evening in the auroral zone at Tromsø. Lower half: Ionospheric $\sigma_H(z)$ and $\sigma_P(z)$ conductivity profiles derived from the observed electron density profiles above and model collision frequency. The electron density profiles are observed by the incoherent scatter radar, EISCAT. (From Brekke and Hall, 1988.)

The current density is determined by the relative velocity between the ions and the electrons, and therefore we find

$$\begin{aligned} \mathbf{j} &= n \cdot e(\mathbf{v}_i - \mathbf{v}_e) = n \cdot e(\mathbf{v}_i - \mathbf{u}_n - (\mathbf{v}_e - \mathbf{u}_n)) \\ &= n \cdot e(\mathbf{v}'_i - \mathbf{v}'_e) = \mathbf{j}' \end{aligned} \quad (7.57)$$

where $\mathbf{j}'$, $\mathbf{v}'_i$ and $\mathbf{v}'_e$ are the current density, the ion and electron velocity, respectively, observed in a reference frame following the neutral gas. By inserting (7.36) and (7.37) into (7.57) we then derive

$$\begin{aligned} \mathbf{j}' &= n \cdot e \left\{ \left(\frac{k_e}{1 + k_e^2} + \frac{k_i}{1 + k_i^2} \right) \frac{\mathbf{E}'}{B} - \left(\frac{k_e^2}{1 + k_e^2} - \frac{k_i^2}{1 + k_i^2} \right) \frac{\mathbf{E}' \times \mathbf{B}}{B^2} \right. \\ &\quad \left. + \left(\frac{k_e^3}{1 + k_e^2} + \frac{k_i^3}{1 + k_i^2} \right) \frac{(\mathbf{E}' \cdot \mathbf{B})\mathbf{B}}{B^3} \right\} \\ &= \sigma_P \mathbf{E}'_\perp - \sigma_H \frac{\mathbf{E}' \times \mathbf{B}}{B} + \sigma_\| \mathbf{E}'_\| \end{aligned} \quad (7.58)$$

The form of (7.58) remains the same as that of (7.56) except that the electric field $\mathbf{E}'$ ($= \mathbf{E}'_\perp + \mathbf{E}'_\|$) is now observed in the reference frame following the neutral gas. Only when $\mathbf{u}_n = 0$ will the current densities given by (7.56) and (7.58) remain the same. For a situation when $\mathbf{u}_n \neq 0$, equation (7.58) is the proper expression for the current.

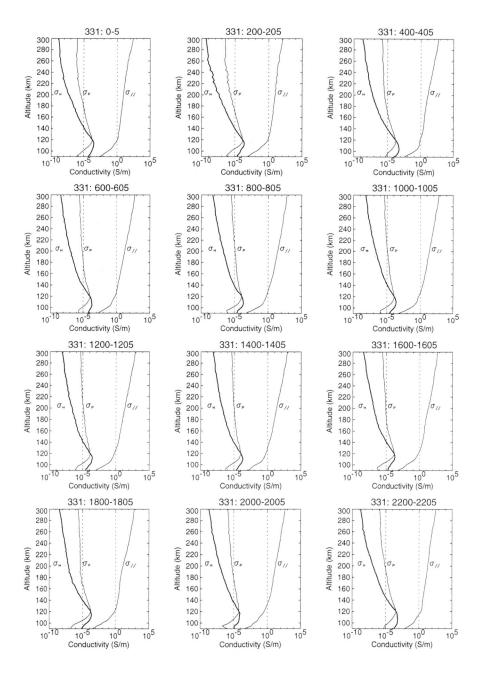

Fig. 7.4. Examples of Pedersen, Hall and parallel conductivity profiles (in Siemens/m) as derived at the auroral zone station Tromsø for 12 different periods on March 31, 1992. (Courtesy Nozawa, 1995.)

7.4 CURRENTS DUE TO GRAVITY AND DIFFUSION

When also allowing for pressure or diffusion as well as gravity forces in the equations of motion for the ions and the electrons, (7.44) and (7.45) should be applied. We then find the following expression for the current density vector

$$
\begin{aligned}
\mathbf{j}'' &= n \cdot e(\mathbf{v}'_i - \mathbf{v}'_e) \\
&= n \cdot e \left\{ \mathbf{G}_i + \frac{k_i}{1+k_i^2} \frac{\boldsymbol{\mathcal{E}}'_i}{B} + \frac{k_i^2}{1+k_i^2} \frac{\boldsymbol{\mathcal{E}}'_i \times \mathbf{B}}{B^2} + \frac{k_i^3}{1+k_i^2} \frac{(\boldsymbol{\mathcal{E}}'_i \cdot \mathbf{B})\mathbf{B}}{B^3} \right. \\
&\quad \left. - \mathbf{G}_e + \frac{k_e}{1+k_e^2} \frac{\boldsymbol{\mathcal{E}}'_e}{B} - \frac{k_e^2}{1+k_e^2} \frac{\boldsymbol{\mathcal{E}}'_e \times \mathbf{B}}{B^2} + \frac{k_e^3}{1+k_e^2} \frac{(\boldsymbol{\mathcal{E}}'_e \cdot \mathbf{B})\mathbf{B}}{B^3} \right\} \\
&\quad - n \cdot e \left\{ \left(\frac{k_i}{1+k_i^2} + \frac{k_e}{1+k_e^2} \right) \frac{\mathbf{E}'}{B} + \left(\frac{k_i^2}{1+k_i^2} - \frac{k_e^2}{1+k_e^2} \right) \frac{\mathbf{E}' \times \mathbf{B}}{B^2} \right. \\
&\quad \left. + \left(\frac{k_i^3}{1+k_i^2} + \frac{k_e^3}{1+k_e^2} \right) (\mathbf{E}' \cdot \mathbf{B})\mathbf{B} \cdot \frac{1}{B^3} \right\} \\
&\quad + n \cdot e \left[\mathbf{G}_i - \mathbf{G}_e + \left(\frac{k_i}{1+k_i^2} \mathbf{G}_i + \frac{k_e}{1+k_e^2} \mathbf{G}_e \right) \times \frac{\mathbf{B}}{B} \right. \\
&\quad \left. + \left(\left(\frac{k_i^2}{1+k_i^2} \mathbf{G}_i - \frac{k_e^2}{1+k_e^2} \mathbf{G}_e \right) \times \mathbf{B} \right) \times \frac{\mathbf{B}}{B^2} \right]
\end{aligned}
\qquad (7.59)
$$

In order to derive (7.59) we have taken advantage of the vector relations

$$(\mathbf{G}_i \times \mathbf{B}) \cdot \mathbf{B} = (\mathbf{G}_e \times \mathbf{B}) \cdot \mathbf{B} = 0 \qquad (7.60)$$

By introducing to (7.59) the expression for $\mathbf{j}'$ given by (7.58) we have

$$\mathbf{j}'' = \mathbf{j}' + \mathbf{j}_G \qquad (7.61)$$

where

$$
\begin{aligned}
\mathbf{j}_G &= n \cdot e \left[\mathbf{G}_i - \mathbf{G}_e + \left(\frac{k_i}{1+k_i^2} \mathbf{G}_i + \frac{k_e}{1+k_e^2} \mathbf{G}_e \right) \times \frac{\mathbf{B}}{B} \right. \\
&\quad \left. + \left(\left(\frac{k_i^2}{1+k_i^2} \mathbf{G}_i - \frac{k_e^2}{1+k_e^2} \mathbf{G}_e \right) \times \mathbf{B} \right) \times \frac{\mathbf{B}}{B^2} \right] \\
&= n \cdot e \left[(\mathbf{G}_i - \mathbf{G}_e)_\parallel + \left(\frac{k_i}{1+k_i^2} \mathbf{G}_i + \frac{k_e}{1+k_e^2} \mathbf{G}_e \right) \times \frac{\mathbf{B}}{B} \right] \\
&= (\mathbf{j}_G)_\parallel + (\mathbf{j}_G)_\perp
\end{aligned}
\qquad (7.62)
$$

$\mathbf{j}_G$ is the current density due to gravity and pressure or diffusion in the absence of an electric field. $(\mathbf{j}_G)_\parallel$ and $(\mathbf{j}_G)_\perp$ are then the components of this current density parallel to and perpendicular to $\mathbf{B}$, respectively.

For the parallel current density $(\mathbf{j}_G)_\parallel$ due to diffusion and gravity forces we have when applying (7.20) and (7.22)

$$(\mathbf{j}_G)_\parallel = n \cdot e(\mathbf{G}_i - \mathbf{G}_e)_\parallel \qquad (7.63)$$

$$= ne\left[\frac{1}{\nu_{in}}\mathbf{g}_\parallel - \frac{\kappa T_i}{m_i \nu_{in}}\left(\frac{\nabla n}{n}\right)_\parallel - \frac{1}{\nu_{en}}\mathbf{g}_\parallel + \frac{\kappa T_e}{m_e \nu_{en}}\left(\frac{\nabla n}{n}\right)_\parallel\right]$$

$$= ne\left[\left(\frac{1}{\nu_{in}} - \frac{1}{\nu_{en}}\right)\mathbf{g}_\parallel - \kappa\left(\frac{T_i}{m_i \nu_{in}} - \frac{T_e}{m_e \nu_{en}}\right)\left(\frac{\nabla n}{n}\right)_\parallel\right]$$

which is essentially the diffusion current along the field line related to the diffusion process at the top of the ionosphere. In the lower region below 300 km or so this current will normally be very weak. The parallel diffusion current will, however, increase in importance at higher latitudes since $\mathbf{g}$, $\mathbf{B}$ and $\nabla n/n$ are all basically vertical there.

For the perpendicular part of the current due to gravity and diffusion we find

$$(\mathbf{j}_G)_\perp = n \cdot e\left(\frac{k_i}{1+k_i^2}\mathbf{G}_i + \frac{k_e}{1+k_e^2}\mathbf{G}_e\right) \times \frac{\mathbf{B}}{B} \qquad (7.64)$$

$$= ne\left[\left(\frac{k_i}{1+k_i^2}\frac{1}{\nu_{in}} + \frac{k_e}{1+k_e^2}\frac{1}{\nu_{en}}\right)\mathbf{g}\right.$$
$$\left. - \kappa\left(\frac{k_i}{1+k_i^2}\frac{T_i}{m_i \nu_{in}} + \frac{k_e}{1+k_e^2}\frac{T_e}{m_e \nu_{en}}\right)\frac{\nabla n}{n}\right] \times \frac{\mathbf{B}}{B}$$

We notice that a gradient in the plasma density perpendicular to $\mathbf{B}$ creates a current density that is perpendicular both to $\mathbf{B}$ and the gradient itself. The current density due to the component of the gravity perpendicular to $\mathbf{B}$ will be very small at high latitudes where $\mathbf{B}$ and $\mathbf{g}$ are close to parallel, but will be important at equatorial regions where $\mathbf{g}$ is almost perpendicular to $\mathbf{B}$. This is also true for the current density due to the gradient in n since this gradient is almost true vertical except at high latitudes close to the auroral arcs and ionospheric irregularities. By a small rearrangement of the equation for $(\mathbf{j}_G)_\perp$ we get

$$(\mathbf{j}_G)_\perp = n \cdot \left[\left(\frac{k_i^2}{1+k_i^2}\cdot m_i + \frac{k_e^2}{1+k_e^2}m_e\right)\mathbf{g}\right. \qquad (7.65)$$
$$\left. - \kappa\left(\frac{k_i^2}{1+k_i^2}T_i + \frac{k_e^2}{1+k_e^2}T_e\right)\frac{\nabla n}{n}\right] \times \frac{\mathbf{B}}{B^2}$$

and we see that since k_e and k_i both increase very rapidly with height (Fig. 7.1b), the current perpendicular to $\mathbf{B}$ due to gravity and diffusion is most important in the collisionless heights. Then we have:

$$(\mathbf{j}_G)_\perp = n\left[m_i\mathbf{g} - \kappa(T_e + T_i)\frac{\nabla n}{n}\right] \times \frac{\mathbf{B}}{B^2} \qquad (7.66)$$

Sec. 7.5]   **Height-dependent currents and heating rates**   329

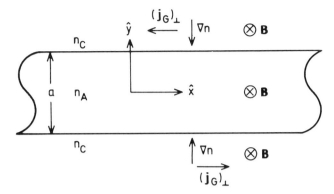

Fig. 7.5. A narrow strip of the xy plane of enhanced density n_A compared to the background density n_C. The strip has a finite width in the y-direction but uniform and unlimited in the x-direction. The magnetic field **B** is parallel to the z-axis. The gradients in n and the current density vectors are indicated.

where we have neglected collisions and m_e with respect to m_i. Equation (7.66) demonstrates that in a situation with no collisions there can be currents flowing in the plasma.

Consider Fig. 7.5 where we have indicated a narrow strip of enhanced density n_A in the xy plane. The enhancement has a finite width, a, in the y-direction but unlimited in the x-direction. The magnetic field **B** is directed in the z-direction. The background density $n_C \ll n_A$. Since ∇n will be pointing towards the strip along the y-axis on both sides of the strip, the current density vector $(\mathbf{j}_G)_\perp$ due to diffusion will be parallel to the strip as indicated in Fig. 7.5, but in opposite directions on the two sides. Such irregularities in the plasma density can therefore create shear currents along the irregularities.

7.5 HEIGHT-DEPENDENT CURRENTS AND HEATING RATES

The main ionospheric currents contributing to observable effects on ground are flowing mainly perpendicular to **B** below 300 km. Especially at high latitudes where the auroral electrojet dominates, it is not expected that the perpendicular currents due to gravity and diffusion will play any important role. In the direction parallel to **B**, however, the situation is similar in the lower ionosphere since $(\mathbf{j}_G)_\parallel$ decreases when the collision frequencies increase. Below 300 km or so, even at high latitudes the contribution to the parallel current $(\mathbf{j}_G)_\parallel$ due to gravity and diffusion will under normal conditions be rather small.

Similar to the analysis done in section 7.2 of the variation by altitude of $\mathbf{v}_i$ with respect to the electric field we will now make an analysis of how the current density vector changes direction by height in the ionosphere. Since the current density $\mathbf{j}$ derived for $\mathbf{u}_n = 0$ and $\mathbf{j}'$ derived for $\mathbf{u}_n \neq 0$ have a similar dependence on $\mathbf{E}$ and $\mathbf{E}'$, respectively, we can study the relationship between $\mathbf{j}$ and $\mathbf{E}$ and infer the derived results to the corresponding relationship between $\mathbf{j}'$ and $\mathbf{E}'$.

As we are mostly interested in the ionospheric currents which can be related to magnetic fluctuations observed on ground in auroral latitudes, we will consider only the current densities perpendicular to $\mathbf{B}$. Furthermore, these currents are predominantly present above 90 km so we can assume, according to Fig. 7.1b, that $k_e \gg 1$. The simplified expression for the current density perpendicular to $\mathbf{B}$ in the auroral ionosphere is then according to (7.54) given by:

$$\mathbf{j}_\perp = \frac{ne}{B} \left\{ \frac{k_i}{1+k_i^2} \mathbf{E}_\perp - \frac{1}{1+k_i^2} \frac{\mathbf{E}_\perp \times \mathbf{B}}{B} \right\} \tag{7.67}$$

By maintaining the same coordinate system as in Fig. 7.2, we find that the angle ϕ this current density makes with the $\mathbf{E}$-field is given by:

$$\tan \phi = -\frac{1}{k_i} \tag{7.68}$$

For this approximation then $\tan \phi \cdot \tan \theta_i = -1$ (see Fig. 7.2). In the absence of the neutral wind the current density therefore is perpendicular to the ion velocity everywhere in the auroral ionosphere. Furthermore, the current density is parallel to the $\mathbf{E}$-field when k_i is very large; that is at the top of the ionosphere, and perpendicular to the $\mathbf{E}$-field when $k_i \ll 1$ which is at the E-region levels and below. The dominant current in the upper ionosphere in the F-region is therefore the Pedersen current, while it is the Hall current that dominates in the E-region.

We also notice that the height-dependent magnitude of the current density is given by:

$$j(z)_\perp = \frac{e}{B} \frac{1}{\sqrt{1+k_i^2}} n(z) E_\perp \tag{7.69}$$

where $n(z)$ is the height-dependent electron density profile. Since $k_i \ll 1$ and the electron density $n(z)$ decreases very rapidly below the E-region peak, the current density below 90 km is often negligible.

We also notice that the height-dependent heat dissipation rate due to the current flowing along an electric field perpendicular to $\mathbf{B}$ (the Joule heat dissipation) is given by:

$$q_J(z) = \mathbf{j} \cdot \mathbf{E} = (\sigma_P(z) \cdot \mathbf{E}_\perp - \sigma_H(z) \mathbf{E}_\perp \times \mathbf{B}/B) \cdot \mathbf{E}_\perp = \sigma_P(z) E_\perp^2 \tag{7.70}$$

where $\sigma_P(z)$ is the height-dependent Pedersen conductivity. We see that the Hall current does not contribute to the heat dissipation. Assuming that the electric field is constant by height and that the neutral wind as well as gravity and diffusion forces can be neglected, the altitude profile of the heat dissipation rate $q_J(z)$ is equivalent to the altitude profile of the Pedersen conductivity except for the constant factor $E_\perp^2$.

If we also include the neutral wind but still neglect the gravity and diffusion forces, the current density vector, $\mathbf{j}'_\perp$, for $k_e \gg 1$ perpendicular to $\mathbf{B}$ will according to (7.58) be given by

$$\mathbf{j}'_\perp = \frac{n \cdot e}{B} \left\{ \frac{k_i}{1 + k_i^2} \mathbf{E}'_\perp - \frac{1}{1 + k_i^2} \mathbf{E}'_\perp \times \mathbf{B}/B \right\} \qquad (7.71)$$

and the angle ϕ' that the current density vector makes with $\mathbf{E}'_\perp$ is given by

$$\tan \phi' = -\frac{1}{k_i} = \tan \phi \qquad (7.72)$$

Since $\mathbf{E}'_\perp$ changes direction by altitude due to the neutral wind, ϕ' is not measured with respect to a fixed direction in space as would be the situation when $\mathbf{u}_n = 0$ and $\mathbf{E}_\perp$ is constant. Furthermore, the height-dependent magnitude of the current density at a height z is given by

$$\begin{aligned} j(z)_\perp &= \frac{e}{B} \frac{1}{\sqrt{1 + k_i^2}} n_e(z) E'_\perp \qquad (7.73) \\ &= \frac{e}{B} \frac{1}{\sqrt{1 + k_i^2}} n_e(z) \left| \mathbf{E}_\perp + \mathbf{u}_{n\perp}(z) \times \mathbf{B} \right| \end{aligned}$$

where $\mathbf{u}_{n\perp}(z)$ is the altitude profile of the neutral wind perpendicular to $\mathbf{B}$. When the neutral wind is present, the current density profile by altitude is a product of the altitude variations in the electron density profile as well as the neutral wind. Since the neutral wind in the regions where most of the current is flowing is often strongly variable by height as well as by time, it is not a straightforward problem to identify the current density profile.

In Fig. 7.6 we notice that the current density has a maximum at about 120 km and that it is mainly westward in the morning and eastward in the evening.

The electric field $\mathbf{E}_\perp$ can be measured by several techniques, and if it is measured at a point in the upper part of the ionosphere, it can be assumed constant at all altitudes down to about 90 km. The neutral wind, however, is a very elusive parameter and extremely difficult to measure simultaneously at different altitudes for considerable length of time. This makes the neutral wind extra-complicated to implement in the models, and usually some averaging assumptions have to be made. Furthermore, since the neutral wind is so variable in the region where most of the current is flowing, altitude profiles with reasonable good height resolution are needed.

For the height-dependent heat dissipation rate we now find, when including the effect of the neutral wind:

$$q'_J(z) = \sigma_P(z)(\mathbf{E}'_\perp)^2 = \sigma_P(z)(\mathbf{E}_\perp + \mathbf{u}_n(z) \times \mathbf{B})^2 \qquad (7.74)$$

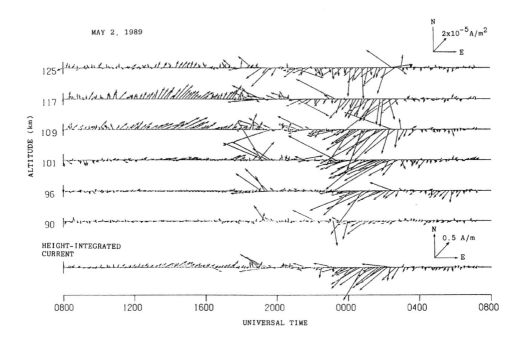

Fig. 7.6. Current density vectors in A/m² observed at 6 different heights in the E-region by EISCAT on May 2, 1989. The bottom panel shows the height-integrated current. The scales are indicated by arrows labelled by proper factors. The northward and eastward directions are also indicated. (From Kamide and Brekke, 1993.)

In the presence of a neutral wind the heat dissipation rate has an altitude profile which can be very different from the altitude profile of the Pedersen conductivity (Fig. 7.7). If the neutral wind induced electric field happens to oppose the electrostatic field at one particular altitude such that:

$$\mathbf{E}_\perp + \mathbf{u}_n(z) \times \mathbf{B} = 0 \tag{7.75}$$

at this altitude, then the heat dissipation will also vanish at the same altitude. In the opposite sense the Joule heating rate may be strongly enhanced at a certain height if $\mathbf{E}_\perp$ and $\mathbf{u}_n(z)$ are enforcing each other.

We notice from Fig. 7.7 that the heating rate per particle, q_n, has its maximum at a greater height than the heating rate itself, and the maximum in the Pedersen conductivity is in general lower than both. In Fig. 7.8 are shown some more examples of heating rate profiles at an auroral zone station (Tromsø).

Sec. 7.5] Height-dependent currents and heating rates 333

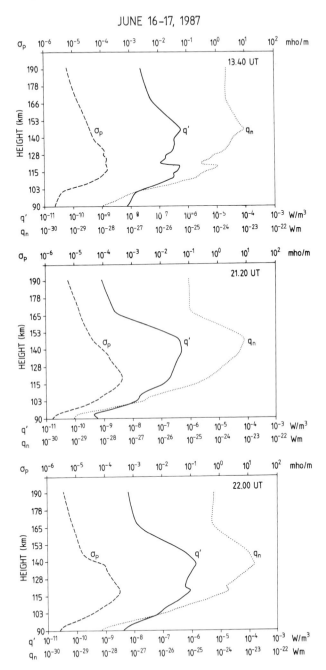

Fig. 7.7. Height profiles of the Pedersen conductivity (σ_P) together with the Joule heating rate (q') and the Joule heating rate per particle (q_n) as obtained by EISCAT measurements at 3 different times on June 16, 1987. (From Brekke et al., 1991.)

Pedersen conductivity, q and q' dated on June 13, 1990 (Ap=119)

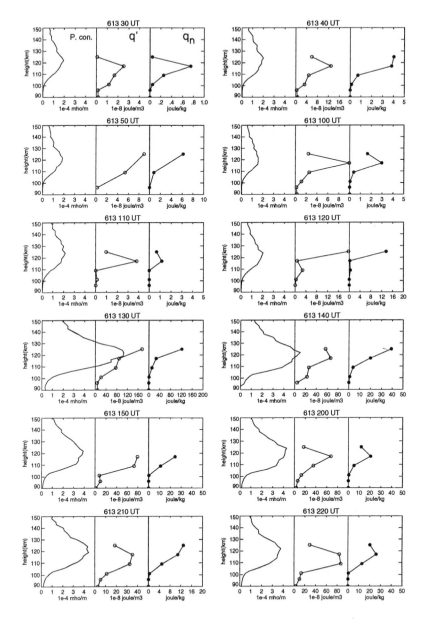

Fig. 7.8. Pedersen conductivity P and Joule heating rate q' profiles as derived by EISCAT at Tromsø on June 13, 1990 in the morning hours. q_n represents the Joule heating rate per particle, as explained in the text. (Courtesy Nozawa, 1995.)

7.6 HEATING DUE TO COLLISIONS

According to (6.82) the rate of energy transfer to the ions due to collisions with the neutrals when it is assumed that $T_i = T_n$, is given by:

$$\dot{Q}_i = \frac{\rho_i \nu_{in}}{m_n + m_i} m_n (\mathbf{v}_i - \mathbf{u}_n)^2 = \frac{m_n m_i}{m_n + m_i} n_i \nu_{in} (\mathbf{v}'_i)^2 \qquad (7.76)$$

where $\mathbf{v}_i$ and $\mathbf{u}_n$ are the ion and neutral velocities observed in the Earth's fixed frame of reference, respectively. $\mathbf{v}'_i$ ($= \mathbf{v} - \mathbf{u}_n$) is the relative velocity between the ions and the neutrals, or equivalently, the ion velocity observed in a frame moving with the neutral wind, m_i and m_n are the ion and neutral masses, respectively, and ν_{in} is the ion–neutral collision frequency. The ion mass density $\rho_i = m_i n_i$ where n_i is the ion particle number density equal to the electron density.

Since our aim here is to obtain the relationship between the rate of mechanical energy input (or frictional interaction) to the ions and the neutrals due to collisions and that of Joule heating, we search for an alternative way of expressing $\dot{Q}_i$. We therefore apply the simplified equation of motions for the ions when neglecting gravity and pressure terms in (7.9)

$$m_i \frac{d\mathbf{v}_i}{dt} = e(\mathbf{E} + \mathbf{v}_i \times \mathbf{B}) - m_i \nu_{in}(\mathbf{v}_i - \mathbf{u}_n) \qquad (7.77)$$

Adding and subtracting the term $e(\mathbf{u}_n \times \mathbf{B})$ on the right-hand side gives:

$$m_i \frac{d\mathbf{v}_i}{dt} = e(\mathbf{E}' + \mathbf{v}' \times \mathbf{B}) - m_i \nu_{in} \mathbf{v}'_i \qquad (7.78)$$

By reorganizing this equation by leaving only terms including $\mathbf{v}'_i$ on the left-hand side, we obtain

$$\Omega_i \mathbf{v}'_i \times \mathbf{B}/B - \nu_{in} \mathbf{v}'_i = \frac{d\mathbf{v}_i}{dt} - \Omega_i \frac{\mathbf{E}'}{B} \qquad (7.79)$$

where the ion gyrofrequency $\Omega_i = eB/m_i$ is introduced. Squaring equation (7.79) and realizing that

$$\mathbf{v}'_i \cdot \mathbf{v}'_i \times \mathbf{B} = 0 \qquad (7.80)$$

we find

$$(\Omega_i^2 + \nu_{in}^2)(\mathbf{v}'_i)^2 = \left(\frac{d\mathbf{v}_i}{dt} - \Omega_i \frac{\mathbf{E}'}{B}\right)^2 \qquad (7.81)$$

The square of the relative velocity $\mathbf{v}'_i$ then becomes:

$$(\mathbf{v}'_i)^2 = \frac{\Omega_i^2}{\Omega_i^2 + \nu_{in}^2} \left(\frac{1}{\Omega_i} \frac{d\mathbf{v}_i}{dt} - \frac{\mathbf{E}'}{B}\right)^2 \qquad (7.82)$$

The rate of mechanical heat exchange can now be rewritten as:

$$\dot{Q}_i = \frac{m_i m_n}{m_n + m_i} n \nu_{in} \frac{\Omega_i^2}{\Omega_i^2 + \nu_{in}^2} \left(\frac{1}{\Omega_i}\frac{d\mathbf{v}_i}{dt} - \frac{\mathbf{E}'}{B}\right)^2 \tag{7.83}$$

In the E-region where dominant portions of the ionospheric current are flowing, we can neglect the collisions between electrons and the neutral gas, and therefore the Pedersen conductivity can be approximately expressed as:

$$\sigma_P = \frac{n \cdot e}{B} \frac{k_i}{1 + k_i^2} = \frac{n \cdot e}{B} \frac{\Omega_i \nu_{in}}{\Omega_i^2 + \nu_{in}^2} \tag{7.84}$$

In terms of σ_P the rate of mechanical heat exchange now becomes:

$$\dot{Q}_i = \frac{m_n}{m_i + m_n} \sigma_P \cdot B^2 \left(\frac{1}{\Omega_i}\frac{d\mathbf{v}_i}{dt} - \frac{\mathbf{E}'}{B}\right)^2 \tag{7.85}$$

Under steady-state conditions ($d\mathbf{v}_i/dt = 0$) we have:

$$\dot{Q}_i^0 = \frac{m_n}{m_n + m_i} \sigma_P (\mathbf{E}')^2 = \frac{m_n}{m_n + m_i} \mathbf{J} \cdot \mathbf{E}' = \frac{m_n}{m_n + m_i} \dot{Q}_j \tag{7.86}$$

where $\dot{Q}_j = \mathbf{J} \cdot \mathbf{E}' = \sigma_P(\mathbf{E}')^2$ is the well-known Joule heating rate due to the ionospheric current density $\mathbf{J}$. Since the steady-state rate of heat input to the neutrals, on the other hand, can be expressed as

$$\dot{Q}_n^0 = \frac{m_i}{m_i + m_n} m_i n_i \nu_{in} (\mathbf{v}')^2 = \frac{m_i}{m_n} \dot{Q}_i^0 \tag{7.87}$$

we have

$$\dot{Q}_i^0 + \dot{Q}_n^0 = \frac{m_n}{m_n + m_i} \dot{Q}_j + \frac{m_i}{m_n + m_i} \dot{Q}_j = \dot{Q}_j \tag{7.88}$$

That is, the Joule heating rate is equal to the total rate of mechanical energy input to the neutrals and the ions due to collisions between the two particle species. It is therefore only for steady-state situations that the Joule heating rate $\dot{Q}_j$ can be set equal to the rate of mechanical energy input due to collisions. If the steady-state condition is not applicable, the result can be quite different, as we soon will see.

Let us return to the equation of motion again and notice that in the ionosphere where the ion velocity is more strongly fluctuating than the neutral velocity:

$$\frac{d\mathbf{v}_i'}{dt} \approx \frac{d\mathbf{v}_i}{dt} \tag{7.89}$$

and that the equation of motion for the ions in a reference frame moving with the neutral wind becomes:

$$m_i \frac{d\mathbf{v}_i'}{dt} = e(\mathbf{E}' + \mathbf{v}_i' \times \mathbf{B}) - m_i \nu_{in} \mathbf{v}_i' \tag{7.90}$$

With reference to (7.89) and introducing (7.90) into (7.85), we can express the rate of mechanical energy input as:

$$\dot{Q}_i = \frac{m_n}{m_n + m_i} \sigma_P \cdot B^2 \left(\frac{1}{\Omega_i} \frac{d\mathbf{v}'_i}{dt} - \frac{\mathbf{E}'}{B}\right)^2 \qquad (7.91)$$

$$= \frac{m_n}{m_n + m_i} \left\{ \sigma_P \cdot (\mathbf{E}')^2 + \sigma_P B^2 \left(\frac{1}{\Omega_i} \frac{d\mathbf{v}'_i}{dt}\right)^2 - \frac{2\sigma_P B}{\Omega_i} \frac{d\mathbf{v}'_i}{dt} \cdot \mathbf{E}' \right\}$$

The rate of energy input now consists of three parts: the first is due to the Joule heating rate, the second is related to inertia motions of the ions in the reference frame of the neutrals, and the last term depends on the relationship between the electric field $\mathbf{E}'$ and the acceleration $d\mathbf{v}'_i/dt$ of the ions in the same reference frame. In the last contribution, if the electric field and the acceleration are in the same direction, then the term represents a maximum negative contribution to $\dot{Q}_i$, while in the opposite sense and when the acceleration is perpendicular to $\mathbf{E}'$ we notice that $\dot{Q}_i$ is larger than the Joule heating rate entering into the ions. When the ions are accelerated up against the electric field, the dynamo action is in play.

7.7 HEATING OF AN OSCILLATING ELECTRIC FIELD

In order to investigate further the relationship between $d\mathbf{v}'_i/dt$ and $\mathbf{E}'$, we will consider the case in which an oscillating electric field is present in the x–y plane perpendicular to $\mathbf{B}$. Expressing this as a circularly oscillating field by:

$$\mathbf{E}' = E_0(\cos\omega t\,\hat{\mathbf{x}} + \sin\omega t\,\hat{\mathbf{y}}) \qquad (7.92)$$

where E_0 is the amplitude, and $\hat{\mathbf{x}}$ and $\hat{\mathbf{y}}$ are the unit vectors for the x- and y-directions, respectively. Owing to this applied electric field, the ion velocity will also oscillate by the same frequency but out of phase with $\mathbf{E}'$. We therefore infer:

$$\mathbf{v}'_i = v_0(\cos(\omega t + \varphi)\hat{\mathbf{x}} + \sin(\omega t + \varphi)\hat{\mathbf{y}}) \qquad (7.93)$$

where φ is the arbitrary phase angle between $\mathbf{E}'$ and $\mathbf{v}'_i$, and v_0 is the amplitude of $\mathbf{v}'_i$.

It is now possible to express the rate of mechanical energy input from (7.91) as:

$$\dot{Q}_i = \frac{m_n}{m_n + m_i}\left\{\sigma_P E_0^2 + \sigma_P\left(\frac{B\omega}{\Omega_i}\right)^2 v_0^2 + \frac{2\sigma_P B\omega}{\Omega_i} v_0 E_0 \sin\varphi\right\} \qquad (7.94)$$

In order to find the relationship between v_0 and E_0 and an expression for the phase angle φ we return to the equation of motion of the ions. Introducing the expressions for $\mathbf{E}'$ and $\mathbf{v}'_i$ and separating the x and y components we find:

$$x:\ v_0(\Omega_i - \omega)\sin(\omega t + \varphi) = \Omega_i \frac{E_0}{B}\cos\omega t - v_0\nu_{in}\cos(\omega t + \varphi) \quad \text{(a)} \quad (7.95)$$

$$y:\ -v_0(\Omega_i - \omega)\cos(\omega t + \varphi) = \Omega_i \frac{E_0}{B}\sin\omega t - v_0\nu_{in}\sin(\omega t + \varphi) \quad \text{(b)}$$

when we introduce $\mathbf{B} = -B\hat{\mathbf{z}}$. Multiplying (7.95a) by $\sin(\omega t + \varphi)$ and (7.95b) by $\cos(\omega t + \varphi)$ and subtracting the resulting equations, we find:

$$v_0 = \frac{\Omega_i}{\Omega_i - \omega} \frac{E_0}{B} \sin \varphi \qquad (7.96)$$

Multiplying (7.95a) by $\cos(\omega t + \varphi)$ and (7.95b) by $\sin(\omega t + \varphi)$ and now adding the resulting equations, we find:

$$\Omega_i \frac{E_0}{B} \cos \varphi - v_0 \nu_{in} = 0 \qquad (7.97)$$

By introducing for v_0 from (7.96) we find:

$$\tan \varphi = \frac{\Omega_i - \omega}{\nu_{in}} \qquad (7.98)$$

We notice that in a steady-state situation (namely, $\omega = 0$), the angle φ reduces to φ_0 as given by:

$$\tan \varphi_0 = \frac{\Omega_i}{\nu_{in}} = \tan \theta_i$$

which is exactly the angle between the ion velocity and the electric field in a reference frame moving with the neutral wind (equation (7.49)). By now introducing the expressions for v_0 and φ from (7.96) and (7.98), respectively, into (7.94), the rate of mechanical energy becomes:

$$\dot{Q}_i = \frac{m_n}{m_n + m_i} \sigma_P E_0^2 \left\{ 1 + \left(\frac{\omega}{\Omega_i - \omega}\right)^2 \sin^2 \varphi + 2 \frac{\omega}{\Omega_i - \omega} \sin^2 \varphi \right\} \qquad (7.99)$$

$$= \frac{m_n}{m_n + m_i} \sigma_P E_0^2 \left(1 + \frac{\Omega_i^2}{\nu_{in}^2}\right) \cos^2 \varphi$$

Applying the relation

$$\cos^2 \varphi = \frac{1}{1 + \tan^2 \varphi}$$

and introducing (7.98) we find:

$$\cos^2 \varphi = \frac{\nu_{in}^2}{\nu_{in}^2 + (\Omega_i - \omega)^2} \qquad (7.100)$$

By inserting this last result into (7.99) we derive:

$$\dot{Q}_i = \frac{m_n}{m_n + m_i} \sigma_P E_0^2 \kappa \qquad (7.101)$$

where

$$\kappa = \frac{\nu_{in}^2 + \Omega_i^2}{\nu_{in}^2 + (\Omega_i - \omega)^2} \qquad (7.102)$$

Sec. 7.7] Heating of an oscillating electric field

We notice that for a given ion gyrofrequency, Ω_i, which is reasonably constant in the ionospheric E-region, κ has a maximum when $\omega = \Omega_i$. This maximum is given by:

$$\kappa_{\max} = 1 + \left(\frac{\Omega_i}{\nu_{in}}\right)^2 = 1 + k_i^2 \qquad (7.103)$$

$\kappa_{\max}$ is presented as function of ν_{in}/Ω_i in Fig. 7.9a showing a strong increase in κ_{max} by a small ratio of ν_{in}/Ω_i. In the ionospheric E-region $\nu_{in} \approx \Omega_i$ at about 125 km or at the peak of the Pedersen conductivity. At this peak $\kappa_{\max}$ is actually twice as large for an electric field oscillating by the ion gyrofrequency as κ would be in steady state ($\omega = 0$). At larger heights in the E-region this factor will be even larger.

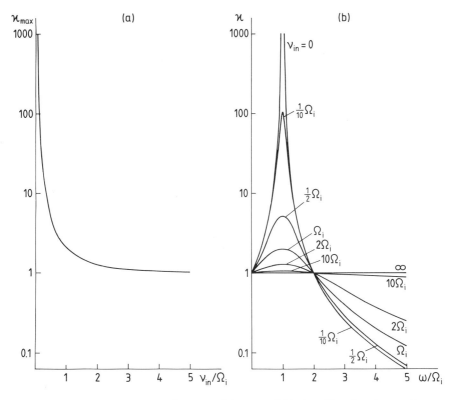

Fig. 7.9. The enhancement factor κ of the rate of frictional heating due to the presence of an electric field oscillating by a frequency ω. (a) The maximum enhancement factor $\kappa_{\max}$ which occurs for $\omega = \Omega_i$ shown for different values of ν_{in}/Ω_i. $\nu_{in} = \Omega_i$ occurs at the peak of the Pedersen conductivity in the auroral ionosphere. (b) The enhancement factor κ for different values of ν_{in} as function of the relative oscillating frequency ω/Ω_i. κ is enhanced for $\omega \leq 2\,\Omega_i$ for all values of ν_{in}. (From Brekke and Kamide, 1996.)

In Fig. 7.9b we have presented κ as function of ω/Ω_i for different values of ν_{in}. We notice that for all collision frequencies there is an enhancement of κ for oscillating frequencies ω between zero and $2\Omega_i$. The enhancement factor is larger the smaller the collision frequency ν_{in}. For $\nu_{in} = 0.1\,\Omega_i$ which corresponds to more than two scale heights above the peak of the Pedersen conductivity the enhancement factor is as high as 100. Even at $\nu_{in} = 0.5\,\Omega_i$ which corresponds to a distance less than a scale height above the peak in the Pedersen conductivity κ is as much as 4 at maximum. Therefore, in a situation of an electric field oscillating by a frequency less than $2\,\Omega_i$ the total rate of frictional heating is larger at all heights than the frictional heating rate in steady state ($\omega = 0$).

From rocket observations of electric field fluctuations in connection with an auroral electrojet it has been demonstrated that considerable power is present in the ionospheric height range between 100 and 120 km in such fluctuating electric fields in the frequency range between 10 and 1000 Hz, within which typical E-region ion gyrofrequencies are included. Such observations are illustrated in Fig. 7.10.

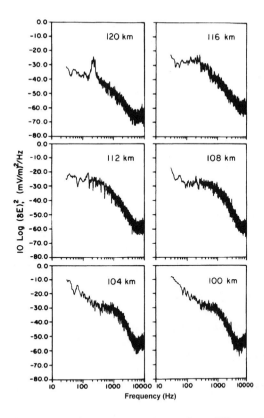

Fig. 7.10. Six electric field spectra observed at different E-region heights from a rocket passing through the auroral electrojet. (From Kelley, 1989.)

7.8 HEIGHT-INTEGRATED CURRENTS

The ionospheric currents perpendicular to the **B**-field in the ionosphere has to be closed either totally in the ionosphere itself or by vertical currents along the magnetic field lines. Since any possible current along the field lines must be of the same order of magnitude as the horizontal current, then:

$$\sigma_P(z) \cdot E_\perp = \sigma_\|(z) E_\| \tag{7.104}$$

Above say 200 km $\sigma_\|(z) > 10^5 \, \sigma_P(z)$, and therefore the parallel **E**-field must be very small compared to the **E**-field perpendicular to **B**,

$$E_\| \approx 10^{-5} \, E_\perp \tag{7.105}$$

The total electric field is for all practical purposes perpendicular to **B**.

For the effects observable on ground it is usually the height-integrated quantities of the currents that account. If these can be derived, they can be compared to the geomagnetic field observations made underneath by ground-based magnetometers.

Since we have shown that the parallel electric field is very small and can be neglected, the height-integrated current density **J** can be derived by height-integrating (7.58) as:

$$\mathbf{J} = \int_{h_1}^{h_2} \mathbf{j}' \cdot dh = \int_{h_1}^{h_2} \sigma_P(h) \mathbf{E}'_\perp(h) dh - \int_{h_1}^{h_2} \sigma_H(h) \mathbf{E}'_\perp(h) \times \mathbf{B} \frac{dh}{B} \tag{7.106}$$

where h_1 and h_2 are the lower and upper integration limits outside which the current is assumed to be zero. A reasonable choice is $h_1 \approx 90$ km and $h_2 \approx 280$ km. Notice that $dh = -dz$. Note that the dimension of **J** is A/m in standard units.

As the perpendicular component of the electric field refers to the magnetic field which makes an inclination angle I with the horizontal plane at the point of reference, $\mathbf{E}'_\perp$ as well as **B** should be properly decomposed into a Cartesian coordinate system where z is in the vertical direction.

At high latitudes where **B** is almost perpendicular to the Earth's surface or equivalently where I is close to 90°, we neglect the inclination of **B** for a moment and assume $\mathbf{B} = B\hat{\mathbf{z}}$ where $\hat{\mathbf{z}}$ is pointing vertically downward. $\hat{\mathbf{x}}$ points positively northward and $\hat{\mathbf{y}}$ completes the orthogonal Cartesian system when counted positive eastward. Then:

$$\begin{aligned}
\mathbf{J}_\perp &= \int_{h_1}^{h_2} \sigma_P(h)(E'_x \hat{\mathbf{x}} + E'_y \hat{\mathbf{y}}) dh - \int_{h_1}^{h_2} \sigma_H(h)(E'_y \hat{\mathbf{x}} - E'_x \hat{\mathbf{y}}) dh \tag{7.107}\\
&= \hat{\mathbf{x}} \int_{h_1}^{h_2} (\sigma_P(h) E'_x - \sigma_H(h) E'_y) dh + \hat{\mathbf{y}} \int_{h_1}^{h_2} (\sigma_P(h) E'_y + \sigma_H(h) E'_x) dh
\end{aligned}$$

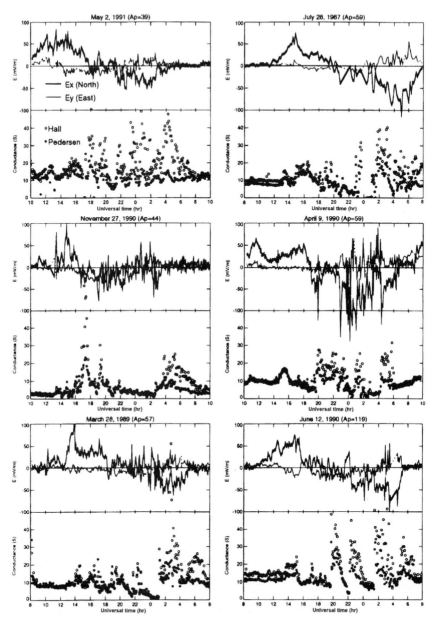

Fig. 7.11. 6 panels showing each in the upper and lower half the time variations of the north and east components of the electric field and the height-integrated Hall and Pedersen conductivities, respectively. The panels are arranged from the upper left corner and downward according to days of increasing A_p index. This index is identified together with the appropriate dates of the EISCAT experiment at the top of each panel. (From Brekke et al., 1995.)

and

$$J_x = \int_{h_1}^{h_2} (\sigma_P(h)E'_x - \sigma_H(h)E'_y)dh \qquad \text{(a)} \qquad (7.108)$$

$$J_y = \int_{h_1}^{h_2} (\sigma_P(h)E'_y + \sigma_H(h)E'_x)dh \qquad \text{(b)}$$

We now neglect the neutral wind, and assume that $\mathbf{E}'_\perp = \mathbf{E}_\perp$. Furthermore, since the magnetic field lines are to be acting like perfect vertical conductors, $\mathbf{E}_\perp$ can be assumed to be independent of height. Therefore:

$$J_x = \Sigma_P E_x - \Sigma_H E_y \qquad \text{(a)} \qquad (7.109)$$

$$J_y = \Sigma_P E_y + \Sigma_H E_x \qquad \text{(b)}$$

where the height-integrated conductivities or conductances are given by:

$$\Sigma_P = \int_{h_1}^{h_2} \sigma_P(h)dh \qquad \text{(a)} \qquad (7.110)$$

$$\Sigma_H = \int_{h_1}^{h_2} \sigma_H(h)dh \qquad \text{(b)}$$

When introducing the expressions for the height-integrated conductivities into (7.106) and assuming that $\mathbf{E}'_\perp(h)$ is constant by altitude we derive for the height-integrated current density:

$$\mathbf{J}_\perp = \Sigma_P \mathbf{E}'_\perp - \Sigma_H \frac{\mathbf{E}'_\perp \times \mathbf{B}}{B} \qquad (7.111)$$

In Fig. 7.11 some typical E-field measurements observed by EISCAT close to Tromsø are presented together with the corresponding conductances. We notice that the field is mainly directed in the northward direction before 2200 UT (2300 LT) and then turns in the southward direction. This turnover occurs at earlier times for more strongly disturbed conditions as indicated by an increasing A_p index. Magnetic midnight in Tromsø is 2220 LT. The largest conductance enhancements, however, are seen around midnight when the field is mainly in the southward direction.

In general, when the field is increased, it is more so in the north–south direction than in the east–west direction and therefore the enhanced field is strongly meridional. When mapped into the equatorial plane, this will correspond to mainly radial fields.

7.9 HEIGHT-INTEGRATED CURRENTS AND MAGNETIC FLUCTUATIONS

If we now assume that the current density we have measured overhead, is representative for the current density in an infinite current sheet, then the magnetic fluctuations we should expect to observe on ground can be derived.

A simple meridional cross-section in the x–z plane is shown in Fig. 7.12. Independent of the position on the ground the contribution to the magnetic field from this current would be according to Ampère's law

$$\Delta B_x = \frac{\mu_0 J_y}{2} \tag{7.112}$$

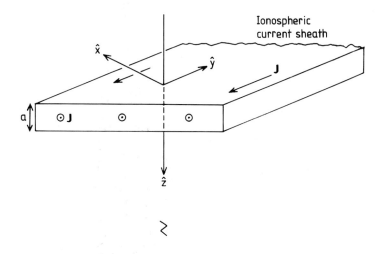

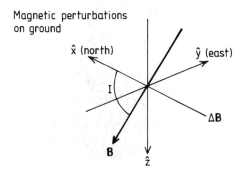

Fig. 7.12. Above: Simplified sketch to illustrate an infinite sheath, carrying a current density **J** (A/m) flowing in the negative y-direction within a thin vertical layer of width a. Below: An illustration of the magnetic field variation $\Delta \mathbf{B}$ observed on ground. It is perpendicular to, and turned anticlockwise with respect to, the current direction.

Sec. 7.9] Height-integrated currents and magnetic fluctuations

If however, the current had an arbitrary direction, then:

$$\Delta \mathbf{B} = \Delta B_x \hat{\mathbf{x}} + \Delta B_y \hat{\mathbf{y}} = \frac{\mu_0}{2} \left(J_y \hat{\mathbf{x}} - J_x \hat{\mathbf{y}} \right) \tag{7.113}$$

For a conducting Earth the factor $1/2$ must be substituted by $3/4$. By therefore comparing the height-integrated overhead current with the simultaneously observed magnetic field fluctuations one could get a feeling for how good the assumptions were. This is done in the bottom panel of Fig. 7.13 where currents derived and integrated overhead Chatanika are compared to the ΔH and ΔD observations at College, Alaska, about 40 km apart. Apart from an arbitrary scaling factor, it is seen that when considering the rather poor time resolution in the current measurements ($\sim$ 30 min) the overall agreement is good as far as the ΔH and J_y components are concerned. The agreement between the ΔD and J_x components are, however, less satisfying, probably illustrating that the latitudinal current is much more homogeneous than the meridional currents in the auroral region. The fact that the D-component is also influenced by field-aligned currents which are not included in the calculations of the overhead currents in Fig. 7.13 may also account for some of this discrepancy.

Also, as we noticed, the two terms in the expression for J_x, the northward component, often subtract each other. The relative uncertainty in subtracting two very similar terms is always greater than if they were added. This may also contribute to some of the discrepancies between J_x and ΔD as observed in Fig. 7.13.

Since the current in the north–south direction is much less than in the east–west current, we will assume as an illustration that

$$J_x = 0 = \Sigma_P E_x - \Sigma_H E_y = 0 \tag{7.114}$$

where we have neglected the contribution to $\mathbf{E}'$ from the neutral wind. We then get the relation:

$$\frac{E_x}{E_y} = \frac{\Sigma_H}{\Sigma_P} \tag{7.115}$$

The ratio between the conductances will then be equal to the ratio between the electric field components. Fig. 7.13 shows an example where this situation is nearly fulfilled since

$$\frac{\Sigma_H}{\Sigma_P} \approx 3 \quad \text{and} \quad \frac{E_x}{E_y} \approx 3$$

The electric field in the meridional direction can therefore be said to be a polarization field. The east–west current now becomes:

$$J_y = \Sigma_P E_y + \Sigma_H \left(\frac{\Sigma_H}{\Sigma_P} E_y \right) = \left(\Sigma_P + \frac{\Sigma_H^2}{\Sigma_P} \right) E_y \tag{7.116}$$

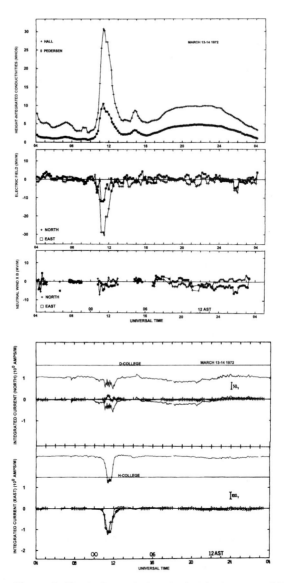

Fig. 7.13. Upper half: Daily variation in height-integrated Pedersen and Hall conductivities, northward and eastward components of the ionospheric electric field and the northward and eastward component of the neutral dynamo electric field observed in the auroral zone at Chatanika for the quiet equinox day of March 13–14, 1972. Lower half: The deduced ionospheric currents from the data in the panels above compared to the observed magnetic field fluctuations observed by a nearby magnetometer. The northward current is compared to the D-component while the eastward component is overlaid on the H-component of the magnetic field, respectively. (From Brekke et al., 1974).

Sec. 7.10]   Equivalent current systems   347

and it is proportional to the east–west electric field component. The proportionality coefficient is an enhanced Pedersen conductance:

$$\Sigma_A = \Sigma_P + \frac{\Sigma_H^2}{\Sigma_P} \qquad (7.117)$$

which can be called the auroral electrojet conductance (the auroral Cowling conductance).

The good agreement observed between the height-integrated current, J_y, derived from the radar observations at Chatanika and the magnetic field fluctuations, ΔH, as observed by a nearby station at College, Alaska, as shown in Fig. 7.13, underlines the significance of the electrojet in the east–west direction. The J_x component is very small in agreement with the polarization condition given above. In the lower part of the upper panel of Fig. 7.13 estimates of the time variations of the $\mathbf{u} \times \mathbf{B}$ term contribution to $\mathbf{E}'$ averaged over the height region of the current are presented. We notice from Fig. 7.13 that the $\mathbf{u} \times \mathbf{B}$ term is insignificant during this event of the electrojet.

7.10 EQUIVALENT CURRENT SYSTEMS

For almost a century magnetic field observations from ground at high latitudes have been used to deduce the so-called equivalent current system. Birkeland first did this to study what he called *the elementary storm* in relation to auroral disturbances which today is called *auroral substorms*. Fig. 7.14 shows two

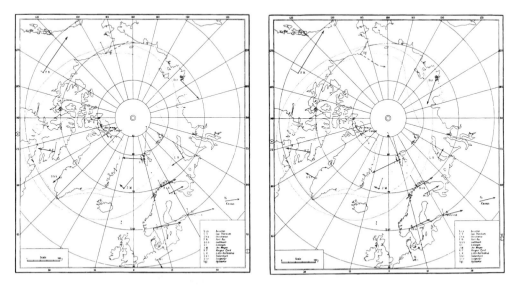

Fig. 7.14. Two of Birkeland's original drawings showing estimates of the equivalent overhead current vectors as derived from the magnetic fluctuation vectors observed on ground at different stations inside and outside the auroral zone. (From Birkeland, 1913.)

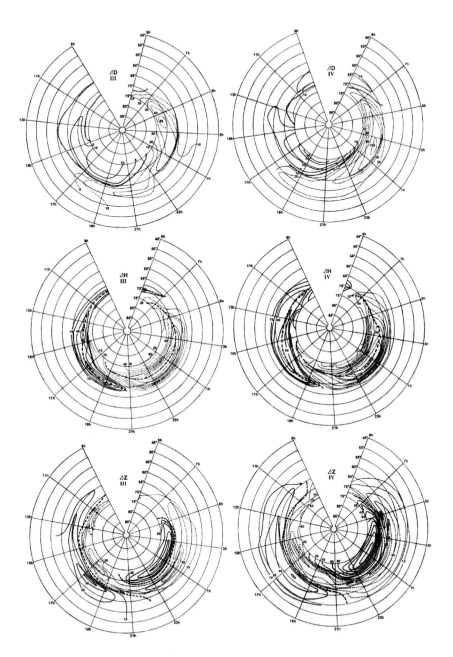

Fig. 7.15. Examples of Harang's original drawings showing contours of constant fluctuations of the magnetic field components (D, H and Z) for two categories of disturbance as drawn on the basis of observations at several stations inside and within the auroral zone. The time indicated is the local time at the 120° geomagnetic meridian. (From Harang, 1946.)

Sec. 7.10] Equivalent current systems 349

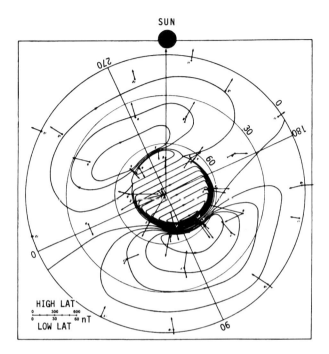

Fig. 7.16. The distribution of magnetic disturbance vectors and equivalent ionospheric current lines which corresponds to the average for many magnetic bays (substorms). (From Silsbee and Vestine, 1942.)

examples of Birkeland's illustrations in which the equivalent current vectors are drawn as arrows at right angles and turned anticlockwise with respect to the observed magnetic field fluctuations observed from different stations in the northern hemisphere.

Harang expanded on Birkeland's method and used high-latitude magnetic records to draw isointensity curves of the magnetic field fluctuations around the polar cap region. He found, as shown in Fig. 7.15, that there is a demarcation line between areas of positive and negative ΔH fluctuations on the nightside. This demarcation line has later been called the Harang discontinuity and is presently an area of great interest. The Harang discontinuity changes position according to the general activity and is the resort for particle precipitation and divergences in the ionospheric electric field, as noticed in relation to Fig. 7.11.

Another classical illustration that has emerged thanks to this method, is due to the work of Silsbee and Vestine (1942) and shown in Fig. 7.16. A large number of magnetic observatories in the northern hemisphere forms the basis for this study of the global current system related to high-latitude magnetic bays or substorms as they usually are called. The horizontal magnetic field fluctuation vector ($\Delta\mathbf{B}$) observed at each station is depicted, and contour lines of constant current densities

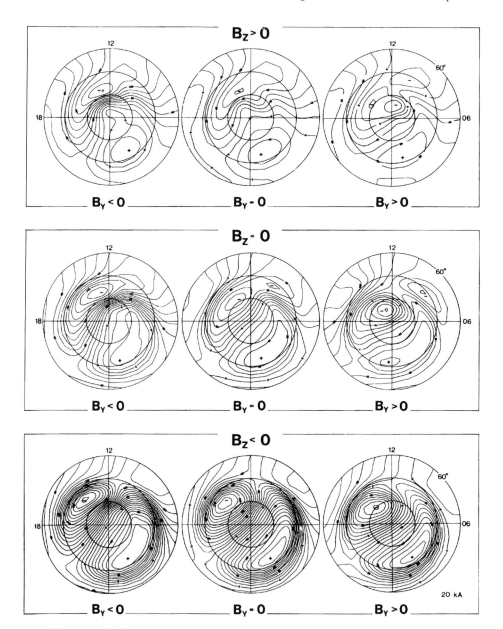

Fig. 7.17. Contours showing the horizontal equivalent currents for different orientations of B_z and B_y of the IMF. A twin vortex system appears to be present above 70° of latitude independent of IMF. The intensity increases and the orientation of the vortices is twisted about 3 hours from the noon–midnight meridian as B_z grows large negative. An additional vortex cell has a tendency to occur on the morning side when B_y is positive. (From Friis–Christensen et al., 1985.)

Equivalent current systems

are drawn perpendicular to these vectors. Two large current cells appear, one rotating in a clockwise sense in the dayside half and the other in an anticlockwise sense in the nightside half, when observed from above the North Pole. These cells are merging together in an intensified westward current, *the auroral electrojet*, above 60° of latitude on the nightside. This electrojet has a maximum close to 0200 local time. The current across the polar cap is fairly uniform and directed from evening towards morning close to the 2000 to 0800 local meridians.

In later years a large number of magnetic observatories have been installed in the Arctic to investigate the high-latitude current systems in more detail, especially those related to auroral substorms. Meridional chains of a multitude of magnetometers have proven extremely useful in this context. Fig. 7.17 demonstrates examples of equivalent current patterns derived from the Greenland chain of magnetometers. The figure is divided up into 9 separate panels depending on the direction of the interplanetary field which has a clear and observable effect on the details in the current pattern system. We recognize for B_z negative a strong similarity with the pattern shown in Fig. 7.16. The nightside current cell is in general penetrating further toward noon than in Fig. 7.16, but the uniform cross-polar cap current from about 2000 to 0800 local time is consistent.

When B_z turns more positive, these current cells have a tendency to deform into a more irregular shape and even break up into several cells. This latter effect appears to be more conspicuous when B_y is positive.

Based on such data it is possible by implementing a model of the ionospheric conductivity to derive the electric potential pattern, the ionospheric currents and the field-aligned currents.

Turning back to the expression for the height-integrated ionospheric current density including the parallel current, but excluding the neutral wind, we have

$$\mathbf{J} = \Sigma_P \mathbf{E}_\perp - \Sigma_H \mathbf{E}_\perp \times \hat{\mathbf{B}} + \Sigma_\| (\mathbf{E} \cdot \hat{\mathbf{B}}) \cdot \hat{\mathbf{B}} \quad (7.118)$$
$$= \mathbf{J}_\perp + \mathbf{J}_\|$$

where $\mathbf{J}_\perp$ and $\mathbf{J}_\|$ are the height-integrated current density vectors perpendicular and along the magnetic field, respectively. Since the divergence of this current must be zero we obtain:

$$\nabla \cdot \mathbf{J} = \nabla_\perp \cdot \mathbf{J}_\perp + \nabla_\| \cdot \mathbf{J}_\| \quad (7.119)$$
$$= \nabla_\perp \cdot \mathbf{J}_\perp + \frac{\partial}{\partial z} J_z = 0$$

where we have split the gradient into a perpendicular and a parallel component. The positive z-axis is counted along the B-field, downward in the northern hemisphere.

Since the current density j_z is given by

$$j_z = \frac{\partial J_z}{\partial z} \quad (7.120)$$

we have

$$j_z = -\nabla_\perp \cdot \mathbf{J}_\perp \quad (7.121)$$

and the vertical current along the **B**-field can be found from the gradient in the perpendicular current. For $\nabla_\perp \cdot \mathbf{J}_\perp$ we have:

$$\nabla_\perp \cdot \mathbf{J}_\perp = \frac{\partial \Sigma_P}{\partial x} E_x + \Sigma_P \frac{\partial E_x}{\partial x} - \frac{\partial \Sigma_H}{\partial x} E_y - \Sigma_H \frac{\partial E_y}{\partial x} \qquad (7.122)$$
$$+ \frac{\partial \Sigma_P}{\partial y} E_y + \Sigma_P \frac{\partial E_y}{\partial y} + \frac{\partial \Sigma_H}{\partial y} E_x + \Sigma_H \frac{\partial E_x}{\partial y}$$

Since $\nabla \times \mathbf{E} = 0$ and

$$\frac{\partial E_x}{\partial y} - \frac{\partial E_y}{\partial x} = 0 \qquad (7.123)$$

the Hall current will not contribute to the divergence of **J** unless Σ_H is anisotropic.

$$j_z = -\left(\frac{\partial \Sigma_P}{\partial x} + \frac{\partial \Sigma_H}{\partial y}\right) E_x - \left(\frac{\partial \Sigma_P}{\partial y} - \frac{\partial \Sigma_H}{\partial x}\right) E_y - \Sigma_P \left(\frac{\partial E_x}{\partial x} + \frac{\partial E_y}{\partial y}\right) (7.124)$$

Even with a constant E-field in the plane perpendicular to the magnetic field we will have a divergence in **J** or a vertical current as long as the conductances are anisotropic. In the neighbourhood of auroral arcs we can expect the conductances to vary quite dramatically in the plane perpendicular to **B**.

7.11 CURRENTS AT THE HARANG DISCONTINUITY

Fig. 7.18 shows a simplified model of the Harang discontinuity which, according to Kamide (1988), serves as a mathematical boundary towards which the electric field converges, poleward ($\mathbf{E}_1$) towards the boundary on the equatorial side, and equatorward ($\mathbf{E}_2$) in the poleward side, respectively. The x-dimension (l) is assumed so small compared to the y dimension (L) of the demarcation region, that $\partial/\partial y$ can be neglected. Also neglecting any east–west (i.e. y) component of the electric field, then the parallel current is given by:

$$j_\parallel = j_z = \frac{\partial \Sigma_P}{\partial x} E_x + \Sigma_P \frac{\partial E_x}{\partial x} \qquad (7.125)$$

Since the Hall current does not contribute to the field-aligned current in this model, the divergence of the Pedersen current is directly connected to the Birkeland current. We notice that:

$$\frac{\partial E_x}{\partial x} = \frac{-E_2 - E_1}{l} = -\frac{2E_1}{l} \qquad (7.126)$$

since $|\mathbf{E}_2| = |\mathbf{E}_1|$. The divergence of **E** then creates a Birkeland current in negative z-direction, i.e. upward from the Harang discontinuity. The Pedersen conductivity which is assumed to have a maximum at the discontinuity as indicated in Fig. 7.18, will contribute by an upward current ($\partial \Sigma_P/\partial x < 0$) on the poleward side and a downward current at the equatorward side. For complete symmetry these currents will exclude each other at the boundary. The currents perpendicular to **B** on both

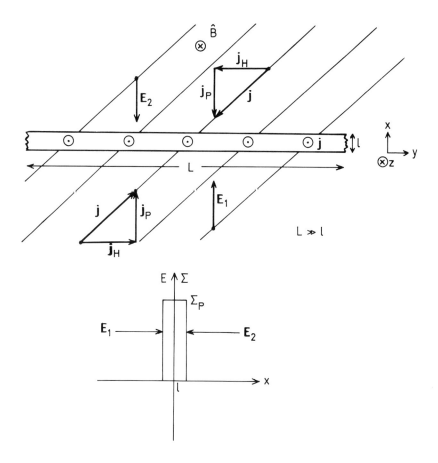

Figure 7.18: A schematic illustration of a simplified model of the Harang discontinuity to show the direction of current flow vectors and E-field converging toward the boundary represented by upward field-aligned currents. An enhancement in Σ_H and Σ_P are introduced across the boundary.

sides of the Harang discontinuity are also shown with an eastward electrojet on the equatorial side and a westward electrojet on the polar side as also indicated by observations.

If, however, the perpendicular current around the Harang discontinuity is divergence-free, it just means that the currents there are part of the current system that closes in the ionosphere, and there is nothing special about the Harang discontinuity. It just marks the area where two current cells diverge. The crucial point is to establish whether there are gradients in the conductivity across the Harang discontinuity. This is still a question not completely solved.

7.12 EQUIVALENT CURRENTS AT DIFFERENT LATITUDES

Let us introduce a Cartesian coordinate system $(\hat{e}_1, \hat{e}_2, \hat{e}_3)$ where

$$\begin{aligned} \mathbf{E}'_\perp &= E_1 \hat{e}_1 & \text{(a)} \\ -\mathbf{E}'_\perp \times \hat{\mathbf{B}}/B &= E_2 \hat{e}_2 & \text{(b)} \\ \mathbf{E}'_\| &= E_3 \hat{e}_3 & \text{(c)} \end{aligned} \qquad (7.127)$$

We then have

$$\begin{pmatrix} J_1 \\ J_2 \\ J_3 \end{pmatrix} = \begin{pmatrix} \Sigma_P & 0 & 0 \\ 0 & +\Sigma_H & 0 \\ 0 & 0 & \Sigma_\| \end{pmatrix} \begin{pmatrix} E_1 \\ E_2 \\ E_3 \end{pmatrix} \qquad (7.128)$$

The conductance tensor has a simple form as long as the electric field is constant by altitude over the height region of integration. In this context the electric field contains the static electric field $\mathbf{E}$ as well as the neutral wind component $\mathbf{u} \times \mathbf{B}$. Since the field changes rather irregularly, a coordinate system with reference to the direction of the total electric field is not really practicable to use from an observational standpoint at the Earth's surface. Let us therefore assume the Earth-fixed Cartesian coordinate system (x, y, z) where x is pointing magnetic northward, y magnetic eastward and z downward towards the Earth's centre. In this coordinate system the magnetic field is given by

$$\mathbf{B} = B(\cos I \, \hat{\mathbf{x}} + \sin I \, \hat{\mathbf{z}}) \qquad (7.129)$$

where I is the inclination angle. The unit vector along $\mathbf{B}$ is:

$$\hat{\mathbf{B}} = \cos I \, \hat{\mathbf{x}} + \sin I \, \hat{\mathbf{z}} \qquad (7.130)$$

For this coordinate system it is assumed that the magnetic dipole axis is antiparallel to the Earth's rotation axis and that the magnetic field is symmetric around this axis. The height-integrated current is now given by (7.118):

$$\mathbf{J} = \Sigma_P \mathbf{E}_\perp - \Sigma_H \mathbf{E} \times \hat{\mathbf{B}} + \Sigma_\| (\mathbf{E} \cdot \hat{\mathbf{B}}) \hat{\mathbf{B}} \qquad (7.131)$$

Here the electric field is assumed constant by height and may include a constant $\mathbf{u} \times \mathbf{B}$ term. In our coordinate system we have:

$$\mathbf{E} = E_x \hat{\mathbf{x}} + E_y \hat{\mathbf{y}} + E_z \hat{\mathbf{z}} \qquad (7.132)$$

$$\begin{aligned} \mathbf{E}_\| &= (\mathbf{E} \cdot \hat{\mathbf{B}}) \hat{\mathbf{B}} & (7.133) \\ &= (E_x \cos^2 I + E_z \sin I \cos I) \hat{\mathbf{x}} + (E_x \cos I \sin I + E_z \sin^2 I) \hat{\mathbf{z}} \end{aligned}$$

$$\mathbf{E} \times \hat{\mathbf{B}} = (E_y \sin I \, \hat{\mathbf{x}} + (E_z \cos I - E_x \sin I) \hat{\mathbf{y}} - E_y \cos I \, \hat{\mathbf{z}} \qquad (7.134)$$

Sec. 7.12] Equivalent currents at different latitudes

$$\mathbf{E}_\perp = \mathbf{E} - \mathbf{E}_\parallel = (E_x \sin^2 I - E_z \sin I \cos I)\hat{\mathbf{x}} + E_y\hat{\mathbf{y}} \qquad (7.135)$$
$$+ (E_z \cos^2 I - E_x \sin I \cos I)\hat{\mathbf{z}}$$

The current density can then be expressed on component form as:

$$J_x = (\Sigma_P \sin^2 I + \Sigma_\parallel \cos^2 I)E_x - \Sigma_H \sin I\, E_y \qquad (7.136)$$
$$+ (\Sigma_\parallel - \Sigma_P)\sin I \cos I\, E_z$$

$$J_y = \Sigma_H \sin I\, E_x + \Sigma_P E_y - \Sigma_H \cos I\, E_z \qquad (7.137)$$

$$J_z = (\Sigma_\parallel - \Sigma_P)\sin I \cos I\, E_x + \Sigma_H \cos I\, E_y \qquad (7.138)$$
$$+ (\Sigma_P \cos^2 I + \Sigma_\parallel \sin^2 I)E_z$$

or on tensor form:

$$\begin{pmatrix} J_x \\ J_y \\ J_z \end{pmatrix} = \begin{pmatrix} \Sigma_P \sin^2 I + \Sigma_\parallel \cos^2 I & -\Sigma_H \sin I & (\Sigma_\parallel - \Sigma_P)\sin I \cos I \\ \Sigma_H \sin I & \Sigma_P & -\Sigma_H \cos I \\ (\Sigma_\parallel - \Sigma_P)\sin I \cos I & \Sigma_H \cos I & \Sigma_P \cos^2 I + \Sigma_\parallel \sin^2 I \end{pmatrix} \begin{pmatrix} E_x \\ E_y \\ E_z \end{pmatrix} \qquad (7.139)$$

At high latitudes where $I \approx 90°$ and $\hat{\mathbf{B}} \approx \hat{\mathbf{z}}$ we have

$$\begin{pmatrix} J_x \\ J_y \\ J_z \end{pmatrix} = \begin{pmatrix} \Sigma_P & -\Sigma_H & 0 \\ \Sigma_H & \Sigma_P & 0 \\ 0 & 0 & \Sigma_\parallel \end{pmatrix} \begin{pmatrix} E_x \\ E_y \\ E_z \end{pmatrix} \qquad (7.140)$$

and the conductance tensor is fairly simple. Especially for $I_x = 0$ we find

$$E_x = \frac{\Sigma_H}{\Sigma_P} E_y \qquad (7.141)$$

and the auroral electrojet is given by

$$J_y = \left(\Sigma_P + \frac{\Sigma_H^2}{\Sigma_P}\right) E_y = \Sigma_A E_y \qquad (7.142)$$

where Σ_A is the auroral electrojet conductance, as already defined. If we, on the other hand, assume that

$$E_\| = \mathbf{E} \cdot \hat{\mathbf{B}} = 0 \tag{7.143}$$

then

$$E_z = -\frac{\cos I}{\sin I} E_x \tag{7.144}$$

and

$$J_x = \Sigma_P E_x - \Sigma_H \sin I\, E_y \tag{7.145}$$

$$J_y = \frac{\Sigma_H}{\sin I} E_x + \Sigma_P E_y \tag{7.146}$$

$$J_z = -\Sigma_P \frac{\cos I}{\sin I} E_x + \Sigma_H \cos I\, E_y \tag{7.147}$$

At high latitudes where $I \approx 90°$, J_z becomes very small, and the vertical current vanishes when $E_\| = 0$. The $E_\| = 0$ constraint, however, enhances and reduces the Hall current in the eastward and westward directions, respectively. The Pedersen current in the horizontal plane is not influenced by this effect. Equations (7.136)–(7.138) are, however, the expressions that should be used in a proper comparison with the X and Y components measured by magnetometers.

At equatorial regions where the inclination angle is zero and $\hat{\mathbf{B}} = \hat{\mathbf{x}}$, we have:

$$\begin{pmatrix} J_x \\ J_y \\ J_z \end{pmatrix} = \begin{pmatrix} \Sigma_\| & 0 & 0 \\ 0 & \Sigma_P & -\Sigma_H \\ 0 & \Sigma_H & \Sigma_P \end{pmatrix} \begin{pmatrix} E_x \\ E_y \\ E_z \end{pmatrix} \tag{7.148}$$

again a simple tensor solution.

Restricting the vertical currents at equator to be negligible, $J_z = 0$, then

$$E_z = -\frac{\Sigma_H}{\Sigma_P} E_y \tag{7.149}$$

and

$$J_y = \left(\Sigma_P + \frac{\Sigma_H^2}{\Sigma_P} \right) E_y = \Sigma_C E_y \tag{7.150}$$

which is the equatorial electrojet and where

$$\Sigma_C = \Sigma_P + \frac{\Sigma_H^2}{\Sigma_P} \tag{7.151}$$

is the Cowling conductivity. The auroral electrojet as well as the equatorial electrojet are therefore results of polarization enforced by the restriction that $J_x = 0$ and $J_z = 0$, respectively.

This last restriction ($J_z = 0$) on the vertical current is usually enforced at low and middle latitudes and not only at equator. This is because the main currents are assumed to be flowing in stratified layers above the ground.

With this more general restriction we find:

$$E_z = -\frac{(\Sigma_\| - \Sigma_P)\sin I \cos I\, E_x + \Sigma_H \cos I\, E_y}{\Sigma_P \cos^2 I + \Sigma_\| \sin^2 I} \tag{7.152}$$

$$J_x = \frac{\Sigma_P \Sigma_\|}{\Sigma_P \cos^2 I + \Sigma_\| \sin^2 I} E_x - \frac{\Sigma_H \Sigma_\| \sin I}{\Sigma_P \cos^2 I + \Sigma_\| \sin^2 I} E_y \tag{7.153}$$

$$J_y = \frac{\Sigma_H \Sigma_\| \sin I}{\Sigma_P \cos^2 I + \Sigma_\| \sin^2 I} E_x + \left(\Sigma_P + \frac{\Sigma_H^2 \cos^2 I}{\Sigma_P \cos^2 I + \Sigma_\| \sin^2 I}\right) E_y \tag{7.154}$$

or on tensor form:

$$\begin{pmatrix} J_x \\ J_y \end{pmatrix} = \begin{pmatrix} \Sigma_{xx} & \Sigma_{xy} \\ \Sigma_{yx} & \Sigma_{yy} \end{pmatrix} \begin{pmatrix} E_x \\ E_y \end{pmatrix} \tag{7.155}$$

where

$$\Sigma_{xx} = \frac{\Sigma_P \Sigma_\|}{\Sigma_P \cos^2 I + \Sigma_\| \sin^2 I} \tag{7.156}$$

$$\Sigma_{xy} = -\Sigma_{yx} = -\frac{\Sigma_H \Sigma_\| \sin I}{\Sigma_P \cos^2 I + \Sigma_\| \sin^2 I} \tag{7.157}$$

$$\Sigma_{yy} = \Sigma_P + \frac{\Sigma_H^2 \cos^2 I}{\Sigma_P \cos^2 I + \Sigma_\| \sin^2 I} \tag{7.158}$$

Since $\Sigma_\| \gg \Sigma_P$ and $\Sigma_\| \gg \Sigma_H$, we find

$$\Sigma_{xx} \approx \frac{\Sigma_P}{\sin^2 I} \tag{7.159}$$

$$\Sigma_{xy} \approx -\Sigma_{yx} \approx -\frac{\Sigma_H}{\sin I} \tag{7.160}$$

$$\Sigma_{yy} \approx \Sigma_P \tag{7.161}$$

except at equator where

$$\Sigma_{xx} = \Sigma_P \qquad (7.162)$$

$$\Sigma_{xy} = -\Sigma_{yx} = 0 \qquad (7.163)$$

$$\Sigma_{yy} = \Sigma_P + \frac{\Sigma_H^2}{\Sigma_P} \qquad (7.164)$$

as before. In the polar regions we find ($I = 90°$):

$$\Sigma_{xx} = \Sigma_P \qquad (7.165)$$

$$\Sigma_{xy} = -\Sigma_{yx} = -\Sigma_H \qquad (7.166)$$

$$\Sigma_{yy} = \Sigma_P \qquad (7.167)$$

As demonstrated in section 5.2 the electron density profile in the ionosphere is strongly dependent on the solar zenith angle χ. Since the ionospheric conductivities at a given height are proportional to the electron density at this height according to (7.55), the conductivity profiles are also strongly dependent on the solar zenith angle. This is the reason why several authors have tried to derive empirical formulas relating the quiet time conductivities to the solar zenith angle. Furthermore the solar irradiation or insulation also varies according to the conditions on the Sun. An often used parameter for monitoring the solar radiation is the radio flux at 2.8 GHz ($\lambda = 10.7$ cm) as discussed in section 1.2. Moen and Brekke (1993) in their empirical model based on electron density measurements by use of the EISCAT system, derived the following formulas for the Hall and Pedersen conductance:

$$\Sigma_H = S_a^{0.53} \left(0.81 \cos \chi + 0.54 \cos^{1/2} \chi\right) \qquad (7.168)$$

$$\Sigma_P = S_a^{0.49} \left(0.34 \cos \chi + 0.93 \cos^{1/2} \chi\right) \qquad (7.169)$$

where S_a is the radio flux index in units of 10^{-22} W m^{-2} Hz^{-1} as presented in section 1.2.

Fig. 7.19a shows some examples of how Σ_P is related to the solar zenith angle for four days in Tromsø when the solar index S_a varies from 74 to 248.

When such models of Σ_P and Σ_H are derived the different values of Σ_{xx}, Σ_{xy}, and Σ_{yy} can be obtained at any observatory around the globe. Fig. 7.19b shows such model calculations of Σ_{xx}, Σ_{xy} and Σ_{yy} as function of latitude between the equator and the poles.

Sec. 7.12] Equivalent currents at different latitudes 359

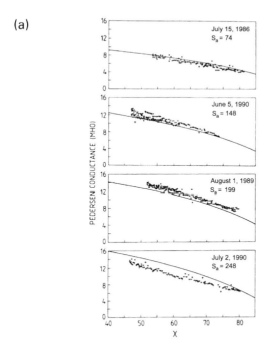

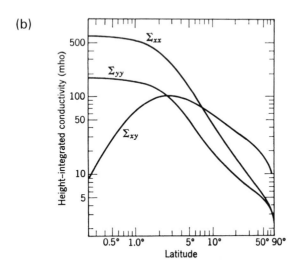

Fig. 7.19. (a) Σ_P versus the solar zenith angle for 4 individual days in Tromsø with different solar flux indices. The observations are compared with an empirical model derived from EISCAT measurements. (After Moen and Brekke, 1993.) (b) Model calculations of the height-integrated conductivities Σ_{xx}, Σ_{yy} and Σ_{xy} as function of magnetic latitude. (From Fejer, 1965.)

7.13 THE S_q CURRENT SYSTEM

Ground-based observed magnetic variations observed at stations around the Earth (Fig. 7.20) can be used to infer electric fields and neutral winds in the upper atmosphere. The magnetic field fluctuations ΔX and ΔY in the x- and y-directions, respectively, are then assumed to be due to overhead equivalent currents in the x,y-plane such that

$$\Delta X \propto J_y \quad \text{and} \quad \Delta Y \propto J_x$$

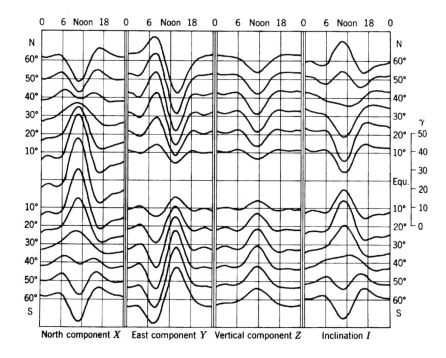

Fig. 7.20. The solar daily variation of the X, Y and Z components and the inclination I at latitudes between 10° and 60° for a sunspot minimum year (1902). (From Chapman and Bartels, 1940.)

Fig. 7.21 illustrates an average global equivalent current system derived as indicated from observations similar to the data shown in Fig. 7.20. This current system is called S_q (solar quiet mean) and is presented as seen from the magnetic equatorial plane at the 00, 06, 12 and 18 local time meridians. The system at the noon meridian is fairly antisymmetric with respect to the two hemispheres. An anticlockwise current vortex is seen in the northern hemisphere with a maximum of about 182×10^3 A at about 30° latitude, and a clockwise vortex is seen in the southern hemisphere with a minimum of about -155×10^3 A close to $-30°$ latitude.

These two vortices are enhancing each other at the magnetic equator, creating an eastward current, the equatorial electrojet. This asymmetry is also maintained at midnight except the rotation directions in the vortices are interchanged and the intensities are about one half compared to daytime. The current patterns in the morning and evening represent the transitions between the night and day, and day and night patterns, respectively.

Since the results presented in Fig. 7.21 were obtained on the basis of magnetic field observations at about 35 stations, all below 60° of latitude, it is clear that the equivalent current system presented at high latitude and in the polar region in Fig. 7.21 does not necessarily represent a true equivalent current system. For this reason special effort has been undertaken to investigate the quiet time high-latitude and polar-cap equivalent current system (S_q^u) (Nagata and Kokubun, 1962). This is particularly important when an understanding of the magnetosphere–ionosphere electrodynamic coupling at high latitude is pursued.

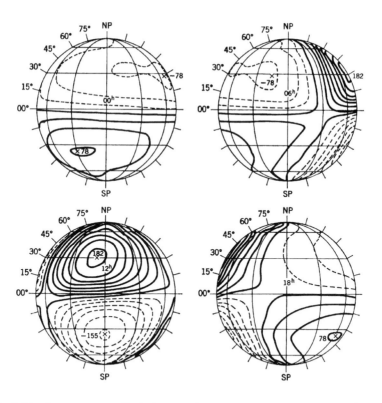

Fig. 7.21. Average mean S_q (solar quiet) current systems during IGY (International Geophysical Year) viewed from the magnetic equator at the 00, 06, 12, and 18 local time meridians. The numbers near the crosses indicate vortex current intensity in 10^3 A. The distance between the current lines corresponds to 2.5×10^3 A. (From Matsushita, 1967.)

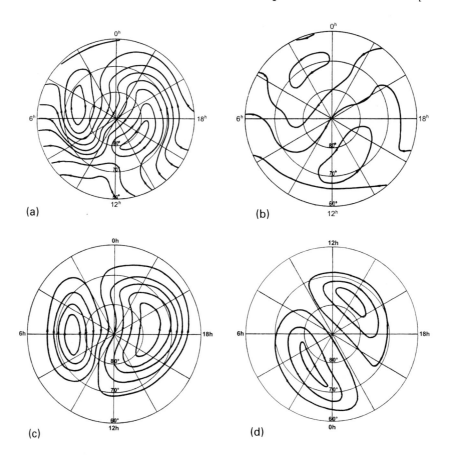

Fig. 7.22. Upper panels: Equivalent current systems of mean daily variations in the north polar region on quiet days: (a) for summer solstice, (b) for winter solstice. The separation between isointensity current lines is 2×10^4 A. Lower panels: Equivalent current systems of the additional geomagnetic variation in the northern polar cap, the S_q^p field: (c) for summer solstice; (d) for winter solstice conditions. The separation between isointensity current lines is 10^4 A. (From Nagata and Kokubun, 1962.)

The solar quiet polar current (S_q^p) patterns, shown in Fig. 7.22 of the northern hemisphere high-latitude region for summer (a) and winter (b) solstice, are based on data from 13 magnetic observatories, all above 60° of latitude. A two-cell pattern is clearly demonstrated in the summer diagram showing the anticlockwise rotation on the dayside and the clockwise rotation on the nightside when observed from above the North Pole. The cross-polar cap current is reasonably uniform and directed close to the 2000 and 0800 local time meridians. The current intensities are, however, more than twice as large in the summer as in the winter season. This current pattern which is based on the same data period as the system shown in

Fig. 7.21, is believed to consist of the extrapolation of the S_q current system at middle latitudes, the so-called S_q^0 system, and a polar cap quiet time system called S_q^p. By subtracting S_q^0 as derived from Fig. 7.21 from the patterns in Fig. 7.22a and b, the special polar cap systems shown in Fig. 7.22c and d, respectively (from Nagata and Kokubun, 1962). These systems are limited to latitudes above 60° and consist of two current cells which depend in magnitude and strength on the season. In the summer season the evening cell which is centred close to the 1800 local meridian is much larger and also stronger than the morning cell which is centred close to the 0600 local meridian. The evening cell is actually extending across the centre of the polar cap and into the morning half. The current flow across the polar cap, however, is fairly uniform and almost parallel to the midnight–noon meridian.

In the winter season (Fig. 7.22d) the two cells are more similar but the axis of symmetry is turned closer to the 2200 to 1000 local meridian.

Traditionally it has been a long discussion about how these different current systems (S_q^0 and S_q^p) are maintained and how they interfere with each other. Since no absolute conclusion can be drawn concerning the closure of the real currents from ground-based magnetic observations alone, it seems like a fertile debate to say at least.

We will show later, however, that progress has been made in this field during the last 20 years due to advanced modelling and new insight from satellite and complementary ground-based observations.

It should be noticed, however, that there is a close resemblance between the S_q^p patterns shown in Fig. 7.22c and d and those current cells at high latitudes shown in Fig. 7.17, in particular for the condition when $B_z = B_y = 0$ which probably represents the average quiet condition rather well, as do the patterns in Fig. 7.22.

It is then believed that the S_q^p current system, contrary to the S_q^0 system, is not driven by the ionospheric dynamo due to solar heat input but rather by external sources related to processes in the magnetosphere and its interaction with IMF. Therefore the S_q^p system is strongly enhanced during geomagnetic disturbances and auroral substorms and does not represent a quiet mean as the index indicates.

By reasonable modelling of Σ_{xx}, Σ_{xy} and Σ_{yy} the electric fields, whether due to a neutral wind or a dynamo process in the magnetosphere, can be solved in a global scale. This method of analysis is based on the atmospheric dynamo theory. Since the electrostatic field $\mathbf{E}_\perp$ of magnetospheric origin or due to polarization charges can be deduced from a potential ϕ, we have

$$\mathbf{E}_\perp = -\nabla \phi \qquad (7.170)$$

If now it is assumed that the neutral wind also is derivable from a potential ψ, then

$$\mathbf{u}_n = -\nabla \psi \qquad (7.171)$$

Ohm's law for the height-integrated current can now be written:

$$\mathbf{J} = -\underline{\underline{\Sigma}} \cdot (\nabla \phi + \nabla \psi \times \mathbf{B}) \qquad (7.172)$$

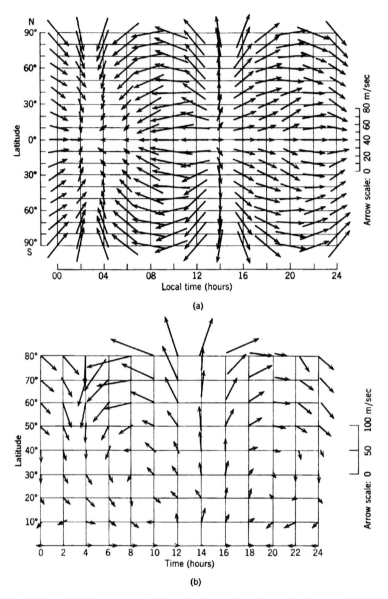

Fig. 7.23. Wind systems deduced from the S_q variations. (a) With curl $\mathbf{u}_n = 0$; (b) with both the rotational and the irrotational part included. (From Maeda and Kato, 1966.)

where $\underline{\underline{\Sigma}}$ is the conductance tensor. In this sense $\mathbf{J}$ must be derivable from a potential χ; so that

$$\nabla \chi = \underline{\underline{\Sigma}} \cdot (\nabla \phi + \nabla \psi \times \mathbf{B}) \tag{7.173}$$

By eliminating ϕ it is possible to solve for ψ when χ is determined by means of the equivalent current system deduced from ground-based magnetic field measurements.

Actually since the neutral wind is not purely irrotational, it must also include a rotational component in terms of a stream function π. The entire velocity field therefore becomes:

$$\mathbf{u}_n = -\nabla\psi - \hat{\mathbf{r}} \times \nabla\pi \qquad (7.174)$$

where $\hat{r}$ is a unit vector radially outward from the Earth. Complete solutions of this problem are shown in Fig. 7.23 in terms of the average ionospheric neutral wind global pattern system. In the upper part of this figure it is assumed that $\pi = 0$ (curl $\mathbf{u} = 0$), while in the lower part the rotational term is included. For the curl-free solution the neutral wind is mainly blowing from the subsolar point at 1400 LT toward the 0300 LT meridian at all latitudes. With the curl indicated, however, a two-cell system is established at both hemispheres and the velocities are increasing toward higher latitudes more strongly than for the curl-free situation.

7.14 THE KAMIDE–RICHMOND–MATSUSHITA (KRM) METHOD

Ground-based observations of the magnetic field fluctuations have, as already mentioned, been widely used to deduce characteristic features of the current sources. Since the magnetic field fluctuations on ground is a compound result of many different sources, externally as well as internally to the Earth, it has been extremely difficult to sort out the contributions from one source with respect to another. Much progress has, however, been achieved in this field throughout the last 20 years as documented by Kamide (1988).

Since our main goal is to find the true current system related to especially the disturbances at high latitudes, it may appear as an insurmountable obstruction when it is realized that the true current system and the Birkeland currents in particular cannot be deduced from the ground-based magnetic measurements only. To understand the current system at high latitudes the Birkeland current is crucial. Thanks to the methods developed lately by the use of advanced computer programmed data handling, new insight in this field has been achieved.

We will follow the outlines of Kamide (1988) in order to present the most advanced methods in this respect.

First of all we will assume that the ground-based magnetic perturbations at high latitudes are due only to ionospheric and field-aligned currents. If one also assumes that the global magnetic perturbation pattern is nearly static with respect to local time, the associated magnetic potential can be directly related to a steady-state current distribution in a thin spherical shell, based on standard procedures from potential theory.

The equivalent current system derived from ground-based magnetometer data is two-dimensional and therefore source-free in the ionosphere. It is assumed to flow in a thin shell at approximately 100 km altitude whose associated magnetic field variations match the external portion of the magnetic variation field $\Delta \mathbf{B}$ observed on the ground. We will here deal with height-integrated quantities and denote this equivalent current $\mathbf{J}_e$. Since this current is toroidal, it can be expressed in terms of an equivalent current function ψ_e such that

$$\mathbf{J}_e = \hat{\mathbf{r}} \times \nabla \psi_e \tag{7.175}$$

where $\hat{\mathbf{r}}$ is a unit vector directed radially outward from the Earth. Furthermore, in the lower atmosphere where negligible currents flow, the magnetic fluctuations must be derivable from a potential χ_B ($\nabla \times \mathbf{B} = 0$) such that

$$\Delta \mathbf{B} = -\nabla \chi_B \tag{7.176}$$

The external potential χ_B must be uniquely related to ψ_e by the mathematical relations inherent in Maxwell's equations (Chapman and Bartels, 1940). Usually ψ_e, the equivalent current function has been estimated by assuming that $\mathbf{J}_e$ is proportional to the horizontal magnetic variation $\Delta \mathbf{B}$ in magnitude but rotated 90° clockwise in direction and by hand-drawn smooth curves of constant intensity on a map parallel to the $\mathbf{J}_e$ vectors (Fig. 7.15).

In order to proceed further from the equivalent current function, additional information about the height-integrated conductivities has to be implemented if an estimate of the parallel currents and a more representative ionospheric current system is to be derived. Detailed information of the conductivities is not generally available, therefore one must fall back on averaged global models. It is here, however, that the basic improvement on this method has been made, because advanced computer programs can, in much greater detail than ever before, account for some of the very local and strongly time fluctuating conductivities. Since the current is intimately linked to the electric field through the conductivities, this method also allows for deduction of the ionospheric electric field pattern.

In order to derive estimates of the horizontal and field-aligned currents on the basis of the equivalent current function ψ_e and a given conductivity model, the following assumptions must be applied (Kamide, 1988):

1. The electric field is electrostatic.

2. Geomagnetic field lines are effectively equipotentials, i.e. there is no parallel electric field.

3. The dynamo effects of ionospheric winds can be neglected.

4. The magnetic contributions of magnetospheric ring currents, magnetopause currents, and magnetotail currents to the equivalent current function can be neglected.

5. Geomagnetic field lines are effectively radial.

The total height-integrated ionospheric current $\mathbf{J}$ must be the sum of the toroidal part $\mathbf{J}_e$ without any source in the ionosphere and a potential part $\mathbf{J}_p$ related to the field-aligned currents and due to a current potential ϕ_c.

$$\mathbf{J}_p = -\nabla \phi_c \tag{7.177}$$

Sec. 7.14]   The KRM method

The total ionospheric current then becomes:

$$\mathbf{J} = \mathbf{J}_e + \mathbf{J}_p = \hat{\mathbf{r}} \times \nabla\psi_e - \nabla\phi_c \tag{7.178}$$

By definition

$$\nabla \cdot \mathbf{J}_e = 0 \tag{7.179}$$

and

$$\nabla \times \mathbf{J}_p = 0 \tag{7.180}$$

The total three-dimensional ionospheric currents $\mathbf{J}$ must be divergence-free as expressed by (7.119)

$$\nabla \cdot \mathbf{J} = \nabla_\perp \cdot \mathbf{J} + \nabla_\parallel \cdot \mathbf{J} = 0 \tag{7.181}$$

where $\nabla_\perp$ and $\nabla_\parallel$ are the gradients perpendicular and parallel to the magnetic field lines, respectively. Since the perpendicular gradient is given by

$$\nabla_\perp \cdot \mathbf{J} = \nabla \cdot \mathbf{J}_p \tag{7.182}$$

in the horizontal plane, and the parallel current density $j_\parallel$

$$j_\parallel = -\nabla_\parallel \cdot \mathbf{J} = \nabla_\perp \cdot \mathbf{J} \tag{7.183}$$

when $j_\parallel$ is counted positive downwards

$$j_\parallel = \nabla \cdot \mathbf{J}_p \tag{7.184}$$

The horizontal current is related to the electric field, $\mathbf{E}$, by Ohm's law which can be expressed as:

$$\mathbf{J} = \underline{\underline{\Sigma}} \cdot \mathbf{E} = \Sigma_P \mathbf{E} + \Sigma_H \mathbf{E} \times \hat{\mathbf{r}} \tag{7.185}$$

where $\hat{\mathbf{r}}$ is pointing radially outwards and antiparallel to $\mathbf{B}$, $\underline{\underline{\Sigma}}$ is the two-dimensional conductivity tensor and Σ_P and Σ_H are the Pedersen and Hall conductances, respectively.

The electrostatic field must be derivable from a potential ϕ and the expression for the height-integrated ionospheric current can be given by:

$$\mathbf{J} = \underline{\underline{\Sigma}} \cdot \mathbf{E} = -\underline{\underline{\Sigma}} \cdot \nabla\phi = -\nabla\psi_e \times \hat{\mathbf{r}} - \nabla\phi_c \tag{7.186}$$

We notice that since $\nabla \times \nabla\phi_c = 0$

$$\nabla \times (\underline{\underline{\Sigma}} \cdot \nabla\phi) = \nabla \times (\nabla\psi_e \times \hat{\mathbf{r}}) \tag{7.187}$$

This gives the relationship between ϕ and ψ_e which shows that when ψ_e is estimated from a global distribution of ground-based magnetic measurements, the electrostatic potential ϕ can be derived, provided that a global conductance model is available.

Again returning to the high latitudes where the height-integrated ionospheric current has a particularly simple form when $\mathbf{B}$ and $\hat{\mathbf{r}}$ are antiparallel, the differential equation between ϕ and ψ_e becomes

$$\nabla \times (\Sigma_P \nabla \phi + \Sigma_H \nabla \phi \times \hat{\mathbf{r}}) = \nabla \times (\nabla \psi_e \times \hat{\mathbf{r}}) \qquad (7.188)$$

If we now assume that the height-integrated conductivities are uniform, then we get:

$$\Sigma_P \cdot \nabla \times \nabla \phi + \Sigma_H \cdot \nabla \times (\nabla \phi \times \hat{\mathbf{r}}) = \nabla \times (\nabla \psi_e \times \hat{\mathbf{r}}) \qquad (7.189)$$

and since $\nabla \times \nabla \phi = 0$, we finally have

$$\Sigma_H \cdot \nabla \times (\nabla \phi \times \hat{\mathbf{r}}) = \nabla \times (\nabla \psi_e \times \hat{\mathbf{r}}) \qquad (7.190)$$

This demonstrates that for a situation with a uniform global conductivity; the electrostatic potential ϕ is simply given by:

$$\phi = \frac{1}{\Sigma_H} \psi_e \qquad (7.191)$$

The global electrostatic potential pattern is then exactly the same as that for the equivalent current function which is derived from ground magnetic perturbations. Furthermore, in this simplified situation only the Hall conductance influences the potential pattern derived from the equivalent current system. This straightforward relationship has been widely used to infer large-scale electric fields from ground magnetic data. This simple relation between ϕ and ψ_e for a global uniform conductivity also demonstrates that for such a situation the Pedersen current does not contribute to the magnetic variations observed on ground (Fukushima, 1969).

We notice, however, that in a more realistic situation, especially at high latitudes where the conductances display large spatial variations at times, a full treatment of the problem including gradients in Σ_H as well as Σ_P has to be achieved. Having obtained the electrostatic potential distribution, the parallel current density is then given by

$$\begin{aligned} j_\parallel &= -\nabla \cdot (\Sigma_P \nabla \phi + \Sigma_H \nabla \phi \times \hat{\mathbf{r}}) \\ &= -\nabla \Sigma_P \cdot \nabla \phi - \Sigma_P \nabla^2 \phi + \nabla \Sigma_H \cdot (\nabla \phi \times \hat{\mathbf{r}}) \end{aligned} \qquad (7.192)$$

Gradients in the Hall as well as the Pedersen conductivity together with the gradients of ϕ will all contribute to the field-aligned currents. In the auroral zone and polar cap, gradients in Σ_P and especially Σ_H are frequently occurring and therefore have a profound influence on the current system at these latitudes.

Fig. 7.24 shows a flow chart diagram of the principles for the so-called KRM (Kamide–Richmond–Matsushita) code (Kamide et al., 1981) used to solve the problem just discussed. A crucial input parameter to the system is a global conductance model, as already mentioned. For derivation of the large-scale current

Sec. 7.14] The KRM method 369

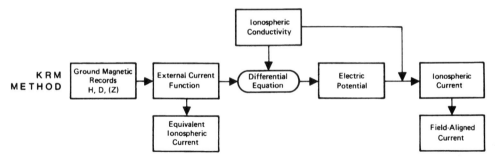

Fig. 7.24. A flow chart diagram showing the principles of the KRM (Kamide–Richmond–Matsushita) method. (From Kamide, 1988.)

and electric potential system the available present-day models are probably sufficient, but for the detailed electrodynamic structure around polar ionospheric substorms the present state of the art probably represents an oversimplification, but is one large step ahead of the earlier models assuming spatially independent conductivities.

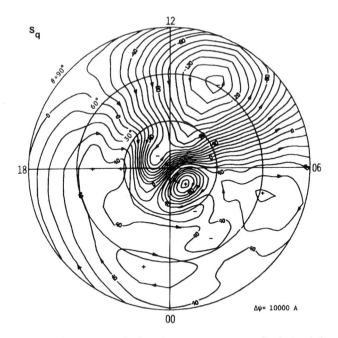

Fig. 7.25. Equivalent external electric current system S_q derived from the northward and vertical components of quiet time geomagnetic fields observed at 40 observatories in summer of 1965. The equivalent current system is presented in polar coordinates with respect to geomagnetic latitude and local time, with a contour interval of 10^4 A. (From Matsushita, 1975.)

Fig. 7.25 shows the equivalent external current system derived for quiet time geomagnetic fields in midsummer calculated from the H and Z perturbations at 40 observatories in the northern hemisphere. The positive and negative signs in the vortex centres represent positive and negative signs in the current function ψ_c, respectively. There are two main features in this diagram: one is a pair of current vortices at low and middle latitudes, and the other is a similar pair at high latitudes. The former characterizes the S_q field at low latitudes (Fig. 7.21). The high-latitude vortices represent the S_q^p (solar quiet polar) field as originally inferred by Nagata and Kokubun (1962).

Fig. 7.26 shows the corresponding estimated electric potential distribution as derived from Fig. 7.25 by implementing an appropriate summertime ionosphere conductance model. The plus and minus signs represent high and low electric potentials, respectively. The electric potential pattern at high latitude is very similar to that of the equivalent current in that both patterns consist of twin vortices whose centra are nearly colocated. This reflects the fact that the quiet summertime polar ionosphere conductivity is fairly uniform. The twin vortices in the electric potential pattern is consistent with the gross features of the convection pattern at high latitudes. An equivalent potential of 21 keV is obtained across the polar cap.

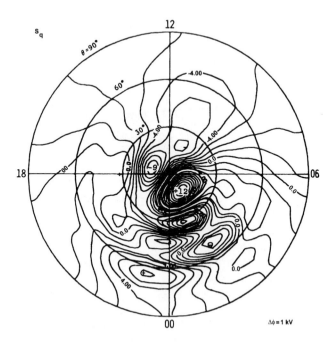

Fig. 7.26. Estimated electric potential distribution (1-kV contour interval) for S_q corresponding to the equivalent current pattern shown in Fig. 7.25 as derived by the KRM method. (From Kamide et al., 1981.)

Sec. 7.14] The KRM method 371

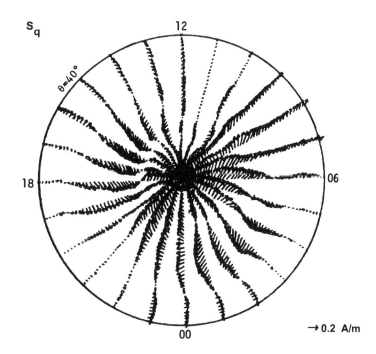

Fig. 7.27. Vector distribution (0.2 A/m vector scale) of the calculated ionospheric currents for S_q corresponding to the data in Fig. 7.25 as derived by the KRM method. (From Kamide et al., 1981.)

Fig. 7.27 illustrates the calculated ionospheric current vectors **J** derived from the same data as used in Figs. 7.25 and 7.26, which can be compared with the equivalent current system in Fig. 7.25. We notice that while the equivalent current at the nightside half in the auroral zone flows predominantly in the east–west direction, the ionospheric current has a considerable northward component in the evening sector and a corresponding southward component in the morning sector. While the westward equivalent current in the morning sector terminates at midnight and turns northward, the ionospheric counterpart penetrates to higher latitudes in the evening side. This reflects the presence of a westward electric field in the high-latitude evening sector, as can be deduced from inspection of Fig. 7.26. Fig. 7.28 illustrates the final step in the KRM procedure, namely the estimated field-aligned currents that represent the divergence of the ionospheric currents shown in Fig. 7.27. Positive and negative signs represent downward or upward currents, respectively. The large-scale current pattern is downward in the poleward half of the auroral belt and upward in the equatorward half on the morning side, the directions are, however, reversed in the corresponding regions on the evening side in agreement with the general picture of these currents as

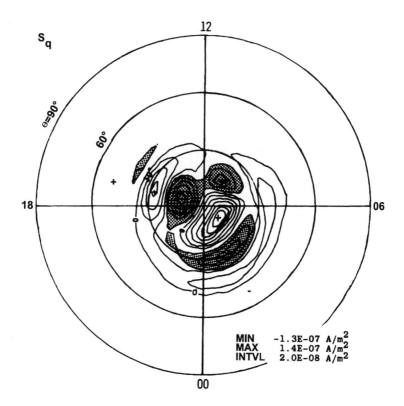

Fig. 7.28. The estimated field-aligned current distribution (2×10^{-8} A/m^2 contour interval) for S_q corresponding to the data presented in Fig. 7.27 as derived by the KRM method. The plus (or minus) sign indicates the downward (or upward) field-aligned current region, and the upward current region is hatched. (From Kamide et al., 1981.)

established from satellite measurements. Probably due to the smoothed conductivity models, however, the parallel currents derived here have a tendency to spread out over larger areas than are obtained from satellite measurements (Kamide et al., 1981).

Figs. 7.29–7.32 show a similar sequence as Figs. 7.25–7.28, but for a maximum phase of an intense substorm in wintertime as obtained on the basis of magnetic observations from 73 observatories in the northern hemisphere. The derived electric potential pattern (Fig 7.30) has been obtained after introducing a suitable model for the enhanced conductivities in the auroral region. This has been represented by a double Gaussian distribution as follows:

$$\Sigma_H = R \cdot \Sigma_P = R \cdot \Sigma_m \exp\left[-\left(\frac{\Delta\theta}{L_\theta}\right)^2\right] \exp\left[-\left(\frac{\Delta\lambda}{L_\lambda}\right)^2\right] \qquad (7.193)$$

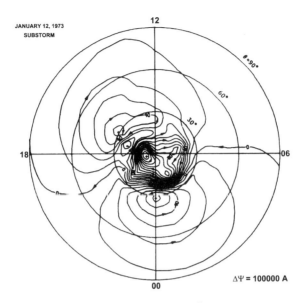

Fig. 7.29. The equivalent current system (10^5 A contour interval) for the maximum phase of an intense substorm at January 12, 1973. The equivalent current function was calculated from H and D component perturbations at 73 northern hemisphere observatories. (From Kamide et al., 1981.)

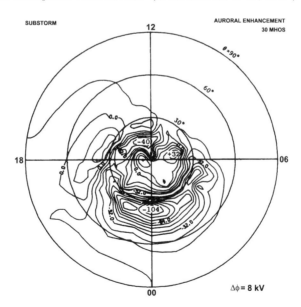

Fig. 7.30. The electric potential distribution (8-kV contour interval) for the substorm shown in Fig. 7.29 as derived by the KRM method. (From Kamide et al., 1981.)

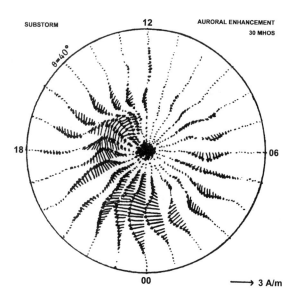

Fig. 7.31. Vector distribution (3 A/m vector scale) of the calculated ionospheric currents corresponding to the data presented in Fig. 7.29 as derived by the KRM method. (From Kamide et al., 1981.)

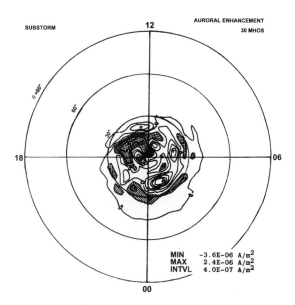

Fig. 7.32. Estimated field-aligned current distribution (4×10^{-7} A/m^2 contour interval) for the substorm shown in Fig. 7.29 as derived by the KRM method. (From Kamide et al., 1981.)

Sec. 7.14] The KRM method 375

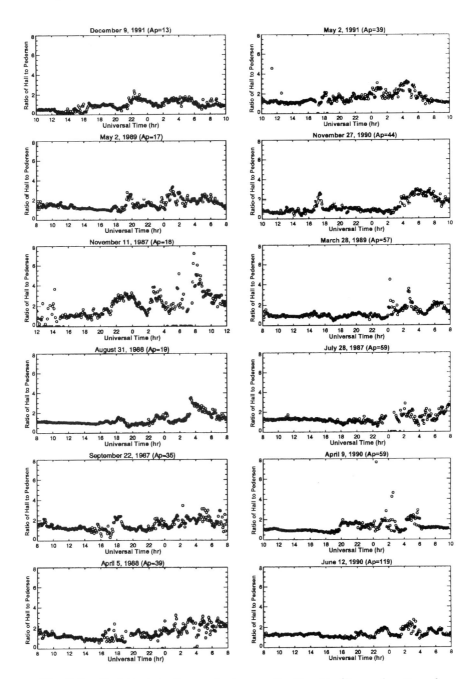

Fig. 7.33. Variations in the conductance ratio $R = \Sigma_H/\Sigma_P$ as function of local time for disturbed days as observed by the EISCAT radar from Tromsø. (Courtesy Nozawa, 1995.)

where $\Delta\theta$ and $\Delta\lambda$ are the angular latitudinal and longitudinal distances from the centre of the enhancement maximum represented by Σ_m, respectively, and L_θ and L_λ are the angular latitudinal and longitudinal width of the enhancement, respectively. R is a factor usually set equal to 2. Clearly the electric potential pattern has a far more detailed structure than the equivalent current function reflecting the complicated relationship between them due to the structured ionospheric conductance. The basic feature of the electric field distribution, however, is a southward field in the late evening and morning hours and a northward field in the afternoon hours at the auroral oval. The intrusion of the southward auroral zone electric field into the evening sector is especially significant and a characteristic feature of auroral substorms.

The ionospheric current as derived from the data presented in Fig. 7.30 is shown in Fig. 7.31. By comparing this current pattern with the equivalent current distribution in Fig. 7.29 the ionospheric westward electrojet is found to be stronger and extending through a broader region than the equivalent current counterpart.

While the gross features of the ionospheric current probably represent the reality rather well, the detailed structure is smeared out. The conductance model only reflects the envelope of fairly strongly structured conductivity pattern, especially is it quite likely that the value of R will vary drastically throughout the precipitation region. Variations in this parameter between 1 and 4 for several hours around midnight is a common feature observed at high latitude (Fig. 7.33).

Finally, the field-aligned currents as derived from the divergence of the current pattern in Fig. 7.29 are illustrated in Fig. 7.32. The current system appears very structured, but the main features still remain; downward currents on the poleward side in the morning sector and on the equatorward side on the evening sector, while the reversed is true for the position of the upward current, i.e. on the equatorward side in the morning sector and on the poleward side in the evening sector.

7.15 POLAR CAP CONDUCTANCE AND CURRENT DISTRIBUTION

In order to demonstrate the intimate relationship between the derived ionospheric current pattern from an electric potential distribution and the choice of the ionospheric conductivity model, we turn to the simplified picture presented in Fig. 7.34.

We assume that there is a constant electric field across the polar cap directed from the high potential area (P) at dawn to the lower potential at dusk (Q). Given the uniform Hall and Pedersen conductances across the polar cap as Σ_H and Σ_P, the ionospheric current across this region is:

$$\mathbf{J} = \Sigma_P \mathbf{E}_P + \Sigma_H \mathbf{B} \times \mathbf{E}_P/B \qquad (7.194)$$

where $\mathbf{E}_P$ is the perpendicular electric field across the polar cap and $\mathbf{B}$ is assumed vertical. Since the vector $\mathbf{B} \times \mathbf{E}_P$ is directed along the midnight–noon meridian, this current will flow from the early morning hours to the early postnoon hours.

Sec. 7.15] Polar cap conductance

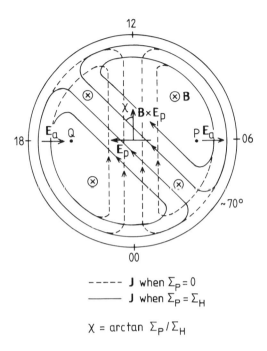

Fig. 7.34. Illustration to show the relationship between the cross-polar-cap current system and the direction of the cross-polar-cap potential when the conductance ratio Σ_P/Σ_H is changed. Dotted lines show the symmetric pattern around the $\mathbf{B} \times \mathbf{E}$ vector when $\Sigma_P = 0$, while the full lines show the pattern tilted at an angle $\chi = \arctan \Sigma_P/\Sigma_H$ when $\Sigma_P \neq 0$ ($\Sigma_P = \Sigma_H$ in the drawing).

Depending on the ratio between the Hall and Pedersen conductances the current will make an angle χ with the $\mathbf{B} \times \mathbf{E}_P$ direction given by

$$\tan \chi = \frac{\Sigma_P}{\Sigma_H} \tag{7.195}$$

For a conductance ratio Σ_P/Σ_H different from zero there will be a finite angle χ. If $\Sigma_P \ll \Sigma_H$, therefore the current is flowing across the polar cap from midnight to noon.

We then notice for the S_q^p summer time current system that the pattern presented in Fig. 7.22c would be in agreement with a dawn–dusk electric field where Σ_P was negligible in relation to Σ_H. In the winter pattern, however, the field direction had to be turned an angle towards the prenoon–premidnight sector, the larger the angle, the higher the ratio of Σ_P/Σ_H. The final results from the KRM analysis also depend on a proper choice of Σ_P/Σ_H.

7.16 MAPPING OF E-FIELDS IN THE IONOSPHERE

We have seen (section 4.4) how electric fields can map downward from the equatorial plane to the ionosphere or vice versa along the magnetic dipole field lines if these are highly conducting. In the ionosphere, however, the conductivity is finite especially perpendicular to the magnetic field lines, and the electric potential between two neighbouring field lines is not necessarily conserved at ionospheric altitudes. Let us adopt a coordinate system where the z-axis is directed vertically upwards from the ground and let x and y be directed southward and eastward, respectively. Let the magnetic field be directed downward so that

$$\hat{\mathbf{B}} = -\hat{\mathbf{z}} \tag{7.196}$$

For an electric field given by:

$$\mathbf{E} = E_x\hat{\mathbf{x}} + E_y\hat{\mathbf{y}} + E_z\hat{\mathbf{z}} \tag{7.197}$$

the ionospheric current density at altitude z will be given by:

$$\begin{aligned}\mathbf{j}(z) &= \sigma_P(E_x\hat{\mathbf{x}} + E_y\hat{\mathbf{y}}) - \sigma_H(E_x\hat{\mathbf{x}} + E_y\hat{\mathbf{y}}) \times \hat{\mathbf{B}} + \sigma_\|(\mathbf{E}\cdot\hat{\mathbf{B}})\hat{\mathbf{B}} \\ &= (\sigma_P E_x + \sigma_H E_y)\hat{\mathbf{x}} + (\sigma_P E_y - \sigma_H E_x)\hat{\mathbf{y}} + \sigma_\| E_z \hat{\mathbf{z}}\end{aligned} \tag{7.198}$$

We now assume that the conductivities are functions of z only. Since the current must be divergence-free we find:

$$\nabla\cdot\mathbf{j} = \sigma_P\frac{\partial E_x}{\partial x} + \sigma_H\frac{\partial E_y}{\partial x} + \sigma_P\frac{\partial E_y}{\partial y} - \sigma_H\frac{\partial E_x}{\partial y} + \frac{\partial}{\partial z}(\sigma_\| E_z) = 0 \tag{7.199}$$

Taking advantage of (7.123) we can reduce (7.199) to read:

$$\nabla\cdot\mathbf{j} = \sigma_P\frac{\partial E_x}{\partial x} + \sigma_P\frac{\partial E_y}{\partial y} + \frac{\partial}{\partial z}(\sigma_\| E_z) = 0 \tag{7.200}$$

Since the electric field must be deduced from a potential ϕ, we obtain:

$$\frac{\partial^2\phi}{\partial x^2} + \frac{\partial^2\phi}{\partial y^2} + \frac{1}{\sigma_P}\frac{\partial}{\partial z}\left(\sigma_\|\frac{\partial\phi}{\partial z}\right) = 0 \tag{7.201}$$

Because of the anisotropy of the conductivity $\sigma_P \neq \sigma_\|$ we will transform the z component by the following substitution (Boström, 1973):

$$dz' = \sqrt{\frac{\sigma_P}{\sigma_\|}}\,dz \tag{7.202}$$

Then we have

$$\frac{\partial}{\partial z} = \sqrt{\frac{\sigma_P}{\sigma_\|}}\frac{\partial}{\partial z'} \tag{7.203}$$

and introducing the geometric mean conductivity

$$\sigma_m = \sqrt{\sigma_P\cdot\sigma_\|} \tag{7.204}$$

Mapping of E-fields in the ionosphere

we have

$$\frac{\partial^2 \phi}{\partial x^2} + \frac{\partial^2 \phi}{\partial y^2} + \frac{1}{\sigma_m} \frac{\partial}{\partial z'} \left(\sigma_m \frac{\partial \phi}{\partial z'} \right) = 0 \qquad (7.205)$$

This equation will be greatly simplified if we assume that σ_m can be represented at least in limited intervals by an exponential function:

$$\sigma_m = \sigma_0 \exp\left(-\frac{z'}{\alpha}\right) \qquad (7.206)$$

where α is a constant scale height. This does not limit the applicability of the solutions to be obtained since any reasonable conductivity profile can be closely approximated by a series of simple exponential functions. The equation now will read:

$$\frac{\partial^2 \phi}{\partial x^2} + \frac{\partial^2 \phi}{\partial y^2} + \frac{\partial^2 \phi}{\partial z'^2} - \frac{1}{\alpha} \frac{\partial \phi}{\partial z'} = 0 \qquad (7.207)$$

which can be solved by the method of separable variables. We will assume that the electrostatic potential is generated at a source level z'_0 and that it has sinusoidal spatial variations with wavelength λ in both the x and y directions. The solution is then given by:

$$\phi = \phi_0 \exp[A(z' - z'_0)] \exp\left[\frac{2\pi i}{\lambda}(x+y)\right] \qquad (7.208)$$

where A is satisfying the equation

$$A^2 - \frac{1}{\alpha} A - \frac{8\pi^2}{\lambda^2} = 0 \qquad (7.209)$$

and

$$A = \frac{1}{2\alpha} \pm \sqrt{\left(\frac{1}{2\alpha}\right)^2 + \frac{8\pi^2}{\lambda^2}} \qquad (7.210)$$

If we let these two roots be denoted by

$$A_1 = \frac{1}{2\alpha} - \sqrt{\left(\frac{1}{2\alpha}\right)^2 + \frac{8\pi^2}{\lambda^2}} \quad (a) \qquad (7.211)$$

$$A_2 = \frac{1}{2\alpha} + \sqrt{\left(\frac{1}{2\alpha}\right)^2 + \frac{8\pi^2}{\lambda^2}} \quad (b)$$

then the complete solution can be expressed as:

$$\phi = \{S_1 \exp[A_1(z' - z'_0)] + S_2 \exp[A_2(z' - z'_0)]\} \times \qquad (7.212)$$
$$\exp\left[\frac{2\pi i}{\lambda}(x+y)\right] \phi_0$$

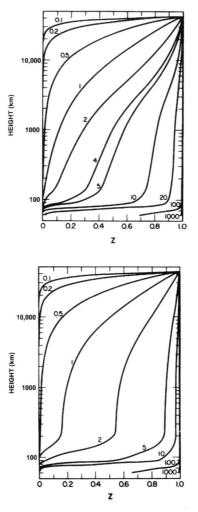

Fig. 7.35. Upper panel: The potential transmission factor (Z) as a function of altitude for daytime conditions. Z is the ratio between the potential at a particular altitude and at the source region. The labels on the curves refer to spatial wavelength (λ) in kilometres. Lower panel: The potential transmission factor (Z) as function of altitude for night-time conditions. (From Reid, 1965.)

where S_1 and S_2 are constants to be determined by the boundary conditions of the problem. In order to make a close approximation to the true physical conditions the ionosphere can be divided below z_0' into a number of slabs in each of which the conductivity σ_m is assumed to obey an exponential law with a fixed scale height α and to be continuous across the interfaces. Examples of such solutions depending on the wavelength λ are shown in Fig. 7.35. We notice that A_1 and A_2 determine how fast the field is damped around z'_0 and that this damping depends on the perpendicular scale length λ as well as the scale height α for the conductivity σ_m. S_1 and S_2 have to be determined from the boundary conditions applied to the problem for each slab (S_{1m} and S_{2m}). For a source region at the top of the ionosphere, $z' - z'_0$, will be negative in the ionosphere and A_2 must be the damping factor while $S_1 = 0$. For a source region in the lower atmosphere, $z' - z'_0$ is positive in the atmosphere and A_1 which is negative, is the damping factor. S_2 is then zero.

From Fig. 7.35 we see that large scale (> 100 km) horizontal electric potentials penetrate from the top of the ionosphere below the E-region almost without damping. For small-scale fields (< 10 km) the damping is severe. It is also noticed that the damping is more severe during night-time conditions, when the conductivities are low. We can also conclude that electric potential variations with characteristic horizontal scale lengths shorter than 1 km can hardly propagate to the E-region from the outer magnetosphere, while those with scale size larger than 10 km easily do.

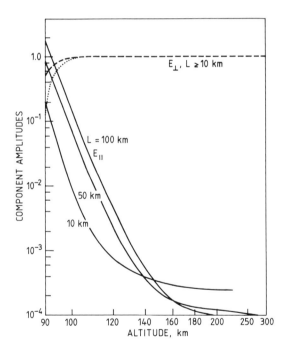

Fig. 7.36. The results of self-consistent mapping of perpendicular ($E_\perp$) and parallel ($E_\parallel$) components of electric fields between 90 and 300 km. The dashed curve ($E_\perp$) and its associated solid curves ($E_\parallel$) for different perpendicular scales (L) are obtained for the boundary conditions $E_\perp$ (300 km) = 1.0 and $E_\perp$ (90 km) = 0.5. The dotted curve shows the change in $E_\perp$ if the lower boundary value is changed to 0.01. (After Park and Dejnakarintra, 1974.)

While the perpendicular electric field component is damped when penetrating to the ionosphere from above, the parallel component is enhanced (Fig. 7.36). A parallel component of the order of 10^{-4} of magnitude of the perpendicular component at 300 km, after penetrating to 90 km will be equal to or larger than the perpendicular component, depending on the scale size. This is partly a consequence of the isotropic conductivities at these lower heights.

For time-varying electric fields the damping relation will be strongly modified and more strongly so for higher-frequency variations. An electric field of characteristic length about 1 km will penetrate to the E-region without much damping for frequencies about 10^{-6} Hz. For 10^{-3}, however, the amplitude is much reduced (Boström, 1974).

Fields of very large scales (100 km) will, as we can see from Fig. 7.35, penetrate well below the ionosphere and all the way down to 20–40 km where they can be observed from balloons. It is not unlikely that electric fields of extreme horizontal dimensions originating in the magnetosphere can affect the vertical electric field at the surface of the Earth which is essential for the current circuit associated with the weather conditions.

7.17 POLARIZATION FIELDS AROUND AN AURORAL ARC

An auroral arc is associated with an area of enhanced electron precipitation and ionospheric conductivities. Associated with the arc there is usually a strong auroral electrojet due to strong electric fields. How the current system is formed and closed around an auroral arc is not well understood since local conductivity enhancements, polarization fields and feedback current loops are all intermingled in a dynamically complicated manner.

We will, however, study a simple model of the current configuration within and close to an auroral arc in order to get some better understanding of the physics taking place (Boström, 1973).

Let the magnetic field be directed vertically downward, the x-axis northward and the y-axis eastward (Fig. 7.37). We will now assume that an electric field with a large (> 100 km) horizontal scale size (in the x–y plane) is penetrating the ionosphere from the magnetosphere above. The field ($\mathbf{E}_0 = E_0 \hat{\mathbf{y}}$) is pointing eastward and will be constant in the whole region also including the auroral arc. The auroral arc is represented by a narrow strip ($\sim$ 10 km) of enhanced electron densities between $x = a$ and $x = -a$. Inside the arc the Hall and Pedersen conductances are given by Σ_H^A and Σ_P^A, respectively. The arc is therefore associated with a thin, uniform layer in the vertical direction. Outside the arc the conductances are small but constant and we denote them by Σ_H^C and Σ_P^C, respectively.

Since the electric field is curl-free it must be continuous across the interface between the background and the arc. This electric field

$$\mathbf{E}_0 = E_0 \hat{\mathbf{y}} \tag{7.213}$$

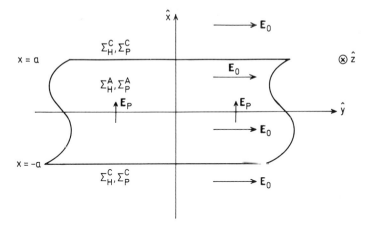

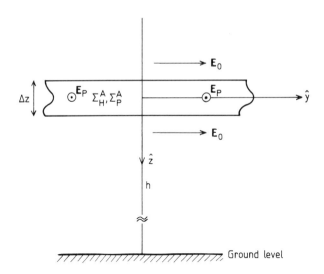

Fig. 7.37. An idealized presentation of the polarization effects associated with an auroral arc at height h above the ground. The arc is assumed to be represented by a slab of enhanced conductivity Σ_P^A and Σ_H^A, within a narrow region in the x-direction ($+a$, $-a$), thickness Δz in the z-direction, and an infinite extension in the y-direction. An applied electric field $\mathbf{E}_0$ in the positive y-direction gives rise to a Hall current which polarizes the arc such that a polarization field $\mathbf{E}_p$ is established in the positive x-direction. Altogether this leads to an enhanced current in the positive y-direction. (After Boström, 1973.)

will drive a Pedersen current

$$\mathbf{J}_{P_0}^A = \Sigma_P^A E_0 \hat{\mathbf{y}} \tag{7.214}$$

within the arc and

$$\mathbf{J}_{P_0}^C = \Sigma_P^C E_0 \hat{\mathbf{y}} \tag{7.215}$$

outside the arc.

This electric field will also drive a Hall current

$$\mathbf{J}_{H_0}^A = -\Sigma_H^A E_0 \hat{\mathbf{x}} \tag{7.216}$$

within the arc, and

$$\mathbf{J}_{H_0}^C = -\Sigma_H^C E_0 \hat{\mathbf{x}} \tag{7.217}$$

outside the arc. This Hall current inside the arc will be larger than the current perpendicular to the arc on the outside of the arc, and charges will accumulate at the interface ($x = \pm a$); positive charges at $x = -a$ and negative charges at $x = +a$. This will again lead to a polarization field perpendicular to the arc in the x–y plane directed parallel to the x-axis. It can be expressed as:

$$\mathbf{E}_p = +E_p \hat{\mathbf{x}} \tag{7.218}$$

This will drive a Pedersen current within the arc given by

$$\mathbf{J}_{P_p}^A = +\Sigma_P^A E_p \hat{\mathbf{x}} \tag{7.219}$$

and a Hall current within the arc given by

$$\mathbf{J}_{H_p}^A = +\Sigma_H^A E_p \hat{\mathbf{y}} \tag{7.220}$$

The current perpendicular to the arc must be continuous as long as charges are not allowed to move along the magnetic field lines. Then

$$J_x^A \hat{\mathbf{x}} = -\Sigma_H^A E_0 \hat{\mathbf{x}} + \Sigma_P^A E_p \hat{\mathbf{x}} \tag{7.221}$$

$$J_x^C \hat{\mathbf{x}} = \Sigma_H^C E_0 \hat{\mathbf{x}} \tag{7.222}$$

If we now assume that $\Sigma_H^C \ll \Sigma_H^A$ such that the current perpendicular to the arc outside the arc can be neglected, then

$$J_x^A = 0 = -\Sigma_H^A E_0 + \Sigma_P^A E_p \tag{7.223}$$

and

$$E_p = +\frac{\Sigma_H^A}{\Sigma_P^A} E_0 \tag{7.224}$$

The polarization field is enhanced with respect to the applied field E_0 by the conductivity ratio Σ_H^A/Σ_P^A. The current along the arc inside the arc is now given by

$$J_y^A \hat{\mathbf{y}} = \Sigma_P^A E_0 \hat{\mathbf{y}} + \Sigma_H^A E_p \hat{\mathbf{y}} \tag{7.225}$$

and finally:

$$J_y^A = \left[\Sigma_P^A + \frac{(\Sigma_H^A)^2}{\Sigma_P^A}\right] E_0 = \Sigma_P^A \left[1 + \left(\frac{\Sigma_H^A}{\Sigma_P^A}\right)^2\right] E_0 \qquad (7.226)$$

and the auroral electrojet is flowing along the applied **E**-field $E_0 \hat{y}$, but it is enhanced with respect to the Pedersen current. The enhancement factor within the brackets can be much larger than 1. The conductivity is now the auroral electrojet conductance

$$\Sigma_A^A = \Sigma_P^A + \frac{(\Sigma_H^A)^2}{\Sigma_P} \qquad (7.227)$$

due to the polarization effect of the Hall current.

If not neglecting completely the current perpendicular to the arc outside the arc but still assuming that the current along the magnetic field line is vanishing, the polarization field could be expressed as:

$$E_p = +\frac{\Sigma_H^A - \Sigma_H^C}{\Sigma_P^A} E_0 \qquad (7.228)$$

and the auroral electrojet will be:

$$J_y^A = \left(\Sigma_P^A + \frac{(\Sigma_H^A)^2}{\Sigma_P^A} - \frac{\Sigma_H^A \Sigma_H^C}{\Sigma_P^A}\right) E_0 \qquad (7.229)$$

The electrojet would then be reduced compared to the ideal case when the current perpendicular to the arc outside the arc is zero.

Finally, allowing for a current vertically upward on the southern edge of the arc would give:

$$J_y^A = \left[\Sigma_P^A + \frac{(\Sigma_H^A)^2}{\Sigma_P^A} - \frac{\Sigma_H^A \Sigma_H^C}{\Sigma_P^A}\right] E_0 - \frac{\Sigma_H^A}{\Sigma_P^A} I_\parallel \qquad (7.230)$$

and the electrojet is even more reduced. By studying the conductivities and electric fields outside and inside the arc, one can deduce to what extent the polarization actually takes place. Marklund (1984) has shown several examples of auroral arcs embedded in different background **E**-field.

7.18 REFERENCES

Birkeland, K. (1913) *The Norwegian Aurora Polaris Expedition 1902–3*, Vols. I and II, H. Aschehoug, Christiania, Norway.

Blixt, M. (1995) Private communication.

Boström, R. (1973) in *Cosmical Geophysics,* Egeland, A., Holter, Ø. and Omholt, A. (Eds.), pp. 181–192, Universitetsforlaget, Oslo.

Boström, R. (1974) in *Magnetospheric Physics*, McCormac, B. M. (Ed.), pp. 45–59, D. Reidel, Dordrecht, The Netherlands.

Brekke, A. and Hall, C. (1988) *Ann. Geophysicae,* **6**, 361–376.
Brekke, A. and Kamide, Y. (1996) *J. Atmos. Terr. Phys.,* **58**, 139–144.
Brekke, A., Doupnik, J. R. and Banks, P. M. (1974) *J. Geophys. Res.,* **79**, 3773–3790.
Brekke, A., Moen, J. and Hall, C. (1991) *J. Geomag. Geoelectr.,* **43**, 441–465.
Brekke, A., Nozawa, S. and Sato, M. (1995) *J. Geomag. Geoelectr.,* **47**, 889–900.
Chapman, S. and Bartels, J. (1940) *Geomagnetism,* Vol. 1, Oxford, at the Clarendon Press.
Fejer, J. A. (1965) *Rev. Geophys.,* **2**, 275.
Friis–Christensen, E., Kamide, Y., Richmond, A. D. and Matsushita, S. (1985) *J. Geophys. Res.,* **90**, 1325–1338.
Fukushima, N. (1969) *Rep. Ionosp. Space Res. Japan,* **32**, 219–227.
Harang, L. (1946) *Geophys. Publ.,* **16**, No. 12.
Kamide, Y. (1988) *Electrodynamic Processes in the Earth's Ionosphere and Magnetosphere,* Kyoto Sangyo University Press, Kyoto, Japan.
Kamide, Y. and Brekke, A. (1993) *Geophys. Res. Lett.,* **20**, 309–312.
Kamide, Y., Richmond, A. D. and Matsushita, S. (1981) *J. Geophys. Res.,* **86**, 801–813.
Kelley, M.C. (1989) *The Earth's Ionosphere. Plasma Physics and Electrodynamics,* Academic Press, New York.
Kunitake, M. and Schlegel, K. (1991) *Ann. Geophysicae,* **9**, 143–155.
Maeda, K. and Kato, S. (1966) *Space Sci. Rev.,* **5**, 57.
Marklund, G. (1984) *Planet. Space Sci.,* **32**, 193–211.
Matsushita, S. (1967) in *Physics of Geomagnetic Phenomena,* Matsushita, S. and Campbell, W. H. (Eds.), Academic Press, New York.
Matsushita, S. (1975) *Phys. Earth Planet. Inter.,* **10**, 299–312.
Moen, J. and Brekke, A. (1993) *Geophys. Res. Lett.,* **20**, 971–974.
Nagata, T. and Kokubun, S. (1962) *Rep. Ionosph. Space Res. Japan,* **16**, 256–274.
Nozawa, S. (1995) Private communication.
Park, C. G. and Dejnakarintra, M. (1974) Paper presented at 5th International Conference on Atmospheric Electricity, Garmisch–Partenkirchen, West Germany, Sept. 2–7, 1974.
Reid, G. C. (1965) *Radio Sci.,* **69D**, 827–837.
Silsbee, H. C. and Vestine, E. H. (1942) *Terr. Mag.,* **47**, 195–208.

7.19 EXERCISES

1. Let the altitude-dependent collision frequency $\nu_{in}(z)$ be expressed as:

$$\nu_{in}(z) = \nu_{in}^0 \exp(-z/H)$$

where ν_{in}^0 and H are constants, and assume that Ω_i and ω in (7.99) are constants. Perform the integration of $\dot{Q}_i$ given by (7.99) from the ground to infinity.

2. Derive (7.54).

3. Indicate the current vector at the 4 different panels in Fig. 7.2.

4. Derive (7.205).

5. Let $\mathbf{E}_0$ in Fig. 7.37 be given by

   $$\mathbf{E}_0 = -E_0 \hat{\mathbf{y}}$$

   and derive the auroral electrojet.

6. An infinitely long line current, I, is situated at a height h above the ground. The current is directed in the eastward (y) direction (Fig. 7.38).

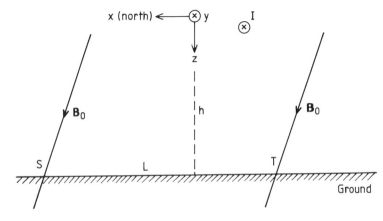

Fig. 7.38.

At two stations S and T separated a distance $L > h$ along the x-axis and perpendicular to the current the variations in the Earth's magnetic field are observed. We will assume that the background quiet time magnetic field $\mathbf{B}_0$ is the same at both stations.

Let the observed components of the magnetic field fluctuations at a certain moment be $(\Delta x_S, 0, \Delta z_S)$ and $(\Delta x_T, 0, \Delta z_T)$ at station S and T, respectively.

(a) Find an expression for the height h of the current when the distance L is known.

Let $L = 200$ km and the observed components:

$\Delta x_T = 1.42 \times 10^{-6}$ tesla $\qquad \Delta z_T = 0.59 \times 10^{-6}$ tesla

$\Delta x_S = 0.65 \times 10^{-6}$ tesla $\qquad \Delta z_S = -0.81 \times 10^{-6}$ tesla

(b) Determine the height and the strength of the current ($\mu_0 = 4\pi \times 10^{-7}$ H/m).

The current is now moving in the x direction with a constant velocity v.

(c) Show that the constant velocity v can be expressed by:

$$v = h \frac{d}{dt}\left(\frac{\Delta z_T}{\Delta x_T}\right)$$

8
The magnetosphere

8.1 THE MAGNETIC FIELD AWAY FROM THE EARTH

Thanks to the satellites it is now possible to observe the geomagnetic field at great distances from the Earth's surface.

Imaging moving out from the Earth in the direction towards the Sun in the equatorial plane measuring the magnetic field, as illustrated in Fig. 8.1 the field will decrease by distance as r^{-3} until about 5 R_e from the centre of the Earth. Between 5 R_e and 8 R_e the field is stronger than expected from the pure dipole field. At about 8 R_e a discontinuity is observed, and further outwards the field is irregular, turbulent and weak. This is confirmed by the observations of the direction of the field which changes abruptly between 8 and 10 R_e. The discontinuity is identified as the magnetopause. On the earthward side is the magnetosphere and on the sunward side is the interplanetary space and solar wind. This would, however, be the situation only at the noon meridian in the equatorial plane. If the satellite moved outwards from the Earth in another meridional plane, it would find the position of the magnetopause at increasing distances from the Earth the larger angle the satellite trajectory is making with the Sun–Earth line (Fig. 8.2). The magnetopause would be found to take the approximate form of a parabola in the equatorial plane with the Earth close to its focal point.

Outside the magnetopause there is a new discontinuity where the turbulence in the field comes to a halt. This is called the shock wave, and outside the shock wave the field is very weak but more regular. The shock wave forms another parabola-shaped curve in the equatorial plane. For other planes making an angle to the equatorial plane the magnetopause and the shock wave will form similar conic sections, and in total the magnetopause makes an approximate paraboloid surface around the Sun–Earth axis with its focal point close to the Earth (Fig. 8.3). The volume inside this paraboloid will be the magnetosphere. The volume between the two paraboloids representing the shock wave and the magnetopause is often called the magnetosheath.

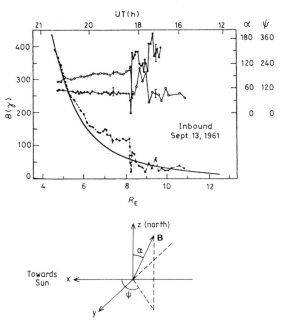

Fig. 8.1. Observations of the Earth's magnetic field strength and direction in interplanetary space as derived from Explorer XII satellite. (From Cahill and Amazeen, 1963.) Also shown at the bottom of the figure is the coordinate system to which the data above are referred. The z-axis is perpendicular to the ecliptic plane and positive northward. The x-axis is directed toward the Sun and the y-axis completes the Cartesian coordinate system. ψ is the azimuthal angle of **B** in the equatorial plane counted positive toward east. α is the poloidal angle observed positive from the positive z-axis.

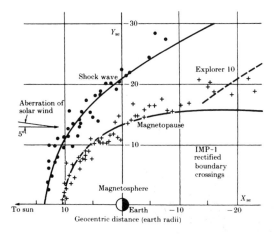

Fig. 8.2. Comparison between observed and predicted positions of the magnetopause and shock wave in the ecliptic or xy-plane. (From Ness et al., 1964.)

Sec. 8.1] The magnetic field away from the Earth

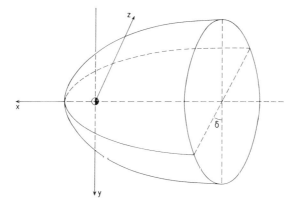

Fig. 8.3. A schematic figure illustrating the shape of the magnetospheric paraboloid. Two cross-sections are shown, one in the ecliptic plane and one making an angle δ with the ecliptic plane. The x, y and z axes are also indicated in agreement with Fig. 8.1.

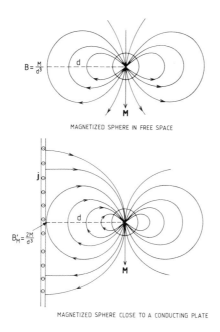

Fig. 8.4. Upper diagram: A perfect magnetic dipole field of a magnetized sphere with magnetic moment **M** in free space. Lower diagram: The magnetic field configuration when a dipole with magnetic moment **M** is placed at a distance d from a conducting plane. The magnetic field will be perpendicular to the plane as if an infinite current of density **j** is flowing in the plane. The magnetic field strength at distance d with a conducting plane present will be twice as great as if the plane was not there.

We also notice from Fig. 8.1 that just inside the magnetopause the field is nearly twice as large as the field expected from a dipole field, and immediately outside the discontinuity the field is close to zero.

The same situation would arise if we had a magnetized sphere close to a conducting surface (Fig. 8.4b). If the magnetized sphere with a dipole moment M were situated in free space (Fig. 8.4a), the magnetic field at a distance d would be

$$B = \frac{\mu_0}{4\pi} \frac{M}{d^3} \tag{8.1}$$

If instead a conducting plate was placed vertically on the magnetic equatorial plane at a distance d, then currents would arise in the plane which would prevent the field from entering behind the plane. This would then enhance the magnetic field on the sphere's side of the plane, and the field strength at the plane would be:

$$B' = \frac{\mu_0}{4\pi} \frac{2M}{d^3} \tag{8.2}$$

It is also interesting to notice that from the outside of the plane where $B = 0$, it would appear as if there were two magnetic dipole moments M in the centre of the sphere with opposite polarities (see Fig. 8.5a). From the sphere's side of the plane it would appear that a magnetic dipole moment parallel to M was situated at a distance equal to $2d$ from the centre of the sphere. Either point of view gives the same result.

The conducting plate must, according to these considerations, be associated with a current. We notice that due to the direction of M this current must be vertical out of the paper (Fig. 8.4b) which will be from dawn towards dusk of the dayside of the Earth.

Let the magnetic field at the surface of the Earth due to the Earth's dipole be B_0 (Fig. 8.5b) and that at a distance d be B_d. Then:

$$B_d = B_0 \left(\frac{R_e}{d}\right)^3 = \frac{\mu_0}{4\pi} \frac{M_0}{d^3} \tag{8.3}$$

Due to the current sheet, however, the field at d will be

$$B'_d = 2B_d = 2B_0 \left(\frac{R_e}{d}\right)^3 \tag{8.4}$$

Due to the same current sheet the field at the Earth's surface will also be enhanced and

$$B'_0 = \frac{\mu_0}{4\pi} \left(\frac{M_0}{R_e^3} + \frac{M_0}{(2d - R_e)^3}\right) \tag{8.5}$$

where the last term is due to the image dipole (Fig. 8.5b). The increase in B_0 at the Earth's surface is then

$$\Delta B_0 = B'_0 - B_0 = \frac{\mu_0}{4\pi} \frac{M_0}{(2d - R_e)^3} \tag{8.6}$$

The magnetic field away from the Earth

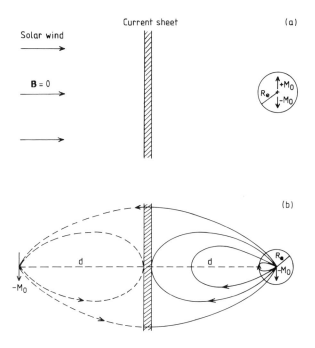

Fig. 8.5. (a) The magnetic field of the Earth that would be produced if the magnetosphere was an infinite conducting plane in front of the Earth represented by a perfect dipole moment at its centre $(-M_0)$. Currents that flow in the plane will produce a field that, seen on the sunward side of the sheet, appears to be due to a dipole $(+M_0)$ at the Earth's centre, and the total field is zero. (b) On the earthward side, however, the field appears to originate at an image dipole $(-M_0)$ at the Earth's distance d on the opposite side of the plane. The field on the sunward side of the plane is zero while on the earthward side close to the plane, it is twice the magnitude it would have without the conducting plane. (After Ratcliffe, 1972.)

If we take $d = 8\ R_e$ as seen in Fig. 8.1, we find that for $M_0 = 7.91 \times 10^{22}$ A m^2

$$\Delta B_0 = 9.06 \times 10^{-9} \text{ tesla} = 9\ \gamma \tag{8.7}$$

which is a negligible increase compared to $B_0 = 30319\ \gamma$. The magnetic field strength $B'_d = 117\ \gamma$ at $d = 8\ R_e$. If the magnetosphere had occurred at $10\ R_e$ as it does in Fig. 8.2, the increase in B_0 at the Earth's surface would be $4.5\ \gamma$, or almost exactly half the magnitude for the sheet at $8\ R_e$. The field strength at d, however, would be close to $60\ \gamma$ in this case. It is believed that under extraordinary conditions in the solar wind the magnetopause has been observed at $4\ R_e$. The expected increase in B_0 at the Earth's surface is then close to $90\ \gamma$.

If the solar wind did not exist, the outer ionosphere would consist of a proton–electron plasma with a concentration determined by diffusion along a magnetic field which could be represented rather well with a dipole field. The ionospheric plasma at the upper strata of the Earth's atmosphere would then merge with the interplanetary plasma at some great distance.

Let us for simplicity assume that the Earth with its magnetic field is embedded in a stationary interplanetary plasma with the same density and temperature as the solar wind plasma. The border between the inner region which we could call the magnetosphere and the outer region which would be controlled by the interplanetary plasma, would be determined by the balance between the magnetic field pressure of the Earth's magnetic field and the thermal pressure of the interplanetary plasma. When we neglect any magnetic field in the interplanetary plasma and the plasma density of the ionospheric plasma at this distance, then we have:

$$\frac{(B_m)^2}{2\mu_0} = n\kappa(T_e + T_p) \tag{8.8}$$

where B_m is the magnetic field strength at this border, n is the plasma density, equal for electrons and protons, and T_e and T_p are the electron and proton temperatures respectively, in the interplanetary medium. The magnetic field is then given by

$$B_m = [2\mu_0 n\kappa(T_e + T_p)]^{1/2} \tag{8.9}$$

and since

$$B_m = B_0 \left(\frac{R_e}{r_m}\right)^3 \tag{8.10}$$

the standoff distance r_m to the border between the two regimes is given by:

$$r_m = R_e \cdot B_0^{1/3} \cdot [2\mu_0 n\kappa(T_e + T_p)]^{-1/6} \tag{8.11}$$

For $n = 5 \times 10^6$ m^{-3}, $T_e = 10^5$ K and $T_p = 4 \times 10^4$ K we find for $B_0 = 3.0 \times 10^{-5}$ tesla

$$r_m = 53 \, R_e \tag{8.12}$$

The extreme limit of a magnetosphere in static equilibrium with an interplanetary thermal plasma would therefore be close to 50 R_e.

Because the solar wind is a highly conducting plasma, the situation is strongly altered. This high conductivity prevents the Earth's magnetic field from penetrating the solar wind plasma, and the solar wind therefore pushes the magnetic field back towards the Earth downstream the wind. Currents are induced in the solar wind plasma as it moves by the Earth, and these currents give rise to new magnetic fields which add to the Earth's dipole field and increase the field at the Earth's surface as if it has been compressed. Moreover, the forces exerted by the Earth's magnetic field on the induced currents cause the wind to alter direction so that it avoids a region surrounding the Earth. The boundary

Sec. 8.1] The magnetic field away from the Earth

between the region where the solar wind continues to flow and the region from which it is excluded, is then the magnetopause.

In the case of a head-on solar wind this pause must be situated at a distance where the kinetic pressure of the solar wind must be balanced by the pressure (energy density) of the magnetic field. Let the number density, molecular mass and velocity of the solar wind be denoted by n, m and v, respectively, then the kinetic pressure in the solar wind assuming elastic collision and reflections of the particles at the magnetopause is:

$$p_k = 2nmv^2 \tag{8.13}$$

while the pressure of the magnetic field is given by

$$p_m = \frac{B_{mp}^2}{2\mu_0} \tag{8.14}$$

where B_{mp} is the magnetic field at the magnetopause. We now know that

$$B_{mp} = 2B_0 \left(\frac{R_e}{r_{mp}^0}\right)^3 \tag{8.15}$$

where r_{mp}^0 is the distance to the magnetopause along the midday meridian from the Earth's centre. For these two pressures being equal at the magnetopause, we will have:

$$2mnv^2 = \frac{2B_0^2}{\mu_0} \left(\frac{R_e}{r_{mp}^0}\right)^6 \tag{8.16}$$

and the distance to the magnetopause is given by:

$$\frac{r_{mp}^0}{R_e} = \left(\frac{B_0^2}{\mu_0 mnv^2}\right)^{1/6} \tag{8.17}$$

For $v = 400$ km/s, $m = m_p$, $n = 5 \times 10^6$ m^{-3} and $B_0 = 3 \times 10^{-5}$ tesla we find:

$$r_{mp}^0 = 9 \cdot R_e \tag{8.18}$$

For $v = 340$ km/s and 570 km/s, r_{mp}^0 is equal to 10 R_e and 8 R_e respectively. We notice that for a point P' at the magnetopause at an angle ϕ from the Sun–Earth line, the distance to the magnetopause from the centre of the Earth is (see Fig. 8.6)

$$r'_{mp} = r_{mp}^0 \cdot (\cos \phi)^{-1/3} \tag{8.19}$$

which to a first approximation gives the position of the magnetopause for $\phi < 90°$. For $\phi = 45°$ we see that ($v = 400$ km/s):

$$r'_{mp} = 10.61\, R_e \tag{8.20}$$

On the other hand, if the solar wind is making an angle ψ with the Sun–Earth line, the stagnation point will not be at the midday meridian but at the angle ψ. The distance to the magnetopause at the midday meridian will then be

$$r''_{mp} = r_{mp}^0 (\cos \psi)^{-1/3} \tag{8.21}$$

Since the direction of the wind changes by time, so will the stagnation point.

As we have already mentioned, the discontinuity in the magnetic field at the magnetopause could be explained by a current sheet perpendicular to the equatorial plane in the dawn–dusk direction. We will consider this current sheet more closely.

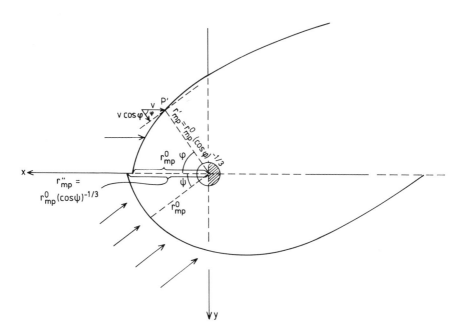

Fig. 8.6. A diagram showing the geometry in the ecliptic plane of the magnetopause at the dayside of the Earth for different directions of the solar wind. The upper half shows the solar wind in a head-on situation at the midday meridian. The lower part shows the solar wind blowing at an angle ψ with respect to the midday meridian.

Let us take an oversimplified model (Fig. 8.7) of protons and electrons with equal concentrations and velocities moving in the negative x-direction towards a uniform magnetic field B perpendicular to their velocities in the positive z-direction. The volume with the magnetic field is divided from the volume without such a field by a plane YY', perpendicular to the xy-plane. A current sheet must then flow in the direction of YY' to produce a magnetic field on the positive x-side which is equal but opposite to the field B_p that would be there without the current sheet present. On the negative x-side the sheet must produce a field that is exactly equal to B_p and the resultant field on this side will be

$$B = 2B_p \qquad (8.22)$$

in the positive z-direction.

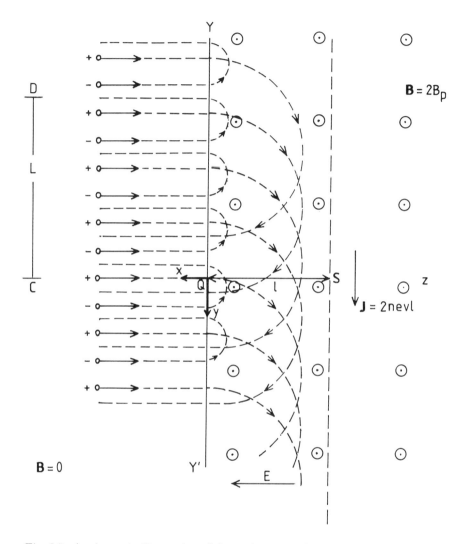

Fig. 8.7. A schematic illustration of the trajectories of protons and electrons when they hit the magnetosheath, here represented by a conducting plane YY' in the yz-plane. On the sunward side of the conducting plane the magnetic field is assumed zero and the velocities of the protons and electrons are **v** and perpendicular to the plane. The protons due to their heavier mass will move in a semicircle with diameter L on the earthward side where the field strength B is twice the value represented by the Earth's dipole moment. The electrons, however, will move in smaller circles, and the motion will be in the opposite direction due to the Lorentz force. The protons will reach to a distance of $l = L/2$ earthward of current boundary YY'. The result is a current density in the positive y-direction given by $J = 2nevl$.

For an infinite current sheet producing a uniform magnetic field strength of B_p the current density has to be:

$$J = \frac{2B_p}{\mu_0} \tag{8.23}$$

If we now consider a charged particle that has crossed the boundary and entered the area with a magnetic field of magnitude $2B_p$, it will follow a circular path with a radius given by

$$r = \frac{mv}{2eB_p} \tag{8.24}$$

and after moving around half a circle, it will return up against the stream again. Since the electrons have much smaller masses than the protons, they will be reflected relatively shortly after they have penetrated into the magnetic field. Let us now consider an area perpendicular to the xy-plane at QS. All particles of mass m that originally are found in the solar wind with velocity v between D and C, will cross the area QS. Again, because the protons have much larger mass, they move in the B-field with larger radius of curvature. Therefore there will be far more protons than electrons penetrating the area QS. The flux of charges (protons) through the area QS per unit time is then

$$\phi_p = 2nevl \tag{8.25}$$

Note that the distance DC is equal to $2l$. This is then the current density J that creates the extra B-field at the magnetopause, and therefore this current can be expressed as:

$$J = 2nevl = \frac{nmv^2}{B_p} \tag{8.26}$$

where we have set r in (8.24) equal to l. Equating J here derived with the one given by (8.23), we find that:

$$\frac{2B_p}{\mu_0} = \frac{nmv^2}{B_p} \tag{8.27}$$

and

$$\frac{(2B_p)^2}{\mu_0} = 2nmv^2 \tag{8.28}$$

which gives us the balance between the magnetic and the kinetic pressures as we originally started out with (equation (8.16)).

We find that the thickness of the current sheet from this simple model is

$$l = \frac{mv}{2eB_p} \tag{8.29}$$

For $m = m_p$, $v = 400$ km/s and $B_p = 117\ \gamma$ at $r_p = 8\ R_e$, the thickness of the sheet becomes

$$l = 17874\text{ m} \approx 18\text{ km}$$

an extremely thin layer on the astronomical scale.

For $r_p = 10\ R_e$ the magnetosheath current would be close to 35 km thick but still a thin layer.

Because the electrons penetrate to a much shorter distance into the magnetic field before they are reflected, there must be an electric field in the positive x-direction which will modify the simple model just described. This E-field will be associated with an $\mathbf{E} \times \mathbf{B}$ motion of the plasma which actually will be opposite to the proton motion but parallel to the electron motion.

8.2 THE MAGNETIC TAIL

If we now consider the magnetic field on the antisunward side of the Earth, we find the magnetopause to be situated at longer and longer distances away from Earth the closer we get to the midnight meridian. Observations with the IMP–1 satellite (Fig. 8.8) of the direction of the field indicate that it is directed in the antisunward direction under the equatorial plane, for $z < 0$. A model of the magnetic field configuration in the midday–midnight meridian plane is shown in Fig. 8.9. At distances larger than 20 R_e on the nightside the field is nearly parallel or antiparallel to the Sun–Earth line depending on whether the field line is below or above the equatorial plane. To sustain such a magnetic field configuration there must be a current in the equatorial plane directed from dawn to dusk across the plane (Fig. 8.10a). For a field strength of about 10 γ as seen by IMP–1 at about 30 R_e, the current density in the sheet must be:

$$J = \frac{2B}{\mu_0} \approx 1.5 \times 10^{-2} \text{ A/m} \tag{8.30}$$

For a current sheet between C and D in Fig. 8.9 approximately 10 R_e wide, the total current will be close to 10^6 A. There is therefore a current in the equatorial plane of the order of 10^6 A to sustain the field configuration close to the equatorial plane.

The magnetic field at the far nightside of the Earth is split into two parts as already mentioned, one above the equatorial plane where the field points towards the Earth and one below this plane where the field points away from the Earth (Fig. 8.10b). This region of the magnetosphere inside the magnetopause is called the magnetospheric tail. We notice that all field lines stretching out in the far tail below the equatorial plane emerge from the southern polar cap, and similarly the field lines directed parallel towards the Sun above the equatorial plane all penetrate the northern polar cap.

It is possible to make an estimate of the cross-section of the magnetic tail when assuming a steady state where the solar wind kinetic and magnetic pressure are

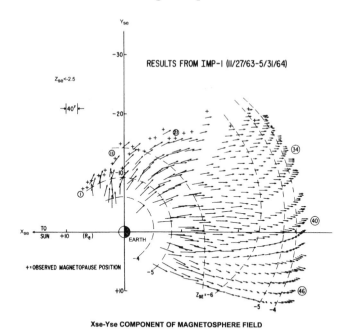

Fig. 8.8. Magnetic field vectors in planes parallel to and below the ecliptic plane. The Z_{SE} value indicates the distance in Earth radii below the ecliptic plane. The arrows represent the magnetic field vectors. The scale of the vectors is indicated by the bar. (From Ness, 1965.)

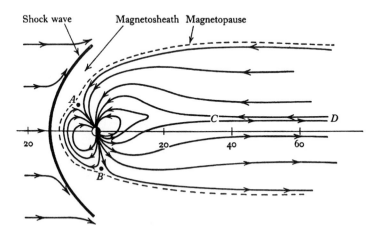

Fig. 8.9. A midday–midnight cross-section of the magnetosphere with the magnetic field directions indicated. Also illustrated is the solar wind and the presence of the magnetosheath, magnetopause and the magnetoshock wave. There are neutral points at A and B and a neutral sheath between C and D. The approximate geocentric distances in Earth radii are also indicated. (From Hess, 1967.)

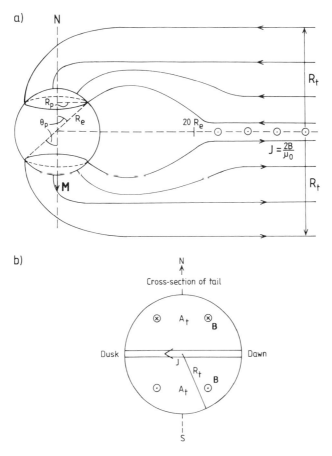

Fig. 8.10. Simplified diagrams showing the geometry of the magnetic field lines: (a) along the day–night meridian in the tail and their connection to the polar cap. The tail is assumed to be an approximate cylinder of radius R_t, (b) and divided by a cross-tail current sheath. The distance from the Earth to the cylindrical tail is set to 20 R_e. The colatitude of the polar cap boundary is indicated by θ_p. The cross-section of the tail shows a cross-tail current from dawn to dusk of density $J = 2B/\mu_0$, where B is the magnetic induction above and below this sheath current.

balanced at the magnetopause. The plasma density in the magnetotail is small except for the regions of the cross-tail current, the plasma sheet, and the kinetic pressure in the tail is therefore small. The pressure balance equation can then approximately be written as

$$\frac{B_t^2}{2\mu_0} = \frac{B_s^2}{2\mu_0} + n_s \kappa (T_p + T_e)_s \tag{8.31}$$

and
$$B_t = [B_s^2 + 2\mu_0 n_s \kappa (T_p + T_e)_s]^{1/2} \tag{8.32}$$

where B_t is the magnetic field in the tail at the magnetopause. B_s and n_s are the magnetic field and particle density in the solar wind, while $(T_e + T_p)_s$ is the sum of the electron and proton temperatures in the solar wind plasma. For solar wind temperatures, $T_e = 10^6$ K and $T_p = 4 \times 10^5$ K, a magnetic field $B_s = 3$ nanotesla and a particle density of $n_s = 5 \times 10^6$ m^{-3}, we find that the magnetic field in the magnetopause will be:

$B_t = 6$ nanotesla

Assuming then that the magnetic field in the upper half of the magnetic tail is uniform and that all field lines merge in the polar cap region within a colatitude θ_p (Fig. 8.10), we find that for magnetic flux conservation

$$\phi = B_p \cdot A_p = B_t \cdot A_t \tag{8.33}$$

B_p is the polar cap magnetic field strength assumed uniform and vertical on the polar cap area A_p. B_t is the magnetotail magnetic field strength assumed uniform perpendicular to the upper cross-tail area A_t. If R_t is the radius of the tail cross-section, then

$$A_t = \frac{1}{2} \pi R_t^2 \tag{8.34}$$

and

$$A_p = \pi R_e^2 \sin^2 \theta_p \tag{8.35}$$

The cross-tail radius is then given by:

$$R_t = (2B_p)^{1/2} \cdot (B_s^2 + 2\mu_0 n_s \kappa (T_e + T_p)_s)^{-1/4} \cdot \sin \theta_p \cdot R_e \tag{8.36}$$

We notice that R_t will decrease when either the solar wind temperature or the solar wind magnetic field increases. With the values used so far for the solar wind parameters and assuming $\theta_p = 15°$, we find for $B_p = 6.0 \times 10^{-5}$ tesla

$R_t = 47 \, R_e$

Therefore an upper estimate of the magnetotail radius is approximately of the order of 50 R_e.

8.3 MAGNETIC FIELD MERGING

The situation described here is obviously not a stationary one since the magnetic field lines are attached to the Earth at least by one end and are forced to rotate by the Earth. The field line on the dayside which are compressed by the solar wind are stretched out on the nightside by the same wind, and actually torn apart.

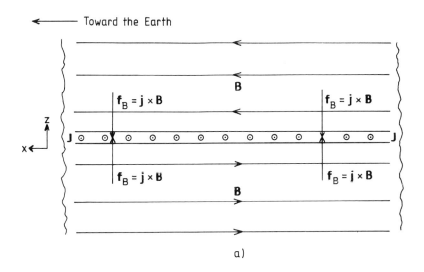

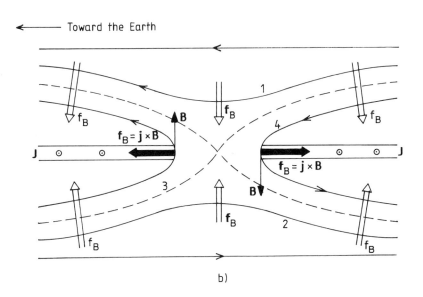

Fig. 8.11. (a) A cross-section of part of the cylindrical magnetospheric tail in the noon–midnight meridian illustrating the necessary current sheath $\mathbf{j}$ in the ecliptic plane, to maintain the magnetic field lines antiparallel above and below this sheath. The magnetic force $\mathbf{f}_B = \mathbf{j} \times \mathbf{B}$ on the plasma is directed toward this sheath from above and below the ecliptic plane. (b) The magnetic field configuration together with the magnetic force is illustrated in relation to a magnetic reconnection (merging) region. The open arrows bring plasma towards, while the filled arrows bring it out of the merging region.

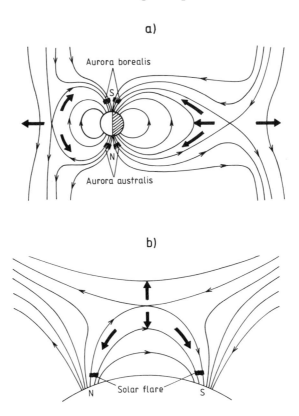

Fig. 8.12. Schematic illustration of magnetic field merging processes taking place in the Earth's magnetosphere creating polar aurora (a), and in the solar atmosphere creating solar flares (b). The thick arrows indicate the plasma flow away from the merging region. The plasma causing the light emissions are propagating along the magnetic field lines towards the magnetic poles. Light emissions occur when the atmospheric density of the Earth or the Sun is high enough that frequent collisions occur between the streaming plasma and the atmospheric particles.

Before many of these field lines return to the day side they have to reconnect or merge again.

Since the magnetic field lines are flowing in the opposite directions in the regions very close to the central current sheet, it is believed that magnetic merging can take place here through reconnection.

The plasma immediately above and below the current sheet is believed to be highly conductive. In the merging process the plasma will therefore be transported toward the current sheet from above and below (see Fig. 8.11a). When the field lines finally reconnect, plasma is jetted partly towards the Earth and partly outward through the tail (Fig. 8.11b).

The only way this merging can take place is when the "frozen-in" concept is invalidated in a small region around the neutral line. In Fig. 8.11b the field lines entering the merging region are numbered 1 and 2 and the double arrows indicate the motion of the plasma attached to the field lines. At the crossing the field lines "cut" and reconnect in a crosswise fashion so that two field lines indicated 3 and 4 appear. As these move away from the merging region earthward and tailward respectively, they carry the plasma along. We notice that there is no magnetic field at the crossing point (the neutral line). It is expected that plasma attached here to field line 3 may be accelerated to considerable energies and transported in a subsequent flow along the magnetic field lines to the lower ionosphere. Similarly the plasma attached to field line 4 may be accelerated and jetted out into interplanetary space. In this process the energy comes from the current which ceases when the neutral sheet disappears.

By cutting field lines we only mean that they can no longer be identified with the plasma motion since the flux through any loop following the plasma motion is not conserved in the vicinity of the neutral line.

It is believed that charged particles forced out of the merging region along the magnetic field lines toward the poles are the source of the polar auroral displays (Fig. 8.12a). Due to the symmetry in the Earth's magnetic field, simultaneous auroral forms may occur in the two hemispheres, *aurora borealis* and *aurora australis* (Figs. 8.12a and 8.13a).

The auroral displays are formed when the energetic plasma (mostly electrons) approaches the Earth and react by collisions with the neutral particles of the upper atmosphere, mainly above 100 km. This light phenomenon is dominated by emissions from neutral species such as atomic oxygen and molecular nitrogen.

Another light phenomenon in the Sun–Earth interaction process, which is strongly related to the aurora, is the solar flare. This is often seen as two parallel luminous bands (Fig. 8.13b) dominated by hydrogen emissions. These bands are often found to be situated between two magnetic poles of opposite polarities (Fig. 8.12b). As in the Earth's magnetosphere, field line merging occurs in a region above the magnetic poles, and energetic plasma is forced along the magnetic field lines toward the solar surface at the magnetic polar regions. When this plasma reaches the solar atmosphere which is so rich in hydrogen, collisions occur and hydrogen emissions are created which can be observed as parallel bands from the ground. Due to these similarities between the polar flares and the aurorae we can claim they are sister phenomena, but there is even a mother–daughter relationship.

Some of the plasma in the merging region is, however, forced outwards into the interplanetary space, and part of it may reach the Earth's magnetosphere. Here it can evidently be forced into a new merging process creating the aurora. In this sense we notice that the plasma processes taking place in the near-Earth environment are almost universal. By learning to understand the merging process in the magnetosphere, we will be able to understand also similar processes at the Sun, the stars and at other celestial bodies in the Universe.

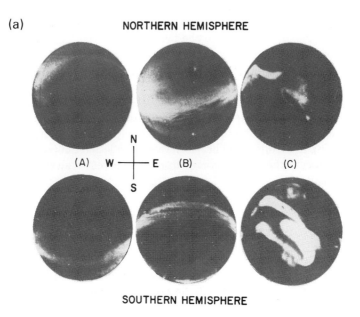

Fig. 8.13. (a) Simultaneous observations of the aurora borealis and the aurora australis at approximately the same magnetic field lines, showing the strong similarity between these two polar light phenomena, thus demonstrating the overall symmetry in the magnetosphere. (b) Solar flare event observed from ground as two approximate parallel luminous bands of H_α emissions. (From Akasofu and Chapman, 1972.)

8.4 SOME MAGNETOHYDRODYNAMIC CONCEPTS

Before we do that, let us study a few points in magnetohydrodynamics which will be indispensable for understanding some of the mechanisms taking place in the magnetosphere.

To describe the magnetohydrodynamic behaviour of plasma we need the continuity equation, the mobility equation, Maxwell's equations and the generalized Ohm's law together with an equation of state.

The continuity equation is represented by

$$\frac{\partial \rho}{\partial t} + \nabla \cdot (\rho \cdot \mathbf{v}) = 0 \tag{8.37}$$

where ρ is the density and $\mathbf{v}$ the velocity of the plasma. The mobility equation is basically the same as for hydrodynamics except that we must include forces that can act on the plasma due to electric currents flowing there

$$\rho \frac{d\mathbf{v}}{dt} = -\nabla p + \rho \mathbf{g} + \eta \nabla^2 \mathbf{v} + \mathbf{j} \times \mathbf{B} \tag{8.38}$$

where p is the pressure, $\mathbf{g}$ the acceleration of gravity, η the coefficient of viscosity and $\mathbf{j}$ is the current density.

The Maxwellian equation will be used in the following form:

$$\nabla \times \mathbf{B} = \mu_0 \mathbf{j} \tag{8.39}$$

$$\nabla \times \mathbf{E} = -\frac{\partial \mathbf{B}}{\partial t} \tag{8.40}$$

$$\nabla \cdot \mathbf{B} = 0 \tag{8.41}$$

$$\nabla \cdot \mathbf{E} = 0 \tag{8.42}$$

where the very high electrical conductivity σ of the plasma has allowed us to neglect the displacement current, and charge neutrality is conserved. The generalized Ohm's law is given by:

$$\mathbf{j} = \sigma(\mathbf{E} + \mathbf{v} \times \mathbf{B}) \tag{8.43}$$

where σ is the conductivity which is assumed isotropic. We can now reduce the number of equations by eliminating the electric field and the current. But first we neglect gravity forces and viscosity so that the mobility equation is reduced to:

$$\rho \frac{d\mathbf{v}}{dt} = -\nabla p + \mathbf{j} \times \mathbf{B} = \mathbf{f}_p + \mathbf{f}_B \tag{8.44}$$

where the motion of the plasma is due either to the pressure force

$$\mathbf{f}_p = -\nabla p \tag{8.45}$$

or the electromagnetic force (Lorentz force)

$$\mathbf{f}_B = \mathbf{j} \times \mathbf{B} \tag{8.46}$$

By eliminating $\mathbf{j}$ from Maxwell's equation we can express the electromagnetic force as

$$\mathbf{f}_B = \frac{1}{\mu_0} (\nabla \times \mathbf{B}) \times \mathbf{B} \tag{8.47}$$

Using the vector identity

$$\nabla B^2 = 2(\mathbf{B} \cdot \nabla)\mathbf{B} + 2\mathbf{B} \times (\nabla \times \mathbf{B}) \tag{8.48}$$

we find that $\mathbf{f}_B$ can be expressed as:

$$\mathbf{f}_B = \frac{1}{\mu_0}(\nabla \times \mathbf{B}) \times \mathbf{B} = -\nabla \frac{B^2}{2\mu_0} + \frac{1}{\mu_0}(\mathbf{B} \cdot \nabla)\mathbf{B} \tag{8.49}$$

The mobility equation can now be written as

$$\rho \frac{d\mathbf{v}}{dt} = -\nabla \left(p + \frac{B^2}{2\mu_0} \right) + \frac{1}{\mu_0}(\mathbf{B} \cdot \nabla)\mathbf{B} \tag{8.50}$$

By applying (8.40), (8.43), and (8.39) we find that:

$$\frac{\partial \mathbf{B}}{\partial t} = \nabla \times (\mathbf{v} \times \mathbf{B}) + \frac{1}{\sigma \mu_0} \nabla^2 \mathbf{B} \tag{8.51}$$

Here we have made use of the fact that σ is isotropic, and taken advantage of the vector identity (2.40)

$$\nabla \times (\nabla \times \mathbf{B}) = \nabla(\nabla \cdot \mathbf{B}) - \nabla^2 \mathbf{B} = -\nabla^2 \mathbf{B} \tag{8.52}$$

The following MHD equations now remain

$$\frac{\partial \rho}{\partial t} + \nabla \cdot (\rho \mathbf{v}) = 0 \tag{8.53}$$

$$\nabla \cdot \mathbf{B} = 0 \tag{8.54}$$

$$\nabla \cdot \mathbf{E} = 0 \tag{8.55}$$

Some magnetohydrodynamic concepts

$$\rho \frac{d\mathbf{v}}{dt} = -\nabla \left(p + \frac{B^2}{2\mu_0} \right) + \frac{1}{\mu_0} (\mathbf{B} \cdot \nabla)\mathbf{B} \tag{8.56}$$

$$\frac{\partial \mathbf{B}}{\partial t} = \nabla \times (\mathbf{v} \times \mathbf{B}) + \frac{1}{\mu_0 \sigma} \nabla^2 \mathbf{B} \tag{8.57}$$

In order to complete the system we need an equation of state which can be either the condition for incompressibility

$$\nabla \cdot \mathbf{v} = 0 \tag{8.58}$$

or the polytropic condition

$$p \cdot \rho^{-\nu} = \text{const.} \tag{8.59}$$

ν is equal to 1 for the isothermal case which is relevant in a plasma with a very high thermal conduction or $\nu = \gamma$ in the situation of an adiabatic process.

We notice that in the mobility equation there are two terms related to the magnetic field $\mathbf{B}$ which both, however, are due to the magnetic force $\mathbf{f}_B$. The first term, $(1/2\mu_0)\nabla B^2$ which is included in the pressure term, is often, due to the analogy, mentioned as the magnetic pressure, while the second term given by $(1/\mu_0)(\mathbf{B} \cdot \nabla)\mathbf{B}$ is called magnetic tension. Therefore any deviation of $\mathbf{B}$ from the force-free case ($\mathbf{j} = 0$) must be balanced by fluid pressure when the fluid is in equilibrium. In a perfect dipole field (Fig. 8.14a), where $\mathbf{j} = 0$ outside the Earth's core, the magnetic forces balance the pressure force everywhere. In a distorted dipole field, such as will occur in the magnetic tail and elsewhere in the magnetosphere (Figs. 8.14b and 8.11), there must be magnetic forces that will in the case of Fig. 8.14b force plasma back towards the Earth. In order to stretch the magnetic field in such a way that $\nabla \times \mathbf{B} \neq 0$, there must be a current out of the plane in Fig. 8.14b inside the magnetic field line. The force due to this current $\mathbf{f}_B = \mathbf{j} \times \mathbf{B}$ will be directed toward the Earth, since $\mathbf{B}$ is pointing vertically upward (northward). This force is equivalent to the magnetic pressure and tension forces, and therefore in a steady state there is a pressure balancing the magnetic forces and directed outward from the Earth due to a pressure gradient directed toward the Earth.

By inspecting the situation in the equatorial tail of the magnetosphere where magnetic field lines are antiparallel above and below the neutral sheet (Fig. 8.11a), we notice that a current must flow across this sheet perpendicular to $\mathbf{B}$ and out of the paper. The magnetic force acting on the plasma ($\mathbf{f}_B = \mathbf{j} \times \mathbf{B}$) will be directed towards the sheet both from above and below. If now merging occurs in the current sheet, then the magnetic force will still be directed toward the sheet away from the merging region. Because the magnetic field will be perpendicular to the neutral sheet within the merging region (Fig. 8.11b), however, the magnetic force will drive the plasma away from the merging region towards the Earth on the earthward side and into planetary space on the tailward side.

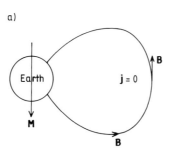

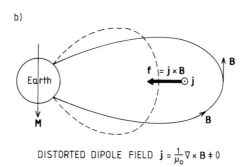

Fig. 8.14. (a) An idealized magnetic field line from a dipole at the centre of the Earth, when no current is flowing between the Earth and the field line. (b) A stretched magnetic field line when a current is present outside the Earth, creating a magnetic force toward the Earth, trying to restore the dipole configuration.

8.5 THE ENERGY FLUX INTO THE MAGNETOSPHERE

There is one more important result coming out of the magnetohydrodynamic theory which is of great importance to the understanding of the physical processes in the magnetosphere, and that is the relationship between thermal and electromagnetic energy. For a magnetic field B in vacuum there is always associated an energy density given by

$$u_B = \frac{1}{2\mu_0} B^2 \qquad (8.60)$$

or identical to the associated magnetic pressure. Within a volume τ there is therefore a total energy

$$U_B = \frac{1}{2\mu_0} \int_\tau B^2 \, d\tau \qquad (8.61)$$

where $d\tau$ is an infinitesimal volume. The time derivative of this energy is given by:

$$\frac{\partial U_B}{\partial t} = \frac{1}{\mu_0} \int_\tau \left(\mathbf{B} \cdot \frac{\partial \mathbf{B}}{\partial t}\right) d\tau \tag{8.62}$$

Now using (8.40) and the relation:

$$\mathbf{B} = \mu_0 \mathbf{H} \tag{8.63}$$

where $\mathbf{H}$ is the magnetic intensity, we find:

$$\frac{\partial U_B}{\partial t} = -\int_\tau \mathbf{H} \cdot (\nabla \times \mathbf{E}) d\tau \tag{8.64}$$

Using the vector identity.

$$\nabla \cdot (\mathbf{E} \times \mathbf{H}) = \mathbf{H} \cdot (\nabla \times \mathbf{E}) - \mathbf{E} \cdot (\nabla \times \mathbf{H}) \tag{8.65}$$

we get

$$\frac{\partial U_B}{\partial t} = -\int_\tau \nabla \cdot (\mathbf{E} \times \mathbf{H}) d\tau - \int_V \mathbf{E} \cdot (\nabla \times \mathbf{H}) d\tau \tag{8.66}$$

Since

$$\mathbf{j} = \nabla \times \mathbf{H} \tag{8.67}$$

and

$$\int_\tau \nabla \cdot (\mathbf{E} \times \mathbf{H}) d\tau = \int_S (\mathbf{E} \times \mathbf{H}) \cdot d\mathbf{s} \tag{8.68}$$

where $\mathbf{s}$ is an area enclosing and pointing out from τ

$$\frac{\partial U_B}{\partial t} = -\int_S (\mathbf{E} \times \mathbf{H}) \cdot d\mathbf{s} - \int_\tau \mathbf{j} \cdot \mathbf{E} \, d\tau \tag{8.69}$$

The time rate of the magnetic energy in a volume τ is equal to the energy flux across the surface into the volume as expressed by the Poynting vector $\mathbf{E} \times \mathbf{H}$ minus the electromechanical work $\mathbf{j} \cdot \mathbf{E}$ done in the volume.

This last term can by the use of Ohm's law (equation (8.43)) be specified into two separate terms. Since

$$\mathbf{E} = \frac{1}{\sigma} \mathbf{j} - \mathbf{v} \times \mathbf{B} \tag{8.70}$$

$$\mathbf{j} \cdot \mathbf{E} = \frac{1}{\sigma} j^2 - (\mathbf{v} \times \mathbf{B}) \cdot \mathbf{j} \tag{8.71}$$

and

$$\frac{\partial U_B}{\partial t} = -\int_S (\mathbf{E} \times \mathbf{H}) \cdot d\mathbf{s} - \int_\tau \frac{j^2}{\sigma} d\tau - \int_\tau \mathbf{v} \cdot (\mathbf{j} \times \mathbf{B}) d\tau \tag{8.72}$$

The second term represents the resistive energy loss inside the volume (the electrical load or the Joule heating rate) while the last term represents the mechanical work done per unit time due to plasma flow against the magnetic force $\mathbf{j} \times \mathbf{B}$ inside the same volume.

8.6 SOME ASPECTS OF THE ENERGY BALANCE

If the Earth was situated in a vacuum, the magnetic field would reach out to infinity without being distorted. Now we have learned that outside the Earth in interplanetary space there is a rarefied plasma penetrated by a magnetic field of solar and cosmic origin and known as the interplanetary magnetic field (IMF). Depending on the magnitude and direction of this magnetic field and the dynamic behaviour of the interplanetary plasma, the Earth's magnetic field will be distorted accordingly. Let us first assume that the interplanetary field, B_z, is directed northward, i.e. perpendicular to the ecliptic plane and parallel to the Earth's magnetic field (Fig. 8.15a). In this situation the magnetic field of the Earth would be embedded in the interplanetary field to form a closed magnetospheric cavity indicated by the dashed line in Fig. 8.15. The surface of this cavity would form an equipotential. Since $\mathbf{E}$ is everywhere perpendicular to this surface and $\mathbf{B}$ is parallel to it, the Poynting vector would be parallel to the surface everywhere and no energy would enter into the magnetospheric cavity except maybe for the two neutral points marked N and N' in Fig. 8.15a.

If, on the other hand, the interplanetary magnetic field was directed southward, the magnetic field of the Earth would open up. A neutral line would form around the equator in a complete symmetric situation. $\mathbf{B}$ is perpendicular to this line, and the electric field would be directed along it. The Poynting flux would therefore point into the magnetosphere and energy could enter.

The situation is, however, much more complicated since the solar wind pushes toward the Earth's magnetic field and deforms it by a compression on the dayside and an elongation into a tail on the nightside. Let us now consider a simplified cylindrical model of the surface of the tail (Fig. 8.16) when there is a southward IMF ($\mathbf{B}_{\text{IMF}}$). The magnetic field in the tail will be directed toward the Earth and against the solar wind stream. The field will make a small angle α with the surface in the northern (upper) half of the tail and far away from the Earth, and also make a similar but negative angle α with the surface in the southern half of the tail in order to merge with the southward IMF. In the solar wind we know that the conductivity is very high so the "frozen-in" concept applies

$$\mathbf{E} = -\mathbf{v}_S \times \mathbf{B} \tag{8.73}$$

where $\mathbf{v}_S$ is the solar wind velocity, and therefore the energy flux (Poynting vector) can be expressed as:

$$\begin{aligned} \mathbf{P} &= \mathbf{E} \times \mathbf{H} = -(\mathbf{v}_S \times \mathbf{B}) \times \mathbf{H} \\ &= -(\mathbf{v}_S \cdot \mathbf{H})\mathbf{B} + (\mathbf{B} \cdot \mathbf{H})\mathbf{v}_S \end{aligned} \tag{8.74}$$

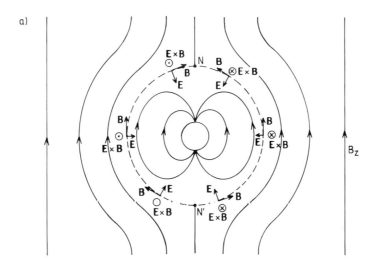

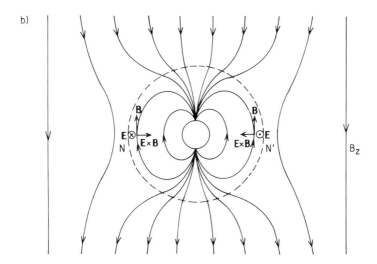

Fig. 8.15. (a) The Earth as an idealized magnetic sphere in an area with a northward directed interplanetary magnetic field ($B_z > 0$), making a closed magnetospheric cavity. The energy flux ($\mathbf{E} \times \mathbf{B}$) is along the cavity surface everywhere. No energy enters the magnetosphere from the outside. (b) The similar situation when there is a southward IMF ($B_z < 0$). The magnetosphere is opened up in the polar regions, and the energy flux is directed inward around the equatorial perimeter of the magnetospheric cavity.

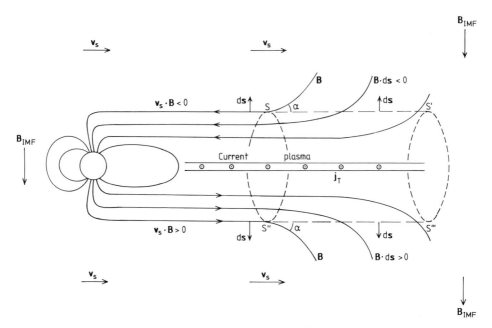

Figure 8.16: An illustration of the interaction between the solar wind and the magnetospheric tail when there is a negative IMF ($\mathbf{B}_{\mathrm{IMF}}$). The solar wind velocity $\mathbf{v}_S$ is parallel and antiparallel to $\mathbf{B}$ on the southern and northern part of the semicylindrical tail lobes, respectively. The surface element $d\mathbf{s}$ is parallel or antiparallel to $\mathbf{B}$ on the same tail lobes, respectively. This result is an energy flux inwards to the tail surface everywhere.

The energy input to the magnetosphere is therefore given by:

$$\phi_E = \int_A \mathbf{E} \times \mathbf{H}\, d\mathbf{s} = \int_A (\mathbf{v}_S \cdot \mathbf{H})\mathbf{B}\, d\mathbf{s} - \int_A (\mathbf{B} \cdot \mathbf{H})\mathbf{v}_S\, d\mathbf{s} \tag{8.75}$$

where A is the surface and $d\mathbf{s}$ a surface element of the magnetospheric tail. Since the solar wind is along the magnetospheric surface, $\mathbf{v}_S$ is perpendicular to $d\mathbf{s}$ everywhere. The last term therefore disappears. The energy flux into the magnetosphere is then given by:

$$\phi_E = \int_A (\mathbf{v}_S \cdot \mathbf{H})\mathbf{B}\, d\mathbf{s} \tag{8.76}$$

We notice that on the northern half $\mathbf{v}_S$ and $\mathbf{B}$ are antiparallel, so $\mathbf{v}_S \cdot \mathbf{H} < 0$, and due to the inclination α the $\mathbf{B}$-field makes with the surface SS', $\mathbf{B} \cdot d\mathbf{s} < 0$, and the energy flux is positive, i.e. into the magnetosphere. On the southern half $\mathbf{v}_S$ is parallel to $\mathbf{B}$, so $\mathbf{v}_S \cdot \mathbf{H} > 0$, and due to the negative inclination α, $\mathbf{B}$ makes with the surface $S''S'''$, $\mathbf{B} \cdot d\mathbf{s} > 0$. The energy flux is therefore directed into the magnetosphere at the southern tail lobe too.

When taking a cross-section of the tail and looking at it towards the Sun, as sketched in Fig. 8.17 when the IMF is southward, we notice that the tail current $\mathbf{j}_T$ must flow from dawn to dusk in order to maintain the open tail. In order to complete the current loop, we let it flow around the surface of the cylinder with a current density $\mathbf{j}_T/2$.

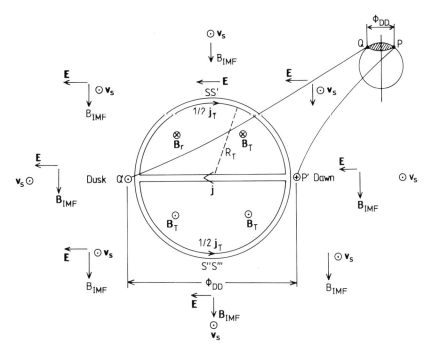

Fig. 8.17. An idealized cross-section of the tail showing the central current sheath from dawn to dusk together with the return currents along the surface of the tail lobes. The electric field associated with the solar wind flow is parallel to the central tail current but antiparallel to the currents at the surface. Thus there is a magnetospheric generator at the surface of the tail driving the currents to the load in the central sheath. A potential ϕ_{DD} is formed between Q' and P', across the tail which is mapped across the polar cap between Q and P.

In the solar wind we notice that the electric field is directed from dawn to dusk everywhere. The current at the surface of the cylinder is therefore antiparallel to **E** and energy is created there. The surface acts as a generator or a dynamo. The cross-tail current, however, will be parallel to the field and acts as a load.

If the IMF turns around, however, the electric field also does, and the generator ceases to work. This is a major difference between the southward-directed and the northward-directed IMF, and it leads to a more closed magnetosphere for northward IMF, as already mentioned.

Applying reasonable magnitudes on $\mathbf{B}_{\text{IMF}}$ ($\approx 5\,\gamma$) and the solar wind velocity (≈ 400 km/s), we can find the potential drop between dawn and dusk (marked P' and Q' respectively) if we assume that the radius of the tail is $R_T \approx 15\,R_e$. The potential is then given by

$$\phi_{DD} = -\int \mathbf{E} \cdot d\mathbf{l} = 2E \cdot R_T = 2v_S B R_T \tag{8.77}$$

$V_{DD} \approx 400$ kV

To understand how the different regions in the magnetosphere are connected to the Earth's ionosphere via the geomagnetic field lines is one of the greatest challenges in ionosphere–magnetosphere physics today. Let us for simplicity assume that this potential ϕ_{DD} between P' and Q' in Fig. 8.17 is exactly mapped to the polar cap between the dawn and dusk points P and Q, then we can give an estimate of the dawn–dusk electric field across the polar cap. If we again let the polar cap be above $75°$ latitude ($\theta_p = 15°$), we find:

$$E_{DD} = \frac{V_{DD}}{2R_e \sin\theta_p} = 8 \text{ mV/m} \tag{8.78}$$

Let us now go back to the force $\mathbf{f}_T$ on the magnetospheric surface due to the tail current $\mathbf{j}_T$ which is generated by the solar wind dynamo and given by:

$$\mathbf{f}_T = \mathbf{j}_T \times \mathbf{B} = \frac{1}{\mu_0}(\nabla \times \mathbf{B}) \times \mathbf{B} \tag{8.79}$$

This force must have a component along the surface (tangential) opposing the solar wind. By using the vector identity

$$\nabla B^2 = 2(\mathbf{B} \cdot \nabla)\mathbf{B} + 2\mathbf{B} \times (\nabla \times \mathbf{B}) \tag{8.80}$$

$$(\nabla \times \mathbf{B}) \times \mathbf{B} = (\mathbf{B} \cdot \nabla)\mathbf{B} - \frac{1}{2}\nabla B^2 \tag{8.81}$$

$$\mathbf{f}_T = \frac{1}{\mu_0}\left[(\mathbf{B} \cdot \nabla)\mathbf{B} - \frac{1}{2}\nabla B^2\right] \tag{8.82}$$

The total force on the volume τ of the magnetospheric cylinder is

$$\mathbf{F}_T = \int_\tau \mathbf{f}_T d\tau = \frac{1}{\mu_0}\int_\tau \left[(\mathbf{B} \cdot \nabla)\mathbf{B} - \frac{1}{2}\nabla B^2\right] d\tau \tag{8.83}$$

where $d\tau$ is a volume element. And finally

$$\mathbf{F}_T = \frac{1}{\mu_0}\int_S \left[(\mathbf{B} \cdot \hat{\mathbf{n}})\mathbf{B} - \frac{1}{2}B^2\hat{\mathbf{n}}\right] ds \tag{8.84}$$

where S is the surface of the tail and $d\mathbf{s}$ is a surface element along the unit vector $\hat{\mathbf{n}}$.

$$d\mathbf{s} = ds\,\hat{\mathbf{n}} \tag{8.85}$$

The last term in (8.84) expresses the magnetic pressure force. Since we are interested in the tangential force only, which acts to stretch the tail, we find when **B** is assumed constant across the surface

$$F_T^t = (F_T \cdot \hat{\mathbf{t}}) = \frac{1}{\mu_0}(\mathbf{B} \cdot \hat{\mathbf{n}}) \cdot (\mathbf{B} \cdot \hat{\mathbf{t}})S \qquad (8.86)$$

where t is a unit vector tangential to S.

$$F_T^t = \frac{1}{\mu_0}(B_n \cdot B_t)S \qquad (8.87)$$

where B_n and B_t are the components of the magnetic field normal and parallel to the surface S, respectively. In a stable situation this force balances the momentum force in the solar wind.

The rate at which the solar wind brings energy into the magnetosphere can now be found:

$$P_S = F_t \cdot v_S = \frac{v_S}{\mu_0}(B_n \cdot B_t)S \qquad (8.88)$$

since v_S is parallel to F_t everywhere. This energy rate can be compared with the energy flux due to the Poynting vector given by:

$$P_E = -\int_S (\mathbf{v}_S \cdot \mathbf{B})\mathbf{H} \cdot ds \qquad (8.89)$$

For reasonable values of v_S, B and S we can find P_E and P_S to be of the order of 10^{11} W.

Probably the most direct interaction taking place between the solar wind and the near-Earth plasma is found above the polar cap regions. In order to gain insight into this, we again give an oversimplified illustration (Fig. 8.18) where the magnetic field lines are entering as parallel and vertical lines to the polar cap and perpendicular to the solar wind. The Earth is viewed from the nightside toward the Sun. In relation to Fig. 8.18 the solar wind will generate an electric field

$$\mathbf{E}_S = -\mathbf{v}_S \times \mathbf{B} \qquad (8.90)$$

which will be directed from dawn to dusk, and since the magnetic field lines are perfect conductors, it will penetrate to the ionosphere and drive a cross-polar Pedersen current (J_{P_c}) in the same direction as the electric field because of the finite resistivity there. The ionospheric current will be limited to a height region h (90–140 km) in which the Pedersen and Hall conductivities can be denoted σ_P and σ_H respectively. We assume for simplicity that these are independent of height. The cross-polar-cap current J_{P_c} will then be:

$$J_{P_c} = \sigma_P h\, E_I = \Sigma_P E_x \qquad (8.91)$$

where we have allowed for some difference in $\mathbf{E}_S$ and $\mathbf{E}_I$ due to the finite conductivity in the ionosphere which effects the mapping. These currents have to be closed in some loops, and due to the high conductivity along the field lines, parallel or Birkeland currents are the likely candidates. How they close in the

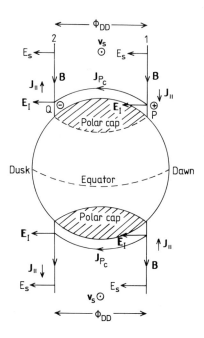

Fig. 8.18. A schematic illustration of the parallel currents penetrating through the ionosphere from the magnetosphere along magnetic field lines (Birkeland currents). They are deflected in a horizontal direction in a thin ionospheric layer between 90 and 150 km.

magnetosphere, magnetosheath or the solar wind is another matter. We notice, however, that these currents enter the polar cap ionosphere on the dawn side and leave it in the dusk side in the northern hemisphere. The opposite situation occurs in the southern hemisphere. These currents are usually referred to as the region 1 currents. Due to the complicated geometry involved (Fig. 8.18) these currents will form curved sheet currents or currents flowing along cylinder-like surfaces.

8.7 CURRENTS IN A COLLISIONLESS PLASMA

Since we are assuming that the magnetospheric and solar wind plasma is collisionless and that the "frozen-in" concept of the magnetic field applies due to the large conductivities along the field lines, one may ask how these regions can act as current sources for currents perpendicular to the B-field at all. In order to obtain a better insight into that, we must go back to the mobility equation of the plasma, and by neglecting the gravity force we get (8.44)

$$\rho \frac{d\mathbf{v}}{dt} = -\nabla p + \mathbf{j} \times \mathbf{B} \tag{8.92}$$

Currents in a collisionless plasma

In order to find the current perpendicular to **B** we make the cross-product with **B**

$$\rho \frac{d\mathbf{v}}{dt} \times \mathbf{B} = -\nabla p \times \mathbf{B} + (\mathbf{j} \times \mathbf{B}) \times \mathbf{B} \qquad (8.93)$$
$$= -\nabla p \times \mathbf{B} - B^2 \mathbf{j}_\perp$$

and

$$\mathbf{j}_\perp = \frac{1}{B^2} \left[\mathbf{B} \times \left(\rho \frac{d\mathbf{v}}{dt} + \nabla p \right) \right] \qquad (8.94)$$

Currents in the magnetosphere and the solar wind can therefore occur perpendicularly to the magnetic field whenever there are spatial variations in the pressure or time variations in the plasma flow itself. Assume for example that the solar wind plasma suddenly is accelerated and that we can neglect pressure forces, then in our simple polar cap model a current $\mathbf{j}_\perp$ will be directed from dawn to dusk. The magnetic force due to this current is

$$\mathbf{f}_{\perp B} = \mathbf{j}_\perp \times \mathbf{B} \qquad (8.95)$$

directed away from the Sun, which is consistent with an acceleration of the solar wind plasma. In this region, however,

$$\mathbf{j}_\perp \cdot \mathbf{E}_S > 0 \qquad (8.96)$$

and the solar wind plasma is acting as a load to the generating mechanism accelerating the plasma, whatever that may be. If it is the ionosphere, energy is actually soaked out of it.

We have now seen that the Earth's magnetic field can merge with the interplanetary field in such a way that it opens up or closes the magnetosphere. Furthermore, the solar wind pressure deforms the symmetric magnetosphere that could otherwise occur in vacuum and stretches the field lines and tears them apart. Since the field lines are anchored in the Earth, they will take part in the Earth's rotation and sweep around in loops which will be stretched and possibly torn apart on the nightside for later to tie together again toward the daytime. The magnetic field lines therefore undergo a steady motion, sometimes disrupted by disturbances in the solar wind, and they will, due to the high conductivity, carry the magnetospheric plasma along. A schematic diagram of this field line in the midday–midnight meridional plane convection is shown in Fig. 8.19 where the interplanetary magnetic field at point 0 starts to merge with the Earth's magnetic field at stage 1. The interplanetary field forced by the solar wind tears the Earth's magnetic field apart and carries it across the poles in stages 2, 3 and 4. In stage 5 the Earth's magnetic field starts on the reconnection process in the tail. At stage 6 the reconnection occurs, and at stage 7 the Earth's field line is tied together again as it is on its way toward a new cycle. Although the picture is illustrated in the meridional plane, the complete process takes place as the field rotates around. At any one moment, however, a snapshot of the meridional plane will show different field lines in different stages.

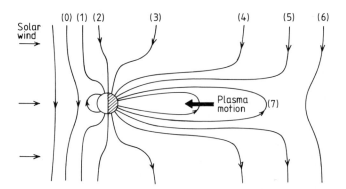

Fig. 8.19. A schematic diagram showing the magnetic field convection in the noon–midnight meridian plane. At point 0 the interplanetary field line starts to merge with a field line from the Earth. At 1 the merging has occurred, and at 2–5 the Earth's field line is drawn back over the pole as the IMF is carried tailward by the solar wind plasma. At 6 the open field line starts to reconnect, and at 7 the Earth's magnetic field line is closed again.

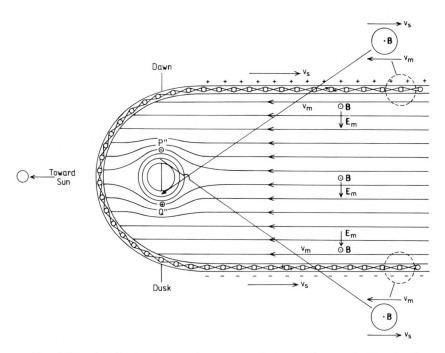

Fig. 8.20. An illustration to show areas of strong shear velocities in the magnetospheric equatorial planes. The shear is such that it leads to vortices to be associated with positive and negative space charges on the dawn and dusk sides of the magnetosheath boundary and on the dusk and dawn side of the inner plasma sheath, respectively.

Sec. 8.7] Currents in a collisionless plasma 421

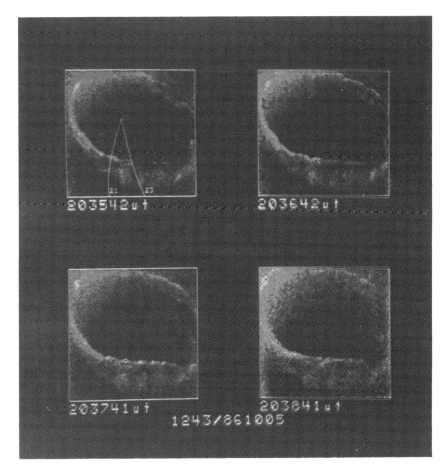

Fig. 8.21. Auroral images from the VIKING satellite showing the auroral oval aligned with approximately equidistant spots of strong auroral emissions. (From Hultqvist, 1988.)

The plasma transport related to this process is indicated by the thick arrow in Fig. 8.19. Of special interest is the return flow toward the Earth in the equatorial plane.

This sunward convection is illustrated in more idealized detail in Fig. 8.20. The plasma flow $\mathbf{v}_m$ is assumed rather uniform in the nightside and is deflected toward the morning and evening flanks to flow out of the magnetosphere and to the magnetosheath on the dayside front of the equatorial plane. The plasma flow in the magnetosphere is of course far more complicated than depicted there, since it certainly is a 3-dimensional one.

We notice that the magnetic field in the equatorial plane is directed upward through the paper and that therefore the "frozen-in" concept must imply an electric field directed from dawn to dusk across the equatorial plane as illustrated in the figure. This plasma flow must, according to what is said above, represent a load to the solar wind dynamo.

There is one dilemma, however, since the magnetosheath is acting as a kind of shielding between the solar wind and the magnetospheric plasma. It is not very well understood how the energy actually can be transferred into the magnetosphere and drive the convection there. There are several suggestions: one refers to viscous interaction between the solar wind and the magnetospheric plasma; another is field line merging and reconnection. Independent of mechanism the processes occurring at the boundary between the solar wind and the magnetospheric plasma create large irregularities and wave-like features there. The interest in magnetospheric sources for wave- and vortex-like large-scale structures has been intensified lately due to recent auroral images obtained from satellites (Fig. 8.21). The remarkable features to be observed are the distinct intensified regions (hot spots) with a fairly constant distance between them along the auroral oval. It is evident that these features must be related to similar characteristic structures in the magnetosphere, but exactly how and where is a question of mapping in a topology that no one knows for certain.

8.8 SPACE CHARGES IN THE MAGNETOSPHERE

The very light spots must also be related to concentrated particle precipitation carrying strong currents. The question is, therefore, where can charges build up and being sustained in such concentrated regions and for such a long time that these high-intensity features can exist?

We have seen that the magnetosheath in the front of the magnetosphere can be very narrow. Let us therefore assume that this sheath is also relatively narrow on the tailward side of the magnetosphere. Because the plasma flow reverses sign across this boundary from being antisunward in the solar wind to becoming sunward inside the magnetosphere, a large gradient will exist in the plasma flow perpendicular to the flow – a region known from the hydrodynamics to be the source for vortices and turbulence. In Fig. 8.20 this is indicated by a turbulent wavy structure.

Let us consider a situation of maintaining a positive space charge q in a highly conducting plasma creating an electric field perpendicular to a magnetic field as illustrated in Fig. 8.22a. In this situation the plasma would have a velocity given by

$$\mathbf{v} = \mathbf{E} \times \mathbf{B}/B^2 \qquad (8.97)$$

which will correspond to a rotation in a clockwise direction around the positive **B**-direction.

On the other hand, if the space charge was negative, the rotation would be in an anticlockwise direction around positive **B** (Fig. 8.22b).

Sec. 8.8] Space charges in the magnetosphere 423

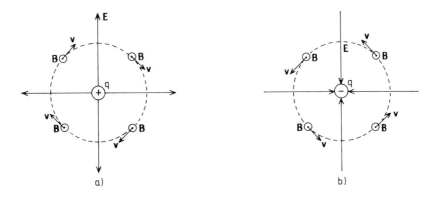

Fig. 8.22. (a) An idealized picture to illustrate the direction of plasma rotation when a positive charge is situated in an area of a magnetic field. The rotation due to the $\mathbf{E} \times \mathbf{B}$ motion is clockwise observed toward the direction of the $\mathbf{B}$ field. (b) When there is a low potential permeated by a magnetic field the rotation direction of the plasma becomes anticlockwise when viewed toward the direction of $\mathbf{B}$.

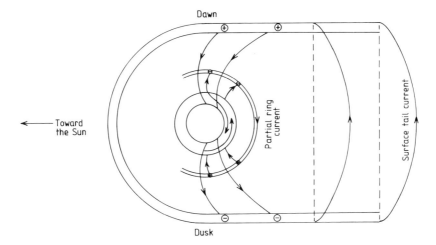

Fig. 8.23. A possible closure of the magnetospheric–ionospheric current system. The space charges in the magnetosheath in the ecliptic plane drive a current along the magnetic field lines to the polar cap, downward on the dawn side and upward on the dusk side respectively. The currents are flowing across the polar region along the poleward boundary of the auroral oval and called Region 1 currents. These currents are probably closed on the magnetospheric surface in the tail as indicated. From the inner magnetosphere similar Birkeland currents are driven from the dusk side toward the dawn side through the lower latitudes of the auroral oval. These currents are denoted Region 2 currents. In the magnetosphere the current is probably closed in the partial ring current as indicated.

If we now go back to Fig. 8.20 and the rather strong velocity shears across the magnetosheath in the tail region, these shears could lead to vortices which on the dawn side would be in the clockwise direction and on the dusk side in the anticlockwise direction. This would then correspond to imaginary space charges, positive on the morning side and negative on the evening side as indicated by the plus and minus signs, respectively. We notice that this would be true, however, if a magnetic field directed out of the paper was prevailing as it is on closed field lines inside the magnetopause at least.

We can therefore associate these vortices with space charges. But space charges cannot exist for very long in the magnetosphere, as they will try to find ways of neutralizing each other. One possibility is as it was indicated for the situation above the polar cap, that the charges neutralize through currents via the ionosphere (Fig. 8.18). The current would then flow into the ionosphere on the morning side and out of the ionosphere on the evening side as illustrated in Fig. 8.23, and this current would be in the same sense as the cross-polar-cap current as discussed earlier and which is again called the region 1 current. In the magnetosphere this current could close as a part of the surface tail current.

There is another area closer to the Earth as indicated in Fig. 8.20 where strong velocity shears are expected to occur. Close to the Earth the plasma will corotate ($r < 6$–$7\ R_e$) while the plasma from the magnetotail will be drifting by the cross-tail electric field. The velocity shears are such that a positive space charge is expected on the dusk side and a negative charge on the dawn side, opposite to what we found in the magnetosheath. These charges can then create currents through the ionosphere from dusk to dawn at a lower latitude than the region 1 current. These currents are related to the so-called region 2 currents and traverse the ionosphere at the equatorward boundary of the auroral oval, in the magnetosphere they are probably closed by the partial ring current to be mentioned later.

8.9 CURRENTS RELATED TO AN AURORAL ARC

Auroral displays associated with strong electron precipitation from the magnetosphere are often appearing as narrow elongated arcs. It is also observed that the plasma motion which can be very strong adjacent to the arc, is changing direction across the arc. In fact, this electron precipitation tends to occur in regions where the flow vorticity has a positive sign. Fig. 8.24 illustrates a cross-section of such an auroral arc directed along the x-axis. The length L of the arc is much larger than the width a. The z-axis is parallel to the vertical magnetic field and pointing into the paper. The y-axis then completes the right-handed coordinate system.

The flow vorticity is defined as the curl of the velocity field

$$\boldsymbol{\xi} = \nabla \times \mathbf{v} \tag{8.98}$$

When $E_z = 0$, the plasma flow velocity $\mathbf{E} \times \mathbf{B}/B^2$ is independent of z, and assuming that the velocity is parallel to the arc everywhere and long the x-axis,

Sec. 8.9]   Currents related to an auroral arc   425

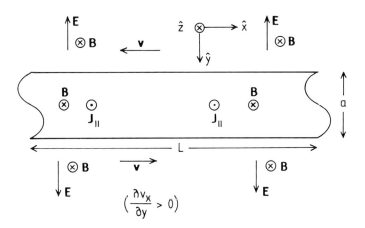

Fig. 8.24. An idealized model of a narrow strip in the ionosphere where precipitation takes place associated with a wind shear across the strip. Vertical currents associated with precipitating electrons are flowing antiparallel to the **B** field and out of the region in the northern hemisphere when the wind vorticity is positive. The strip is then associated with a high potential as also the **E**-field is directed away from the strip.

the vorticity has only a z-component

$$\xi = \frac{\partial}{\partial y}\hat{y} \times (v_x\hat{x} + v_y\hat{y}) = -\frac{\partial v_x}{\partial y}\hat{z} = \xi_z\hat{z} \tag{8.99}$$

as there are no variation in the z and x direction of **v**, and the v_z component is zero. Again the parallel current density is given by (remember $\mathbf{B} = B\hat{z}$):

$$j_\| = -\nabla_\perp(\underset{=\perp}{\Sigma} \cdot \mathbf{E}) \tag{8.100}$$

When assuming uniform conductances, we obtain

$$j_\| = -\Sigma_P \nabla_\perp \cdot \mathbf{E} = -\Sigma_P \frac{\partial E_y}{\partial y} \tag{8.101}$$

since $\partial/\partial x \equiv 0$. When the velocity is due to the $\mathbf{E} \times \mathbf{B}$ drift, we find that

$$\mathbf{E} = -\mathbf{v} \times \mathbf{B} = -v_x B\hat{y} + v_y B\hat{x} \tag{8.102}$$

and

$$\frac{\partial E_y}{\partial y} = -\frac{\partial v_x}{\partial y} B \tag{8.103}$$

since **B** is independent of y. Finally

$$j_\| = \Sigma_P B \frac{\partial v_x}{\partial y} = \Sigma_P B \xi_z \tag{8.104}$$

The current parallel to the magnetic field, i.e. downward on the northern hemisphere, is therefore associated with a negative vorticity. Precipitating electrons, however, are associated with currents out of the ionosphere or antiparallel to **B** in the northern hemisphere. Therefore this current is related to a positive vorticity as stated above.

We also notice from Fig. 8.24 that the electric field will point away from the arc on both sides as if there is an excess positive charge in the arc. The current then tends to neutralize the charge either by positive charges leaving the ionosphere or electrons precipitating into the ionosphere.

8.10 HIGH-LATITUDE CONVECTION PATTERNS

Because of the high conductivity along the magnetic field lines, any potential that is created between them in the magnetosphere will map down to the ionosphere where they will give rise to an electric field which sets the ionospheric plasma

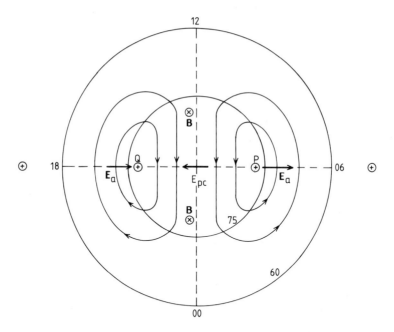

Fig. 8.25. An oversimplified model of the high-latitude ionospheric convection pattern showing a symmetric two-cell system with clockwise rotation on the dusk side and anticlockwise rotation on the dawn side. The plasma convection is thought to be driven by the dawn–dusk electric field $\mathbf{E}_{pc}$ across the polar cap. The return flow takes place along narrow belts at lower latitudes in the auroral oval. The points P and Q are related to high and low potentials, respectively, thus a poleward electric field, $\mathbf{E}_a$, occurs in the dusk side and an equatorward electric field in the dawn side of the auroral oval, respectively.

Sec. 8.10]   **High-latitude convection patterns**   427

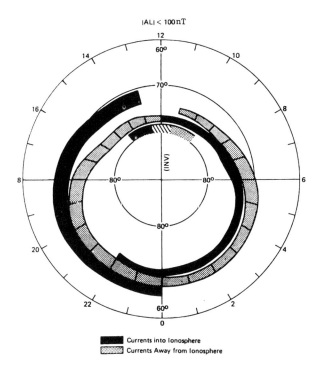

Fig. 8.26. The average pattern of field-aligned currents in the high-latitude region when the IMF is southward. The currents into and out of the ionosphere are indicated by different symbols, thus showing currents out of the ionosphere in the evening and into the ionosphere in the morning associated with the region 1 current. The equatorial currents into the ionosphere in the evening and out of the ionosphere in the morning are associated with the region 2 currents. The field-aligned currents at very high latitude at noon (the cusp currents) are strongly dependent on the B_y-component of the interplanetary magnetic field (IMF). (From Ijiima and Potemra, 1976.)

adrift. As already mentioned, in order to understand this mapping in full, we need a detailed knowledge of the topology of the magnetosphere which we are not in the possession of today.

An idealized representation of the plasma flow in the high-latitude and polar-cap ionosphere is shown in Fig. 8.25.

The dawn–dusk directed polar cap electric field $\mathbf{E}_{P_c}$ which corresponds to the cross-tail magnetic field $\mathbf{E}_m$ or the solar wind electric field $\mathbf{E}_{DD}$, drives the ionospheric plasma from the dayside across the polar cap to the nightside. In the nightside this plasma flow diverges eastwards and westwards along a narrow latitudinal belt associated with the auroral oval. On the morning side these two flow branches converges again to complete the flow circuit. A flow reversal takes place between 70° and 80° of latitude on the dawn and dusk side. Since the rotation is clockwise around the positive magnetic field direction in the dawn sector at P in the figure (Fig. 8.25), a positive potential occurs here and the contrary takes place in the dusk sector at Q. The electric field, $\mathbf{E}_a$, in the auroral oval will therefore be directed poleward in the dusk side and equatorward in the dawn side. This picture has emerged from years of observations and study by means of ground-based and satellite techniques. The picture portrayed in Fig. 8.25 is of course a considerable oversimplification, but it helps in getting the main frame of reference under control. Due to the great variability in the solar wind, the interplanetary field and their interaction with the magnetosphere, this two-cell convection model in the polar cap is highly distorted and may be broken up into a large number of local cells and turbulence.

Another result which has emerged from satellite measurements is the presence of the field-aligned or Birkeland currents at any time. An overview of the main results from these observations is shown in Fig. 8.26. The spiral-formed areas around the polar cap indicates the region for the average position of the Birkeland currents flowing into and out of the ionosphere. The inner ring is termed region 1 current, while the outer ring is denoted region 2 current. There is, however, an area of overlap around midnight. Similarly there is another area of confusion at midday. For the dawn and dusk sides the situation appears well behaved.

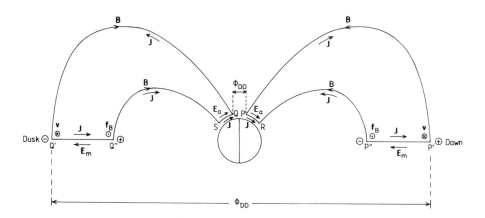

Fig. 8.27. A possible synthesis of ground-based and satellite observations of the ionospheric–magnetospheric current system and electric fields. The cross-section is in the xz-plane along the dawn–dusk meridian seen from the tail side of the Earth (toward the Sun). The magnetospheric dynamo drives the currents in the dusk to dawn direction in the ecliptic plane. These currents continue to the ionosphere as Birkeland currents (currents parallel to B). In the ionosphere they are deflected in the horizontal direction as Pedersen currents along the auroral zone electric field. The magnetic force $\mathbf{f}_B$ in the ecliptic plane is opposing the convection velocity $\mathbf{v}$. The electric potential ϕ_{DD} in the magnetosphere between the dawn and the dusk sector is projected along the magnetic field line across the polar cap between the points P and Q on the dawn and dusk side, respectively.

Sec. 8.10] High-latitude convection patterns

If we now try to make a synthesis of our understanding of the dawn–dusk current system between the ionosphere and magnetosphere, we refer to the diagram sketched in Fig. 8.27. Here the Earth is seen from the nightside with the midnight meridian marked as the diameter. The Birkeland currents in the dawn–dusk meridian are in accordance with the satellite observations illustrated in Fig. 8.26. In the ionosphere we know from especially balloon and radar observations that the auroral zone electric field $\mathbf{E}_a$ is directed northward in the dusk sector and southward in the dawn sector, i.e. from dusk to dawn. Since the ionosphere has a finite conductivity, the Birkeland currents will mainly shorten in the ionosphere by Pedersen currents, i.e. along the E-fields, northward in the dusk sector and southward in the dawn sector, respectively. In the magnetosphere the closure of the loop is not well established, but we indicate here that it is in radial in the dusk–dawn direction. This will correspond to currents closing between P', Q' and P'', Q'' in Figs. 8.17 and 8.20, respectively. We notice that when the dusk to dawn auroral zone electric field is mapped out to the magnetosphere, they will be reversed and directed from dawn to dusk in agreement with the polar cap and cross-tail electric fields $\mathbf{E}_{Pc}$ and $\mathbf{E}_m$ respectively. In the magnetosphere therefore the electric field and currents are antiparallel and thus forming a generator, while they are parallel in the ionospheric load.

In the tail region of the magnetospheric equatorial plane the convection is driven towards the Sun by the dawn–dusk electric field. This is indicated by the arrows $\mathbf{v}$. Since the magnetic field is directed perpendicular to the equatorial plane, the magnetic force

$$\mathbf{f}_B = \mathbf{j} \times \mathbf{B} \qquad (8.105)$$

due to the closure currents in the magnetospheric plane will be directed tailwards and against the plasma flow. If therefore the plasma flow for some reason is decelerated so that $d\mathbf{v}/dt < 0$, the term $\mathbf{B} \times d\mathbf{v}/dt$ in (8.95) will be directed from dusk to dawn and increase the current $\mathbf{j}$ in the loop. If, on the other hand, the plasma motion is accelerated, the current in the loop will decrease. This is the effect of the magnetospheric generator. If the acceleration was so severe that the magnetospheric cross-tail current came to a halt, we could expect short-circuiting of the circuit by penetration of electrons down to the auroral oval at low latitude (R) in the dawn sector and at high latitude (Q) in the dusk sector, or outflow of positive charges from the ionosphere could be expected here. Similarly, positive charges could be expected to penetrate into the auroral oval at low latitudes (S) in the dusk side and at high latitude (P) on the dawn side, or conversely, negative charges could be soaked up from the ionosphere at these places.

We will now leave the magnetosphere–ionosphere interaction and these rather speculative reflections concerning the current systems for a while. It should be clear that the magnetosphere with all its complexity is not only open for intrusion by the solar wind but also to the imagination of mankind.

8.11 HIGH-LATITUDE CONVECTION AND FIELD-ALIGNED CURRENTS

We have indicated that the polar cap and auroral zone convection in reality is quite different from the simplified picture in Fig. 8.26 and that it will change according to the direction of the IMF. In Fig. 8.28 a synthesis of the high-latitude daytime convection patterns as derived from satellite observations are presented for a negative B_z component or a southward IMF, but for different directions and magnitudes of the B_y component. When $B_y = 0$ we recognize the two-cell structure. When $B_y < 0$, or IMF is directed eastward, the eastern or dawn side convection cell shrinks and retreats to the polar cap, while the dusk side cell intrudes the eastern half of the dayside auroral oval. For a positive B_y or B_y directed westward, the dusk side convection cell covers the whole polar region while the dawn side cell keeps in place but shrinks.

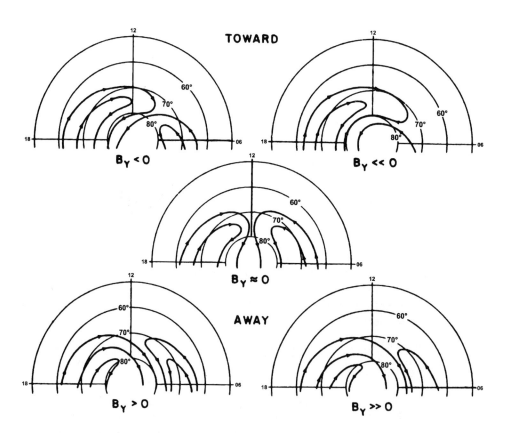

Fig. 8.28. Schematic representation of the dayside high-latitude convection pattern showing its dependence on the IMF when B_z is south. (From Heelis, 1984.)

Sec. 8.11] **High-latitude convection and field-aligned currents** 431

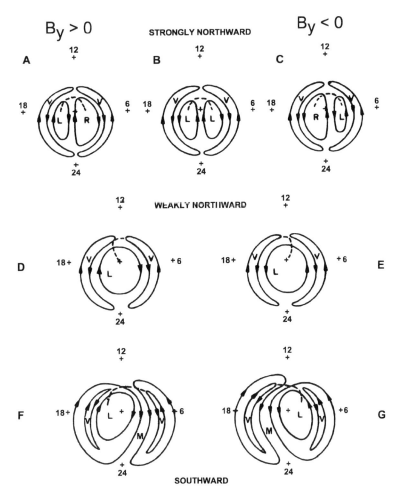

Fig. 8.29. Schematic northern hemisphere polar-cap convection patterns for various orientations of the IMF. The left column (i.e., A, D and F) is for $B_y > 0$, the centre column (i.e., B) has $B_y = 0$, and the right column (i.e., C, E and G) is for $B_y < 0$. The top row (i.e., A, B and C) is for strongly northward IMF ($B_z > 0$); the middle row (i.e., D and E) for weakly northward IMF, and the bottom row (i.e., F and G) for southward IMF. Viscous cells are marked with a "V", merging cells are marked with an "M" and lobe cells with an "L". (From Reiff and Burch, 1985.)

In Fig. 8.29 are the high-latitude convection patterns illustrated for different conditions of B_y and B_z. The most characteristic difference between southward and northward IMF in these patterns is that several convection cells occur for $B_z > 0$. At $B_y \simeq 0$ there are two closed cells inside the polar cap, fairly symmetric around the noon meridian and situated between the more extended and permanent dawn and dusk cells. For $B_y < 0$ the dusk side polar cap cell increases while the opposite is true for $B_y > 0$, then the dawn side polar cap convection cell dominates.

When B_z is close to zero or weakly northward, a circumpolar convection cell covers the polar cap between the dawn and dusk cells. The flow also changes direction when the y-component changes sign from clockwise when $B_y > 0$ to anticlockwise when $B_y < 0$.

When the IMF is southward, the convection pattern is dominated by an asymmetric two-cell system where the dawn cell is enhanced for $B_y < 0$, and for $B_y > 0$ the dusk side is largest.

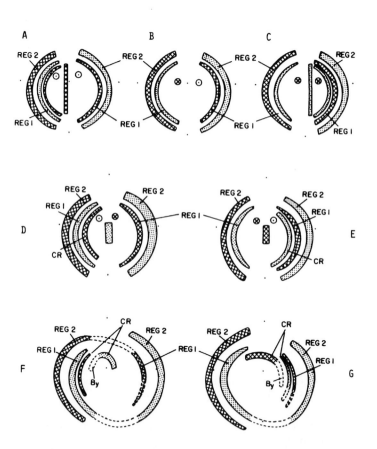

Fig. 8.30. Schematic Birkeland current patterns to correspond to the convection patterns in Fig. 8.29. Upward currents are shown dotted; downward currents are shown cross-hatched. The predominant near-noon cusp region (CR) current is dependent on B_y: outward for $B_y < 0$ and inward for $B_y > 0$, as observed. The pair of sheet currents (CR and B_y) in F and G which occur on the dawn (dusk) side for $B_y < 0$ (> 0) are more narrow than drawn here and may simply appear as a brief reversal in the region 1 currents. (From Reiff and Burch, 1985.)

For $B_y > 0$ there is also a narrow cell inside the dusk cell, which displays the same rotation as the main cell, and similarly there is a small cell inside the dawn cell when $B_y < 0$. It is not clear whether this is a separate cell or whether it simply represents a flow reversal. It is indicated in Fig. 8.29 that the different cells have different origins. The labels V, M and L represent viscous cells, merging cells and tail lobe cells, respectively.

In general, we notice that the area taking part in the high-latitude convection is larger for IMF southward than IMF northward, possibly a result of the more efficient energy transfer from the solar wind to the magnetosphere when IMF < 0.

Finally, in Fig. 8.30 the patterns of Birkeland (field-aligned) currents in the high-latitude ionosphere corresponding to the ionospheric convection pattern shown in Fig. 8.29 are presented. We notice that the region 1 and region 2 currents with opposite flow directions on the dawn and dusk side are always present with the region 2 current flowing on the equatorward side upward on the dawn side and downward on the dusk side. The area covered by these current systems, however, are expanded for southward IMF with respect to IMF positive. When B_y is equal to zero, the region 1 and region 2 current patterns are fairly symmetric with respect to the midnight–noon meridian when IMF $\gg 0$. We also notice a pair of field-aligned currents observed close to the dayside cusp in this situation. When B_y is different from zero, this cusp current pair appears as more complicated current structures. For example, when $B_y > 0$ and $B_z \gg 0$, the downward cusp current is distributed partly along the midnight–noon meridian and partly along a high-latitude circle on the dusk side. The upward current in the cusp, however, is seen as approximately two parallel line currents. For $B_y < 0$ this line current pair close to the cusp changes sign as well as the current along the noon–midnight meridian, while the upward current, polar cap current, is found at high latitudes on the dawn side.

For $B_y \approx 0$ and especially $B_y < 0$, the Birkeland current pattern appears even more structured and fragmented.

8.12 REFERENCES

Akasofu, S.-I. and Chapman, S. (1972) *Solar–Terrestrial Physics,* Oxford, at the Clarendon Press, London.

Cahill, L. J. and Amazeen, P. G. (1963) *J. Geophys. Res.,* **68**, 1835–1843.

Heelis, R. A. (1984) *J. Geophys. Res.,* **89**, 2873–2880.

Hess, W. N. (1967) *The Radiation Belt and the Magnetosphere,* Blaisdell.

Hultqvist, B. (1988) *IRF Scientific Report 196,* Kiruna.

Ijiima, T. and Potemra, T. A. (1976) *J. Geophys. Res.,* **83**, 5971–5979.

Ness, N. F. (1965) *J. Geophys. Res.,* **70**, 2989–3005.

Ness, N. F., Scearce, C. S. and Seek, J. B. (1964) *J. Geophys. Res.,* **69**, 3531–3569.

Ratcliffe, J. A. (1972) *An Introduction to the Ionosphere and the Magnetosphere,* Cambridge, at the University Press.

Reiff, P. H. and Burch, J. L. (1985) *J. Geophys. Res.,* **90**, 1595–1609.

8.13 EXERCISES

1. Assume different values for v in (8.17) and derive r_{mp}^0.

2. Derive (8.51) and discuss the implications of the "frozen-in" concept.

3. Derive (8.72).

4. Discuss (8.74) in terms of different directions on IMF.

9

The aurora

9.1 AN HISTORICAL INTRODUCTION

No one knows where and when the aurora was first observed. Most certainly it has embellished part of the dark night sky and hurled between the stars as long as man has lived on Earth; its apparition and scene, however, might have changed.

Some people think they find traces of auroral allegories in the Bible or in mythological deliberations such as in the cast of the Norse mythology. Others believe they can recognize the auroral motif in ancient stonewall carvings, as for example in India.

In old China it is believed to have been called "flying dragons", in Scotland "The Merry Dancers", in ancient Greece "Chasmata" and in Norway just "Northern Lights" or "Weather Lights".

Among several cultures in the Arctic regions it has been fairly common to relate the aurora to the realm of the death. In Scandinavia, for example, it was believed in some areas that the aurora was the spirit of dead maidens, and among Eskimos in Greenland it was said that it was the spirit of stillborn children playing ball. Others believed it was related to the eternal fight between those who were killed in war or by other brutal means.

In historic time the aurora has been observed as far south as Greece and Italy in Europe and in the northern part of Japan, and in modern times in the southern states of USA.

As in so many other areas of science auroral research had a fumbling start at the end of the Middle Ages. From pamphlets distributed in Germany as late as the 17th and 18th centuries, the aurora was depicted as fighting hordes (Fig. 9.1) and imaginative monsters in the sky, believed to be bad omens, of war, pestilence, famine and fire. It was often seized as an opportunity by both the clerical and the secular power to menace the people to pay their tithes or taxes.

Individuals such as the celebrated Danish astronomer Tycho Brahe (1546–1601), however, who meticulously observed the dark sky, night after night from his Uraniborg, also noticed a few auroral displays without contemplating further about their cause. He did have a few notes (Fig. 9.2), however, which in later

Fig. 9.1. A fantastic illustration of an aurora observed in Middle Europe February 10, 1681. The Sun is just dipping below the horizon to the right, and the aurora is represented by burning castles and an advancing troop of cavalry. The river at the front is the Danube. (From Réthly and Berkes, 1963.)

times have been used to study the occurrence of aurora in history.

As the late medieval time faded away with all its wars and depressions not favourable for auroral recording and in a period of possible low auroral activity, nothing much was reported about the aurora between Brahe's notes and the beginning of the 18th century. As a matter of fact, the phenomenon was considered as new in many academic circles when strong auroral displays occurred in March, 1716, which could be observed over large areas in Europe. When looking back on the large amount of papers which were devoted to these events in the early 18th century, it is not an exception to find articles dealing with the question of the age of the aurora. Notable was the small group of scholars in Scandinavia almost offended by contentions from more renowned scientists in the central parts of Europe that the aurora was a new heavenly display marking the entrance to a new century. The Norwegian historian Schøning (1722–86), for example, wrote a treatise in two parts covering 120 pages in the Proceedings of the Royal Norwegian Academy in 1760, carrying the title: "The age of the Northern Lights proven with testimonies from ancient writers". He strongly refuted the absurdity of claiming that the aurora was a new phenomenon since he could prove that the Vikings had already known this celestial apparition which they so descriptively called the "northern light". The facts that Schøning referred to could be found in the ancient

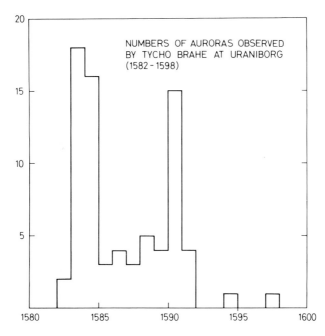

Fig. 9.2. Number of auroras observed per year between 1582 and 1597 by Tycho Brahe. (From Brekke and Egeland, 1983.)

annals from about 1230 AD, called "The King's Mirror". Here it was mentioned as a phenomenon often observed by people sailing to Greenland, although the author himself lived in the middle part of Norway. This statement has remained a puzzle for generations. Since the aurora at our time is probably more often observed in Norway than in Greenland, it is surprising that a Norwegian author talked about it as if he had not seen it. This can, however, be understood by the close connection between the appearance of the aurora and the configuration of the Earth's magnetic field as mentioned in section 4.5.

In "The King's Mirror" attempts were also made to explain the cause of the phenomenon. Elements like a big fire encircling the Earth, radiation of stored sunlight in ice and snow and reflection of solar rays by snow crystals in the air, were ingredients in the scholarly comprehension of the aurora among the Norse population.

Sir Edmond Halley (1656–1742) who was afraid of having to leave this Earth without having had the pleasure of observing the aurora, the only known light phenomenon in the sky that he had not seen, by the way. When he finally experienced the gigantic display of March 1716, he immediately sat down and wrote about his observations. He also launched a theory far more advanced than the almost 500 years old allegations forwarded by the author of "The King's Mirror". Halley took advantage of his detailed knowledge about the Earth's magnetic field and proposed that due to the appearance of the auroral display with its arched form encircling the northern sky, it had to be linked in some way to the geometry of this field. He turned the matter somewhat upside-down and proposed that so-called "magnetic effluvia" streamed out of pores in the ground, and when it reached up into the atmosphere and mingled with the moisture there, aurora was formed.

The modern auroral science is rooted in scholarly discussions in a part of the world where the aurora was seldom seen. It was, however, in these circles that the phenomenon was to become known by its international name, the "aurora borealis", a misleading appellation which has nothing to do with the beautiful goddess of dawn. Since the appearance of the aurora, however, is dominated by its crimson-like colour when it is observed in Paris or Rome, the name is quite adequate. For a northerner, however, used to observing the phenomenon with its spectrum of colours and wild, wavy performance, the name is rather unseemly.

9.2 THE HEIGHT OF THE AURORA

In the early days of science the aurora was considered to be a meteorological phenomenon, and it was often mentioned as a "meteoron". Traditionally in Norway the aurora was often called "weather light", and all around in the country there were weather signs related to the aurora. The northern lights were prognostic as far as weather was concerned, and interestingly enough, the local weather signs around in the country with respect to the aurora were quite similar.

As the repressive Middle Ages in Europe with its hostile attitude towards individual and progressive science faded away after the turmoil following the Reformation and the subsequent Thirty Years' War (1618–1648), the interest for the auroral phenomenon grew, as can be substantiated by the increasing numbers of auroral recordings into the beginning of the 18th century.

The height of the phenomenon became an important issue. Rather conflicting records appeared, where the aurora was said to have been observed below the peaks of the mountains and even as far down as to the surface of the sea. Others maintained that it was a phenomenon appearing in the upper strata of the atmosphere. How far the atmosphere reached above the ground, however, was still an unsolved problem around the turn of the 18th century, although Halley had shown evidence for an exponentially decaying pressure with height – in principle, an atmosphere of infinite extension.

Already the famous French astronomer Gassendi had on one occasion as early as 1621 estimated the height of the aurora to be about 850 km as observed from Paris. A few individual measurements of the auroral altitude were reported from places like Paris, Geneva, Copenhagen and St. Petersburg between 1730 and 1750, giving values varying from 200 to 1000 km; most of the events observed, however, were reported to be situated above 600 km. From Uppsala in Sweden the well-known physicist Anders Celsius (1701–44) and his successors reported on 13 observations of the auroral altitude made by triangulation to be between 380 and 1300 km with a mean height of 760 km.

Sec. 9.2] The height of the aurora

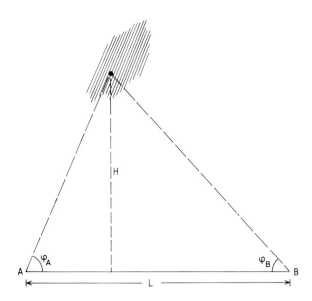

Fig. 9.3. A schematic illustration of the principle for auroral triangulation. Two observers A and B at a known distance L are simultaneously observing a point in an auroral form at the elevation angles φ_A and φ_B, respectively. By simple geometric calculations the height can then be derived.

To perform a precise triangulation (Fig. 9.3) of an auroral form in the early part of the 18th century was not an easy task. Because of the real height of the aurora, which is more than 100 km above the Earth's surface, two observers on ground had to be separated by a substantial distance in order to obtain reliable observations. Furthermore, these observers had to observe exactly the same point in the auroral form at exactly the same time, and this was more or less impossible without communication.

The problem of the height of the aurora could therefore not be solved before the invention of the telephone and the photographic technique, and as this took place toward the end of the 19th century, the Norwegian mathematician Carl Størmer (1874–1957) was ready for a lifelong enterprise in auroral triangulation. From his more than 40,000 auroral pictures he derived the height of 10,000 forms, the statistical distribution of which is presented in Fig. 9.4.

The maximum in the distribution is found to be close to 100 km, none is observed below 70 km, and only about 6.5% of the auroral forms are found above 150 km.

Although improvements have been made to Størmer's method, thanks to TV-cameras and modern automation, these results have proven to be a frame of reference for most of the discussions concerning the height of the aurora. There are indications, however, that thin auroral layers occurring quite frequently below 100 km may have escaped Størmer's careful analysis of photographic plates.

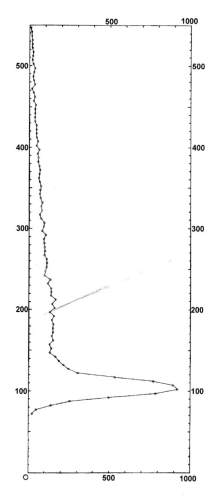

Fig. 9.4. The distribution of 20,000 observations of the height of auroral forms. (From Størmer, 1955.)

9.3 THE OCCURRENCE FREQUENCY OF THE AURORA

When the solar cycle was discovered in 1844 by Schwabe, it was soon realized that the auroral occurrence is strongly related to this cycle. The fact that the auroral displays appeared to agitate the magnetic needle was already reported in 1741 by Hiorter and Celsius, and in 1826 Ørsted was able to explain this relationship by an electric discharge along the auroral arc. A close relationship between the variations in the Earth's magnetic field and the solar cycle was established by the middle of the 19th century by Sabine and Wolf. Then the foundation was made for auroral research to become an intimate part of planetary science implying an electromagnetic coupling between the Sun and the Earth.

As early as 1733 the French astronomer de Mairan, in a beautiful treatise entitled "Traité Physique et Historique de l'Aurore Boréale", introduced a hypothesis that the aurora was the result of an interaction between the atmospheres of the Sun and the Earth. Lacking the knowledge of electromagnetism, he leant upon the forces of gravity only. His basic idea was brought to light again toward the end of the 19th century by the theory that particles from the solar atmosphere were a source of the aurora. This was highlighted by the work of Birkeland who proposed that the aurora was caused by cathode rays (electrons) streaming out from the Sun and guided by the magnetic field towards the polar atmosphere of the Earth.

One early graph showing the high correlation between the number of auroras reported per year in Scandinavia between 1761 and 1877 and the annual sunspot number, is due to Tromholt (1851–1896) and is shown in Fig. 9.5. The correlation is good especially until 1865, after that, however, the number of the auroral observations appears to increase in disproportion. This small discrepancy serves as a warning to people employing statistics of visual auroral sightings in scientific work, because they are the subject of many error sources. In this example the apparent increase in auroral occurrence after 1865 is more a result of Tromholt's dedicated work to encourage people to observe the phenomenon (he had about 2000 correspondents), and also the fact that the report of occurrence of aurora now became a part of the meteorological observations at many weather stations in Scandinavia, than a genuine increase in the auroral occurrence.

A well-known and much used catalogue of auroral observations is due to Rubenson, the director of the Swedish Meteorological Institute in the last part of the 19th century. The annual numbers of recordings are shown in Fig. 9.6. Again the increase of auroral observations towards the end of the 19th century is evident with respect to the earlier part of the time series. Although the Rubenson catalogue represents a meticulous work in collecting newspaper articles, notes, diaries, meteorological journals, etc. for almost 200 years, the statistics have little scientific value in a quantitative sense because it does not represent a homogeneous data set. When studying the auroral activity from annals and old suspect notes, one also need to know something about the activity of the individual recorders and this is often the most difficult part of the task. The likelihood of observing an aurora and getting it written down on paper for later use was far higher toward the end of the 19th century than at the beginning of the 18th. The fact that the aurora has a tendency to appear in relation to the variations in the solar cycle cannot be denied, but to express this in a quantitative manner by, for instance, a correlation coefficient, is far from well established.

On an annual basis it appears that the aurora has a tendency to occur more frequently at equinoxes than solstices (Fig. 9.7), even when corrected for cloudiness and number of dark hours per night. The reason for this is not quite clear but is most likely due to a seasonal difference in the electromagnetic coupling between the solar wind and interplanetary field with the Earth's magnetic field and magnetosphere.

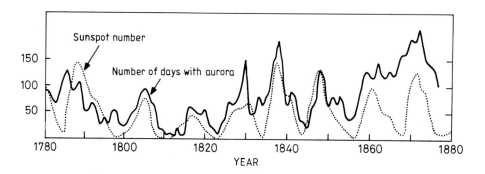

Fig. 9.5. The annual numbers of aurora recorded in Scandinavia between 1780 and 1877 according to a survey made by Sophus Tromholt (1851–1896). These numbers are compared with the annual sunspot number for the same period. (After Brekke and Egeland, 1983.)

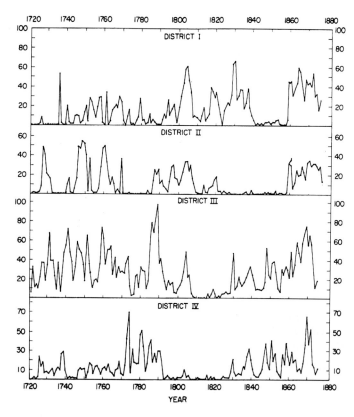

Fig. 9.6. The annual numbers of aurora recorded in Sweden for the period 1720 to 1875 according to a data catalogue prepared by Rubenson (1879, 1882).

Sec. 9.4] The global distribution of the aurora 443

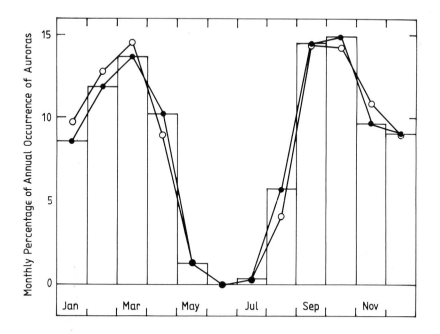

Fig. 9.7. Average monthly percentage of auroral observations according to data compiled in Sweden (dots) from 1720 to 1877 (Rubenson, 1882) and data compiled in Southern Finland (circles) between 1748 and 1843 (Hällström, 1847). (After Nevanlinna, 1995.)

In relation to the disturbances on the Sun, which have a tendency to repeat themselves with a period determined by the rotation period of the Sun as seen from the Earth (close to 27 days), the aurora frequently shows a 27-day periodicity.

9.4 THE GLOBAL DISTRIBUTION OF THE AURORA

The existence of an auroral zone encircling the polar regions were envisaged as early as 1833 by the German Muncke. In 1860 the American physicist Elias Loomis had collected enough data on a global scale to make an attempt to draw the geographical position of the auroral zone. In 1881, however, the Swiss physicist Hermann Fritz published a book called "Das Polarlicht" in which he, on the basis of a large set of global data, was able to draw so-called isochasms in the northern hemisphere, i.e. lines where the aurora occurs with equal frequency at midnight.

Fig. 9.8 gives a reproduction of Fritz's drawing. The numbers labelled on the isochasms indicate the number of nights per year when the aurora occurs. At the maximum zone this is, according to Fritz, more than 100 nights. In Rome and Madrid, however, it is less than one. On the average it corresponds to about one auroral display over Madrid per solar cycle.

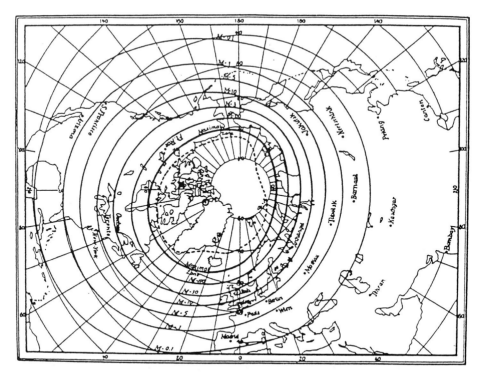

Fig. 9.8. Fritz's map of isochasms showing the occurrence frequency of auroral displays at midnight for different places on the Earth. The labels marked on the curves show the number of nights per year when the aurora can be observed. The geomagnetic pole is marked by a small circle in the northwestern corner of Greenland. The magnetic pole is shown by a cross in Northern Canada. The maximal zone within which the aurora can be observed more than 100 nights per year, is located roughly at 67° in European sector. (From Fritz, 1881.)

When we follow the maximum zone around the polar region, we notice that it follows approximately the geomagnetic latitude close to 67° and touches the northern part of Scandinavia, crosses Novaya Zemlja and passes north of Siberia and reaches the northern part of North America. Greenland and Iceland, however, are in the poleward side of the maximum zone where the auroral occurrence is less.

The mild climate in Scandinavia and the relatively easy access to this part of the world has, because of the position of the auroral zone, given these countries an advantage in ground-based auroral research substantiated by the many Scandinavian pioneers within this field of science.

Since the early 18th century, however, the notion has been put forward that an auroral arc formed a part of a luminous ring encircling the global pole. As a matter of fact, this concept was used as a reference for some height estimates of the aurora based on measurements by one observer only (Fig. 9.9). Assuming that the aurora

Sec. 9.4] **The global distribution of the aurora** 445

was a complete ring at a fixed height encircling the pole, observations from one point of the elevation and azimuth angles to the apex of the auroral arc together with the azimuth angles to the points where the arc went below the horizon, was in principle sufficient to derive the height of this ring if the radius of the Earth was known. The method was first used in 1731 and later on in the 18th and 19th centuries. In 1881, however, Nordenskiöld argued that the centre of the ring was not the geographic pole but rather the geomagnetic pole, at that time determined by Gauss to be in the northern corner of Greenland. Fig. 9.10 is an illustration representing the concept of the auroral ring as redrawn from Nordenskiöld's book "The Vega Expedition".

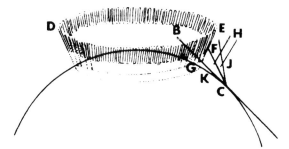

Fig. 9.9. A schematic illustration to show the concept of the triangulation of auroral heights from one point only, assuming the auroral arc is part of a complete circumpolar ring. (From Hansteen, 1827.)

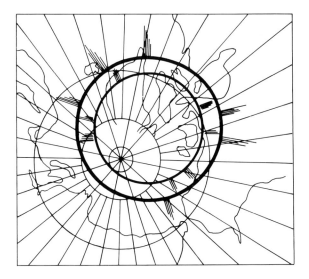

Fig. 9.10. An illustration from the book "The Vega Expedition" by A.E. Nordenskiöld, showing the auroral ring centred at the geomagnetic pole as defined by Gauss. (From Nordenskiöld, 1880–81.)

Fig. 9.11. Examples from Birkeland's terrella experiments showing the two luminous rings around a magnetized fluorescent sphere bombarded by cathode rays or electrons, thus supporting his own theory of the aurora. (From Birkeland, 1913.)

At the entrance of the 20th century when Birkeland set up his laboratory experiments to prove his theory that the aurora was caused by cathode rays (electrons) streaming out of the Sun and caught by the Earth's magnetic field, he was able to produce two illuminated rings around the terrella poles supporting the old concept of auroral rings (Fig. 9.11).

During the International Geophysical Year 1956–57 this theory was put to test when a large array of all-sky cameras was installed in the auroral zones of the Arctic and Antarctic regions. When analysing all-sky films from this network of stations, Feldstein was able to show (Fig. 9.12) that indeed the aurora at any instant forms an annular belt around the geomagnetic pole which was to become known as *the auroral oval*.

This oval has a fixed position with respect to the Sun. It is compressed somewhat on the sunward side where it on the average reaches down to approximately 78° geomagnetic latitude, and stretched out toward the antisunward side where it reaches down to approximately 67° geomagnetic latitude on the average. As a station on Earth at auroral latitudes rotates around the geographic pole in the course of a day, it can be partly inside, partly poleward and partly equatorward of the oval. A station at very high latitude, however, may be poleward of the oval throughout the whole day. This relative motion between an observer at high latitude on ground and the auroral oval also explains why the auroral zone is centred at 67° geomagnetic latitude.

This oval, however, is not static as it expands and contracts with respect to the situation on the Sun and the potential distribution across the polar cap, the dawn–dusk potential. Fig. 9.13 shows the variation of the polar cap area inside the auroral oval as function of this dawn–dusk potential indicated by the equivalent solar wind electric field. When the potential grows, the oval expands to lower latitudes as it also widens in latitudinal width.

Sec. 9.4] The global distribution of the aurora

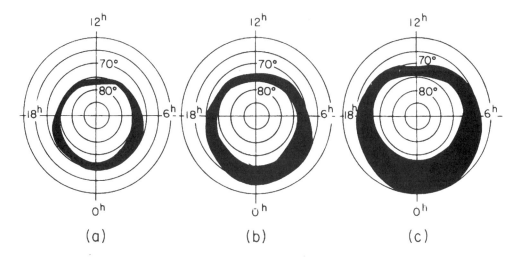

Fig. 9.12. The auroral oval for different levels of disturbance. (a) For quiet conditions. (b) For medium disturbance condition, and (c) for strongly disturbed conditions. (From Starkov and Feldstein, 1967.)

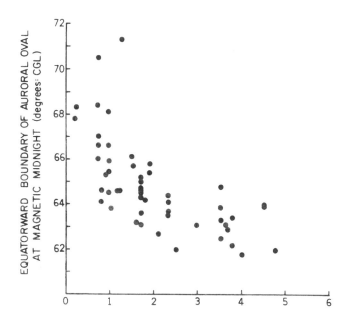

Fig. 9.13. Location of the equatorward border of the auroral oval versus the maximum 30-min average $\mathbf{B}_S \cdot \mathbf{V}_S$ value during six hours preceding the time of each auroral observation. $\mathbf{B}_S$ and $\mathbf{V}_S$ are the solar wind magnetic field and velocity, respectively. (From Kamide, 1988.)

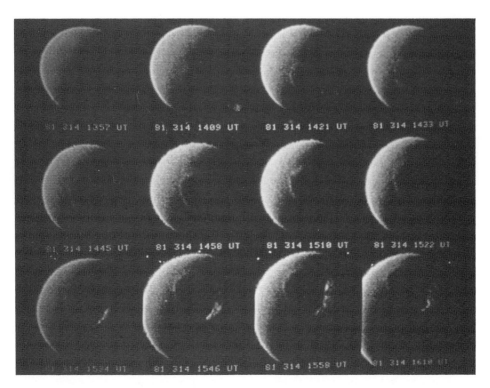

Fig. 9.14. A series of images of the northern auroral oval prior to and during an isolated, modest substorm on November 10, 1981. The dayglow and auroral oval are seen primarily in the emissions of atomic oxygen at 130.4 nm. (From Frank and Craven, 1988.)

In recent years it has been possible to image this oval in its total extent from space, and Fig. 9.14 shows a few examples of the oval. It is becoming evident from these images that the oval is always present although it changes its shape and may sometimes be bifurcated by auroras crossing the polar cap and sometimes appears as a spoked ring when torches of aurora stretch toward the pole from the main luminous oval. Furthermore, it is important to notice that the aurora is not only a night-time phenomenon, as we may think when watching it by eye from the ground, but when observed in the ultraviolet from space as these figures show, the auroral process is always active in the day as well as at night.

9.5 THE AURORAL APPEARANCE

As already mentioned, when the aurora is observed at lower latitudes, i.e. when the oval has expanded equatorward, it appears red, and that is how the phenomenon got its name, the northern dawn – aurora borealis. At auroral latitudes the phenomenon exposes a multitude of colours, often mingled together in such a drapery of beauty that it may be hard to conceive that the aurora seen from Madrid is the same as the one seen from Tromsø.

Auroral spectroscopy was introduced by Ångström who in 1866 observed the most prominent auroral emission, the yellow–green emission, and determined its wavelength to be 5577 units, the unit known as the ångström at a later time. Nearly 60 years, however, had to pass before one could find an explanation for this emission. Then it was revealed that it was due to a transition in atomic oxygen from the metastable excited state ^1S to the lower ^1D state. It is still a puzzle why this emission is so strong in the aurora, because the metastable state makes it very vulnerable to collisions as these are very frequent at auroral heights. The green line emission is centred around 100 km.

Auroral spectra have been obtained in great amount by many different observers. Because the spectra are so extremely variable due to varying auroral conditions, the study of auroral spectra is a whole scientific field by itself. The different emissions are due to different excited atmospheric atoms, molecules and ions. Careful studies of the intensities and height variations of these emissions have therefore increased our knowledge considerably of the composition and temperature in the upper atmosphere above 90 km. As a matter of fact, before the 1960s when rockets and incoherent scatter radars were brought into use in ionospheric science, our main knowledge about the composition of the upper polar atmosphere resulted from optical auroral research.

Fig. 9.15 shows some examples of typical spectra for medium to bright auroras in the visual part of the spectrum.

The observed features in the spectrum are almost all due to lines and bands of neutral or ionized N_2, O, O_2 and N, roughly in the same order of importance. Some of the most outstanding lines and bands are indicated in the spectrum such as the oxygen emissions at 5577 Å and 6300 Å and the N_2^+ 1N band with a maximum at 4278 Å. Some of the most characteristic emissions together with their emitting source are listed in Table 9.1.

The auroral emissions are then due to transitions between energy states in atoms and molecules. We will not go into detail here about the many possibilities for transitions in atoms and especially molecules, but will only indicate that the wavelength of the light emission due to a transition between two energy states, E_1 and E_2, can be expressed as

$$\lambda = \frac{hc}{E_2 - E_1}$$

In the reverse direction a light emission with this wavelength can be absorbed by an atom or a molecule to leave the particle in an excited state ($E_2 > E_1$). For the atmospheric species of interest here, we have seen that the ionization potential is of the order of $E_I = 15$ eV. Therefore the different emissions in the aurora are usually associated with wavelengths larger than the one corresponding to this energy:

$$\lambda_I = \frac{E_I}{hc} = 824 \text{ Å} = 82.4 \text{ nm}$$

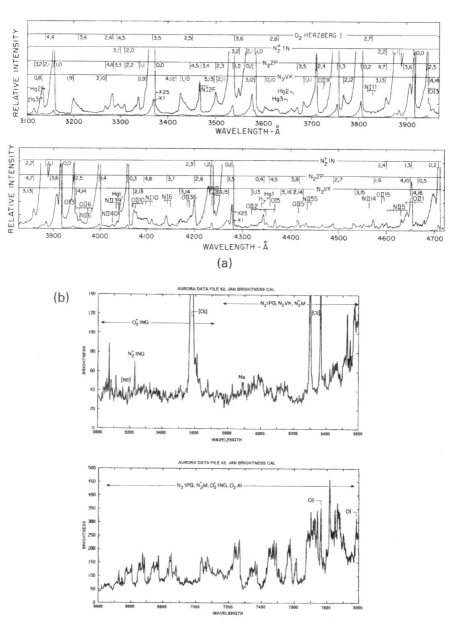

Fig. 9.15. (a) Auroral spectrum between 3100 and 4700 Å. The two different curves are at the gains shown. Bands and atomic features present or possibly present are indicated. The ordinates are relative intensity uncorrected for atmospheric transmission. (From Vallance Jones, 1974.) (b) Auroral spectrum in the visible wavelength region 5000–7000 Å showing a diversity of spectral emission features, some of which are indicated by their assumed sources. (From Rees, 1989.)

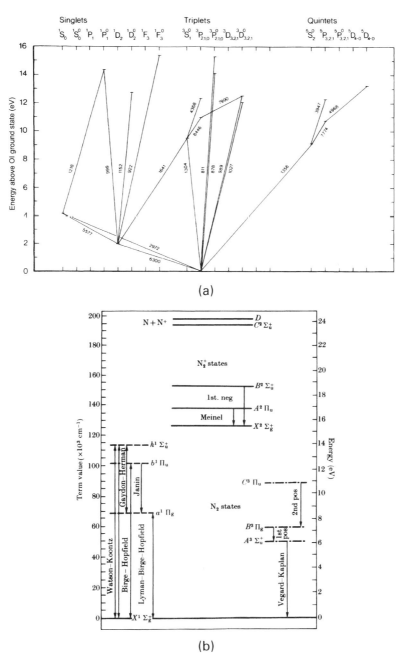

Fig. 9.16. (a) Energy levels in atomic oxygen related to some of the most characteristic auroral emissions such as 6300 and 5577 Å. (From Rees, 1989.) (b) Energy levels in molecular nitrogen related to some of the auroral emissions such as 4278 and 3914 Å. (From Akasofu and Chapman, 1972.)

Table 9.1. Characteristic auroral emissions and their emission source (upper state). The lifetime and dominant quenching particle for each upper state is also given

Emission (in Å)	Upper state	Lifetime	Dominant quenching particle, Q
$\lambda 5577$	$O_I{}^1S$	0.7 s	O_2
$\lambda 2972$			O
$\lambda\lambda 6300$–64	$O_I{}^1D$	110 s	N_2
$\lambda\lambda 7319$–30	$O_{II}{}^2P$	5 s	N_2
$\lambda\lambda 3727$–9	$O_{II}{}^2D$	2.9 h	N_2
$\lambda\lambda 10395$–404	$N_I{}^2P$	12 s	N_2
$\lambda 3466$			O_2
$\lambda\lambda 5199$–201	$N_I{}^2D$	26 h	O_2
			O
			NO
$\lambda 5755$	$N_{II}{}^1S$	0.9 s	O_2, O
$\lambda 6584$	$N_{II}{}^1D$	246 s	O_2, O
N_2(LBH)	$a^1\pi_g$	0.14 ms	
N_2(VK)	$A^3\Sigma_u^+$	~ 2 s^n	O
O_2(Atm.)	$b^1\Sigma_g^+$	12 s	N_2
O_2(IR Atm.)	$a^1\Delta_g$	60 mn	O_2
N_2^+(1N)	$B^2\Sigma$	70 ns	$N_2 + O_2$
N_2^+(M)	$A^2\pi$	14 μs	N_2
N_2(1P)	$B^3\Sigma$	6 μs	N_2
N_2(2P)	$C^3\pi$	50 ns	O_2
O_2^+(1N)	$b^4\Sigma$	1.2 μs	N_2

Fig. 9.16b shows the energy levels of different auroral emissions from O, N_2 and N_2^+. The characteristic auroral green line is found to correspond to an upper level with energy of about 4.2 eV, while the oxygen red line at 6300 Å and the N_2^+ 1 neg. emission at 4278 Å correspond to upper states with energies close to 2 eV and 18.75 eV, respectively.

9.6 AURORAL PARTICLES

It has been known for a long time that energetic electrons and protons penetrating the polar atmosphere from above are the ultimate cause of the auroral emissions. Not until rockets equipped with instruments to catch the incoming electrons and to measure their energies could be launched into the auroral forms, was it known what the energy spectrum of the precipitating electrons were. Fig. 9.17 shows a few examples of energy spectra of auroral electrons. Typically the number of electrons with energies above a few tens of keV are falling rapidly, while there is

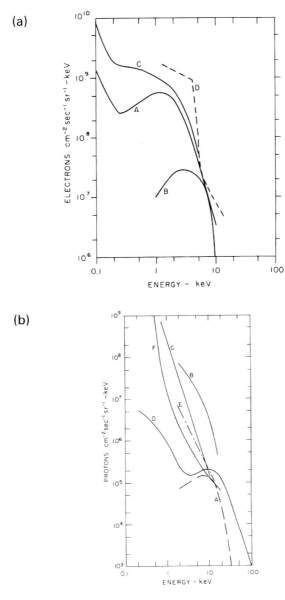

Fig. 9.17. Energy spectra of precipitating particles related to different auroral forms. (a) Electron spectra observed by satellites above the auroral oval. A: Early evening auroral oval; B: Auroral oval early morning; C: Broad weak homogeneous arc; D: Auroral oval very disturbed conditions and strong aurora at midnight sector. (b) Proton spectra observed in rocket flights over aurora. A: Early evening proton aurora; B: Medium strong aurora; C: Post break-up aurora; D: Recovery phase aurora; E: Recovery phase aurora; F: Break-up aurora. (From Vallance Jones, 1974.)

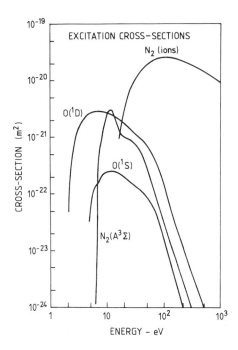

Fig. 9.18. Excitation and ionization cross-sections, for electron impact of some characteristic states of importance in aurora. (After Vallance Jones, 1974.)

a tendency for a peak in the number of electrons at a few keV. The number of electrons increases strongly towards lower energy, below, say, a couple of 100 eV.

Based on such energy spectra of the incoming particles it would be possible to derive the auroral emission spectra if all cross-sections for the various transitions between the numerous energy states of atoms and molecules in respect of these energies were known. Since this is extremely difficult to obtain in an exact manner, one has to lean on only a few typical cross-sections, some of which are shown in Fig. 9.18 as function of the energy of the precipitating particles.

We notice that the cross-sections are of the order of 10^{-20}–10^{-15} cm^2 or 10^{-24}–10^{-19} m^2 in the energy range between 1 eV and 1 keV. Table 9.2 presents the maximum cross-section as well as the typical energy at this maximum for a few important energy states of atomic and molecular species in the auroral atmosphere.

When comparing the energy at the maximum of the different cross-sections with the energy spectra of the penetrating auroral particles, it becomes clear that there is a high-energy tail of particles which must be decelerated before they can become efficiently active in the ionization and excitation processes. A 1 keV particle has to go through a whole series of secondary, tertiary and higher-order collisions before it finally has its energy reduced below the average 15 eV to excite an atom. During this process about 300 ion pairs have been formed due to one auroral particle with

Auroral particles

1 keV energy at the top of the ionosphere. We notice here that the maximum cross-section for the O(^{1}S) and O(^{1}D) states are 0.25×10^{-21} and 0.28×10^{-20} m^2, respectively, and that the maxima occur at roughly the same energy. Let us then assume a monoenergetic electron beam of intensity I_∞ at the top of the atmosphere, which will be degraded as it penetrates downward due to collisions with atomic oxygen. The maximum production of the two states will occur when (see section 5.2)

$$\sigma \cdot n_m \cdot H = 1 \tag{9.1}$$

where σ is the cross-section, n_m the density of the target atoms at maximum, and H the scale height. When assuming H to be constant for the density of the target atoms, then

$$n = n_0 \exp(-z/H) \tag{9.2}$$

Here n_0 is the density of the target atoms at $z = 0$ which is a reference height. The height of maximum production is then

$$z_m = +H \ln(\sigma \cdot n_0 H) \tag{9.3}$$

Since $\sigma(\mathrm{O}^1\mathrm{D}) = 10 \cdot \sigma(\mathrm{O}^1\mathrm{S})$, we find that

$$z_m(\mathrm{O}^1 D) = z_m(\mathrm{O}^1 S) + H \cdot \ln 10 \tag{9.4}$$

The scale height at about 100 km where the production of the ^{1}S state has a maximum, is about 7.0 km. The maximum in the production of the O(^{1}D) state should then be about 16 km higher according to this simple calculation. In reality the height difference is greater than this.

Table 9.2. Some representative data on cross-sections in the auroral atmosphere

Excited state	σ_{max} (m^2)	E_{max} (eV)
O(^{1}S)	0.25×10^{-21}	10
O(^{1}D)	0.28×10^{-20}	5.6
N$_2$(A$^3\Sigma$)	0.28×10^{-20}	10
N$_2$(B$^3\pi$)	0.11×10^{-19}	12
O$_2$(a$^1\Delta$)	0.85×10^{-21}	6.5
O$_2$(b$^1\Sigma$)	0.20×10^{-21}	6
N$_2^+$ 1N(0,0)	0.17×10^{-20}	100
O$_2^+$ 1N(1,0)	0.43×10^{-21}	100

456   The aurora   [Ch. 9

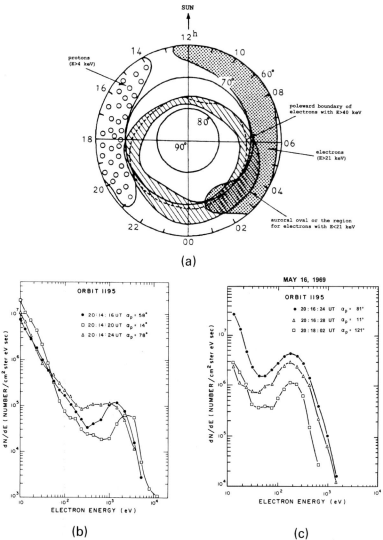

Fig. 9.19. (a) Characteristic precipitation zones at high latitude. The typical auroral oval is associated with electrons with energy less than 21 keV. Equatorward of the oval on the evening side is the proton precipitation zone corresponding to protons of energy above 4 keV. On the morning side there is an equatorward zone of high energetic electrons (> 21 keV). The poleward boundary of trapped electrons of energy above 40 keV is illustrated by a dotted line. (From Tohmatsu, 1990). (b) Typical electron spectra observed in the polar cap at three different pitch-angles from the ISIS–I satellite. (From Winningham and Heikkila, 1974.) (c) Typical electron spectra observed in the dayside oval at three different pitch-angles from the ISIS–I satellite. (From Winningham and Heikkila, 1974.)

Sec. 9.8]   The energy deposition profiles of auroral particles   457

9.7 PRECIPITATION PATTERNS OF AURORAL PARTICLES

There are other ways by which auroral particles can be mapped in addition to rocket probes launched through auroral forms. During the deceleration process X-rays are formed in the auroral ionosphere which can be observed from balloons at lower heights. These X-ray emissions then give information of the precipitation above the balloon height. From particle detectors in satellites the particles can be mapped before they hit the atmosphere below, and from the energy and direction of the particle movement the precipitation zone can be deduced.

From the vast data sets now available on auroral particle precipitation it has been found, in addition to the visual auroral oval, that there are other zones of distinct precipitation (Fig. 9.19). The central region of the average oval appears to correspond to the poleward boundary of trapped electrons with energy greater than 40 keV. This tends to be an important clue with respect to the origin of the auroral particles since trapped electrons have to be on closed field lines. The central region of the average oval therefore appears to be the demarcation between open and closed field lines. Equatorward of the auroral oval in the late morning side and partly covering the oval at the early morning sector is a pattern of precipitating electrons with energy higher than 21 keV. On the equatorward side of the oval in the evening sector there is a zone of precipitating protons with energies higher than 4 keV. The auroral oval in itself appears to be closely related to the precipitation zone of electrons carrying energies of less than 21 keV.

We have demonstrated that the energy spectra of auroral electrons as measured by rockets have a strong decaying slope with increasing energy and that they sometimes display an enhancement at a few keV. One can imagine these spectra as composed of two parts, where one forms the background spectrum, fairly well represented by an exponential or a power law distribution. The other part, however, represents a quasi-monoenergetic beam with a beam width of a few keV. It is evident that these spectra must be replicas of the energy source processes, and it is a problem of great interest among auroral scientists how to find the ultimate source of the auroral particles.

Fig. 9.20 shows a comparison between an electron spectrum observed in the dayside auroral oval and a similar spectrum observed in the magnetosheath by a satellite at $L = 20$. The scales are in relative units in order to enhance the similarity between the two. This should indicate that the magnetosheath is a likely candidate for being a reservoir to auroral electrons.

9.8   THE ENERGY DEPOSITION PROFILES OF AURORAL PARTICLES

We have already discussed the stopping cross-section for a charged particle penetrating the atmosphere and the distance it has to penetrate before it is thermalized, expressed as the range. The stopping cross-section for protons and electrons in O, O_2 and N_2 are illustrated in Fig. 9.21, derived partly from theory and partly from laboratory experiments.

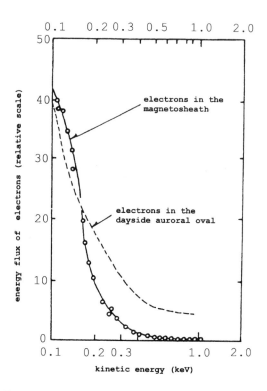

Fig. 9.20. Energy spectrum from the dayside auroral oval of auroral particles related to the energy spectrum of electrons in the magnetosheath at 19.8 R_e. (From Sharp and Johnson, 1974.)

Based on such a cross-section, a neutral atmosphere model and monoenergetic beams of auroral particles penetrating the atmosphere from above, it is in principle possible to derive the energy deposition profile for each individual energy. Fig. 9.22 demonstrates such profiles derived for energies between 2 and 20 keV. As expected, the peak of the deposition rate profile occurs at a lower height when the energy increases. The peak energy deposited per unit length in the atmosphere, however, increases with energy. Table 9.3 gives the maximum energy deposition per unit length and the height of this maximum as function of the initial energies used in Fig. 9.22. As can be noticed, a 10 keV particle has a maximum deposition rate at about 105 km. Since the aurora is found to occur most frequently in this height region, one can conclude, also based on the rocket measurements of auroral particles, of course, that the typical energy of these particles is below 10 keV.

In this context the range–energy relation $R(\varepsilon_0)$ is often introduced. $R(\varepsilon_0)$ gives the depth of penetration in a particular medium as function of the incident particle energy ε_0. The unit of R is typically given in g/cm^2. The important quantity is not the particle path length but rather the matter traversed by the

Sec. 9.8] The energy deposition profiles of auroral particles

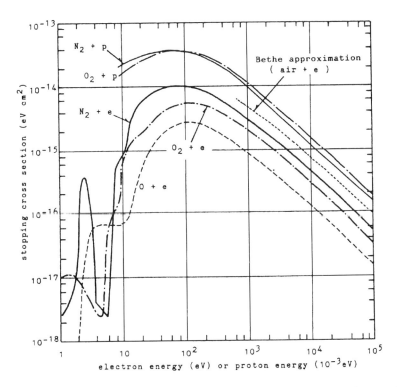

Fig. 9.21. Stopping cross-sections of electrons and protons in N_2, O_2 and O. (From Tohmatsu, 1990.)

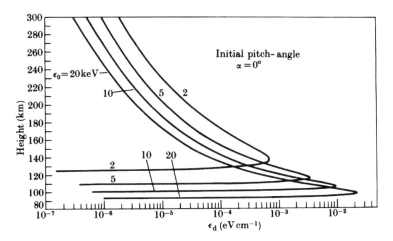

Fig. 9.22. The height profile of the energy deposition by electrons in the CIRA (1965) mean atmosphere normalized to one incident electron. (From Berger et al., 1970.)

Table 9.3. Height (H_m) at which the energy deposition peaks, and the maximum energy deposition (ε_l) per unit length according to the energy (ε_0) of the precipitating particle

ε_0 (keV)	H (km)	ε_l (eV m^{-1})
20	97	2.22
10	105	0.97
5	115	0.34
2	136	0.07

particle along its path, and this is given by:

$$N(l) = \int_0^l n(s)ds \qquad (9.5)$$

where l is the point of interest and $n(s)$ is the particle density of the matter along the path. For vertical incidence the corresponding function will be

$$N(h) = \int_h^\infty n(z)dz \qquad (9.6)$$

where h is the height of interest. For protons in air, the range–energy relation is given by (Rees, 1989):

$$R(\varepsilon_0) = 5.05 \cdot 10^{-6} \, \varepsilon_0^{0.75} \text{ g cm}^{-2} \qquad (9.7)$$

for ε_0 incident proton energy between 1 keV and 100 keV. The corresponding range–energy relation for incident electron energies between 200 eV and 50 keV is given by (Rees, 1989):

$$R(\varepsilon_0) = 4.30 \cdot 10^{-7} + 5.36 \cdot 10^{-6} \, \varepsilon_0^{-1.67} \text{ g cm}^{-2} \qquad (9.8)$$

When the incident energy ε_0 is known, $N(h)$ is calculated by adjusting the integration limit h until the stopping altitude h_s is found, given by:

$$N(h_s) = R(\varepsilon_0) \qquad (9.9)$$

The corresponding stopping altitudes for typical electron and proton incident energies are shown between 40 and 130 km in Fig. 9.23.

Now applying a library of all the individual cross-sections for ionization in the auroral atmosphere one can derive ionization profiles for monoenergetic particle beams as well as composite particle spectra. Fig. 9.24 shows typical ionization profiles derived for monoenergetic electron beams with energies between 2 and 100 keV normalized to the incident electrons per cm^2 and presented in units of ion pairs per cm^3 s. The shape of the profiles are very similar to the profiles for the energy deposition rate indicating that the ionization per energy is constant and independent of height. Of course, the electron beams are not monoenergetic,

Sec. 9.8]   The energy deposition profiles of auroral particles   461

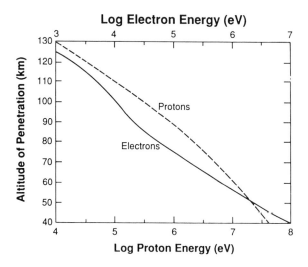

Fig. 9.23. Stopping altitude for electrons (1–10^4 keV) and protons (10–10^5 keV) in the case of vertical incidence to the atmosphere. (From Luhmann, 1995.)

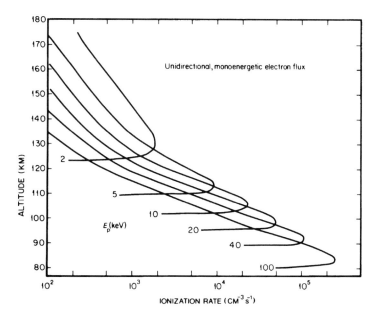

Fig. 9.24. Ionization profiles in the Earth's atmosphere of unidirectional and monoenergetic electron beams. The labels on the curve indicate the energy in keV. The ionization rate is given in ion pairs cm^{-3} s^{-1}. The primary electron flux is 10^8 erg cm^{-2} s^{-1} or 10^5 J m^{-2} s^{-1}. (From Rees, 1989.)

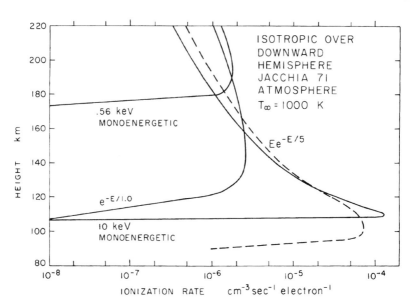

Fig. 9.25. Calculated ion production rates for electrons of various initial energy spectra. E is in keV. The particles are assumed to have an isotropic distribution over downward hemisphere. (From Vallance Jones, 1974.)

but having developed a library of such ionization profiles for a variety of energies within the energy range of interest, composed energy spectra can be used either from measurements or theoretically and more "realistic" ionization profiles can be derived. In Fig. 9.25 are two ionization profiles derived from such composed spectra represented by the Maxwellian function

$$\phi = \phi_0 \varepsilon \exp^{-\varepsilon/\varepsilon_0} \qquad (9.10)$$

where ε_0 is the characteristic energy and ϕ_0 is a normalized flux. The characteristic energies used are 1.0 and 5.0 keV, respectively. These profiles are compared with the ionization profiles derived from monoenergetic electron beams with 0.56 and 10 keV, respectively. The composite spectra produce a broader profile with respect to height, especially do they reach down to lower altitudes due to the high energetic tail in the spectrum.

Having then a library of excitation cross-sections for a number of reactions adequate for the auroral processes involving primary and secondary electrons, it is then possible to estimate the production profiles as function of height. In Fig. 9.26 we show some excitation rate profiles in units of $cm^{-3} s^{-1}$ as function of height for the ^1D and ^1S states in atomic oxygen as well as the $B^2\Sigma_u^+$ in molecular nitrogen, corresponding to the upper state of the 6300 Å, 5577 Å and 4278 Å emissions, respectively. The $N_2^+(B^2\Sigma_u^+)$ and the $O(^1S)$ state peak approximately at the same height, while the ionized nitrogen profile falls off very rapidly by height above the peak, partly due to the relatively strong decrease in the density of N_2 compared

Sec. 9.9] Deriving energy spectra from electron density profiles 463

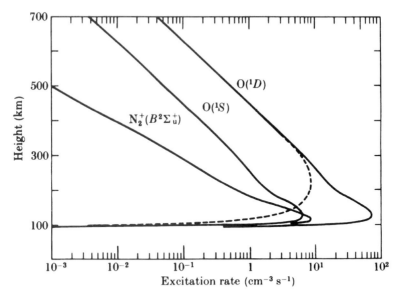

Fig. 9.26. The height distribution of the rate of excitation of the $N_2^+(B^2\Sigma_u^+)$, $O(^1S)$ and $O(^1D)$ states for primary auroral electrons of spectral form $J_0 \exp(-\varepsilon/5000)$ in units of cm^2 s ster eV, where ε is in eV. $J_0 = 10^3$ $(cm^2\ ster\ s\ eV)^{-1}$. The dashed curve also takes deactivation into account. (From Kamiyama, 1966.)

to the density of O above 150 km. The energy spectrum used in these calculations has an e-folding energy of 5 keV.

9.9 DERIVING ENERGY SPECTRA FROM ELECTRON DENSITY PROFILES

In the reversed sense it is possible to retrieve the energy spectrum of the precipitating auroral particles if the ionization profile can be measured by, for example, a rocket probe or an incoherent scatter radar. If we assume that a steady state is valid, then the electron production at a height z in the E-region is approximately given by:

$$q_0(z) = \alpha(z) \cdot (n_e(z))^2 \qquad (9.11)$$

where α is the recombination coefficient and $n_e(z)$ is the electron density as function of height. An observed electron density profile therefore can, to a first approximation, when transport terms are neglected and steady state is assumed, yield the ion-pair production profile. Applying now a library of ionization profiles $q_i(\varepsilon_i, z)$ derived for a large variety of monoenergetic beams of energy ε_i, a combination of these profiles should yield the observed ion production profile. Then assuming that

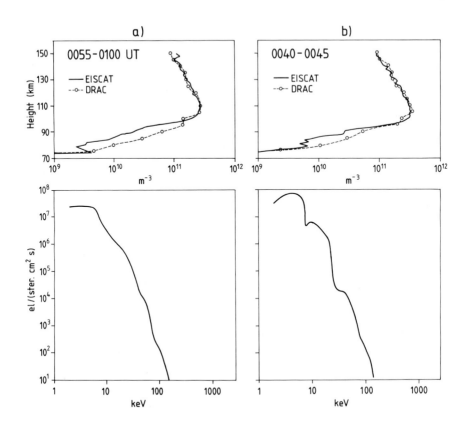

Fig. 9.27. Upper panels: Night-time electron density profiles observed by EISCAT on August 14, 1985 at 00:40 to 00:45 UT and 00:55 to 01:00 UT respectively. Also included (dashed line) are the synthetic profiles as calculated by using the corresponding electron spectra shown in the bottom panels. Bottom panels: Energy spectra derived by following the procedure described in the text from the corresponding measured electron profiles in the upper panels. (From Brekke et al., 1989.)

we can form a linear combination of $q_i(\varepsilon_i, z)$ at each discrete height z_j, we have

$$q(z_j) = \sum_{i=1}^{N} a_i \cdot q_i(\varepsilon_i, z_j) \qquad (9.12)$$

where the energy range is $\varepsilon_1 - \varepsilon_N$ and the number of heights can be chosen arbitrarily. The factor a_i now represents the weight with which the different monoenergetic energy beams contribute to the total profile.

By making a least squares fit to the deduced ionization profile $q_0(z_j)$ by means of the observed electron densities at the discrete heights z_j, we try to minimize the sum of the squared differences:

Sec. 9.9]   Deriving energy spectra from electron density profiles   465

$$\delta = \sum_{j=1}^{M}[q_0(z_j) - q(z_j)]^2 \tag{9.13}$$

$$= \sum_{j=1}^{M}[q_0(z_j) - \sum_{i=1}^{N} a_i q_i(\varepsilon_i, z_j)]^2$$

where M is the number of heights between the lower height z_1 and the upper height z_M. By minimizing δ with respect to the coefficients $\{a_i\}$ we derive the spectrum. Fig. 9.27 shows an observed electron density profile during an auroral event together with the steady-state ion production profile and finally the energy spectrum of the electrons derived as just described. The spectrum is rather similar to spectra observed in the auroral forms directly by rocket probes. Therefore measured electron density profiles can be used at least as a first approximation to derive the energy spectra of auroral particles from the ground in a quasi-continuous manner. The method can actually be improved by allowing for time variations in n_e. This is certainly more appropriate since there are rather dramatic time variations in n_e during the auroral displays.

An auroral electron emits electromagnetic radiation when it is deflected by Coulomb collisions with atmospheric molecules. This radiation, called bremsstrahlung, represents a flat continuum in the spectral range ν given by

$$\nu < \frac{\varepsilon}{h}$$

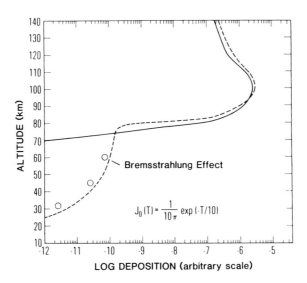

Fig. 9.28. Example of an energy-deposition profile for an incident electron spectrum of the form indicated including the contribution from absorbed bremsstrahlung photons. (From Luhmann, 1995.)

where ε is the energy of the penetrating particle. The X-ray region of this spectrum is especially important since an X-ray photon at a given energy is able to penetrate to lower altitudes than its parent charge particle. Therefore the bremsstrahlung X-rays can produce ionization much lower down in the atmosphere than the stopping altitude of primary electrons. Fig. 9.28 represents model calculations of the ionization profiles produced by penetrating electrons and their accompanying X-rays when the primary particles have an e-folding energy spectrum with a characteristic energy of 5 keV. While the primary and secondary particles are stopped above 90 km where they ionize the atmospheric constituents, the X-rays may reach below 30 km where they create ionization that can be of importance for the fair weather electric circuit.

This also means that auroral X-ray emissions can be monitored at balloon heights, and balloons therefore represent an alternative platform to study the spatial and temporal variations of auroral precipitation. Caution has to be taken, however, since X-rays at balloon heights are mainly formed by the high energetic tail of the auroral particles ($\varepsilon > 25$ keV) while the auroral forms are mainly due to particles with energy less than 10 keV, and it is not well known how particles at these different energy ranges are related to each other.

Since the auroral X-rays contribute so strongly to the ionization of the D-region, auroral events are often found to be related to fairly strong galactic cosmic radio noise at 20–40 MHz.

9.10 EXCITATION PROCESSES IN THE AURORA

One of the best understood auroral emissions is related to the excitation of the N_2^+ ion, more specifically to the $B^2\Sigma_u^+$ state which has a maximum cross-section of excitation close to 100 eV (see Fig. 9.19). It is most often monitored by the (0,0) or the (0,1) emission bands at 4278 Å (427.8 nm) or 3914 Å (391.4 nm), respectively. Because these are spontaneous emissions, the radiation occurs at the incidence of the primary and secondary particles (within 10^{-7} s). Approximately 1 photon at 3914 Å appears, according to laboratory experiments to be produced per 50 ion pairs formed in the atmosphere by the auroral particles. Therefore the numbers of photons resulting from an incident electron with an initial energy ε_0 can be calculated to be

$$\eta(3914 \text{ Å}) = 0.02 \cdot \frac{n(N_2)}{n} \cdot \frac{\varepsilon_0}{W} \tag{9.14}$$

where W is the mean ionization energy equal to about 35 eV. $n(N_2)$ and n are the number densities of the nitrogen molecules and the total number density in the atmosphere at the height of the emission, respectively. Since $n(N_2) \approx 0.80\, n$, we find that

$$\eta(3914 \text{ Å}) \approx 0.5 \cdot \varepsilon_0 \tag{9.15}$$

when ε_0 is given in keV. There is therefore a close relationship between the emission rate η of the N_2^+ emissions and the energy ε_0 of the precipitating particle. Note that $\eta(3914 \text{ Å})/\eta(4278 \text{ Å}) = 3.3$.

Excitation processes in the aurora

The most predominant emission in the high-latitude aurora is the yellow–green line at 5577 Å. The line is known to be due to the ^1S–^1D transition in the atomic oxygen, where the ^1S state has a lifetime against radiation of approximately 0.75 s. The transition is therefore so-called forbidden and leaving the oxygen atom long enough in the excited ^1S state that it can be quenched by collisions with other gas particles in the atmosphere. How the excitation occurs, is still not clear, and much work has been contributed in the attempt to solve the problem.

One likely candidate, however, is by direct impact of secondary and higher-order electrons produced during the stopping process of an incoming high-energetic auroral electron. This process can be illustrated by the following reaction equation:

$$O(^3P) + e \to O(^1S) + e \tag{9.16}$$

Another possible source for excitation of the $O(^1S)$ state is by dissociative recombination with an O_2^+ ion:

$$O_2^+ + e \to O(^1S) + O \tag{9.17}$$

or an energy transfer from an excited N_2 molecule as follows:

$$N_2(A^2\Sigma) + O \to O(^1S) + N_2 \tag{9.18}$$

Both of these last processes will have a delay time with respect to the initial ionization and excitation process by the auroral primary and secondary particles. If the reaction rates for the two processes are $k_{O_2^+}$ and k_{N_2}, respectively, then the time constants involved will be

$$\tau_{O_2^+} = \frac{1}{k_{O_2^+} \cdot [O_2]} \tag{9.19}$$

$$\tau_{N_2} = \frac{1}{k_{N_2} \cdot [N_2]} \tag{9.20}$$

respectively. If these last two processes are effective in producing the $O(^1S)$ state, it should show up as a delay in the 5577 Å emission compared to the 4278 Å emission which is spontaneously emitted at the incidence of excitation by the primary and secondary particles.

When observing the 5577 Å emission carefully in comparison with the 4278 Å emission, for example, in so-called pulsating aurora which often displays quasi-regularly sinusoidal variations in intensity, one often finds a phase shift between the two time series. This is shown in Fig. 9.29a where the 4278 Å emission is leading the other in time. Whether this time delay is only due to the time constant against radiation of the $O(^1S)$ state or a combination of that with a delayed production due to recombination of O_2^+ and energy transfer of N_2, is still a question of uncertainty. Probably all the agents contribute.

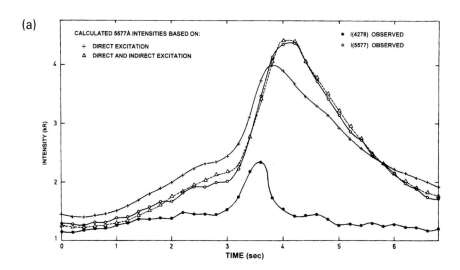

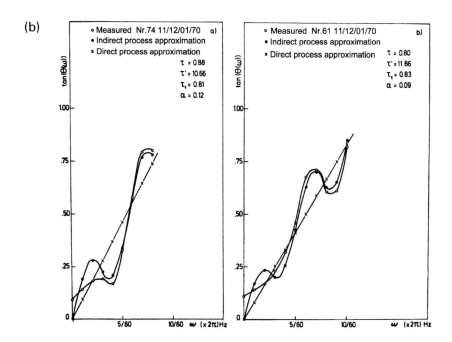

Fig. 9.29. (a) Intensity variations of 5577 Å and 4278 Å in a pulse of a pulsating aurora showing a phase shift between the two emissions. Also indicated are theoretical estimates of the 5577 Å emission depending on different excitation sources. (b) The tangent of the phase angle φ between the 5577 Å and 4278 Å as a function of the frequency ω as measured in pulsating aurora. (From Henriksen, 1974.)

Now let the effective lifetime against radiation for the $O(^1S)$ state be τ. Then the time variation of the 5577 Å emission ($I_O(t)$) can be expressed by:

$$\frac{dI_O(t)}{dt} = k\,I_N(t) - \frac{I_O(t)}{\tau} \tag{9.21}$$

$I_N(t)$ is the intensity of the 4278 Å emission which is proportional to the production of the excited states. Therefore it is also proportional to the production of the $O(^1S)$ by the proportionality constant k, if direct electron impact is the only source. If now the 4278 Å emission exhibits a sinusoidal variation, then we denote:

$$I_N(t) = I_{N_0} \cdot \exp(i\omega t) \tag{9.22}$$

Due to the phase lag of the 5577 Å emission we express

$$I_O(t) = I_{O_0} \exp(i(\omega t + \varphi)) \tag{9.23}$$

and find:

$$\frac{1}{\tau} + i\omega = k\,\frac{I_{N_0}}{I_{O_0}}(\cos\varphi + i\sin\varphi) \tag{9.24}$$

which gives $\tan\varphi = \omega\tau$.

Given the frequency ω of the pulsations, it is therefore in principle possible to derive the effective lifetime τ of the upper state of the excited atom related to the 5577 Å emission. If one performs a cross-spectral analysis of the time signals and derive φ for each frequency component, then there is a linear relationship between ω and $\tan\varphi$ where the slope in the line is represented by the lifetime τ (Fig. 9.29b).

9.11 THE QUENCHING PROCESS

When a so-called forbidden transition occurs, its upper state is metastable and has a lifetime against radiation which is much longer than the lifetime against spontaneous emission which is of the order of 10^{-7} s. The metastable states of major interest in the aurora are $O(^1D)$, $O(^1S)$ and $N_2(A^3\Sigma_u^+)$ having time constants corresponding to 110, 0.75 and 2 seconds, respectively.

For an emission due to a transition between an upper energy level E_m and a lower level E_n, the probability of radiation is expressed by the Einstein factor A_{nm}. The lifetime for the energy state E_m against a spontaneous transition is then given by

$$\tau_m = \frac{1}{A_{mn}} \tag{9.25}$$

Now there may be several lower levels E_n to which a transition can take place, and each individual transition will have its own Einstein coefficient. The final lifetime for the energy state m can then be expressed as:

$$\tau_M = \left(\sum_n A_{mn}\right)^{-1} \tag{9.26}$$

for all $E_n < E_m$. If, on the other hand, there is a number density N_m of excited states of energy E_m, then the volume emission rate of a photon with wavelength λ corresponding to the transition $E_m - E_n$ will be

$$\eta(\lambda) = N_m \cdot A_{mn} \tag{9.27}$$

If now the upper state with energy E_m is metastable, it may be deactivated by a collision before it can emit a photon and the volume emission rate will be reduced. Let the metastable atom or molecule in the excited state E_m be denoted by $X(m)$, and let it be deactivated by collision with another unspecified agent M. Then this collision can be expressed by:

$$X(m) + M \to X + M^* \tag{9.28}$$

where X is in a lower state or the ground state compared to $X(m)$, and M^* is in any arbitrary state. If this process has a reaction rate denoted by k^q, then the loss of $X(m)$ species per unit time due to these collisions is

$$L_q(N_m) = k^q \cdot N_m \cdot [M] \tag{9.29}$$

where the square brackets mean number densities as usual. In equilibrium the number of excited states produced per unit time $\eta(N_m)$ must equal the number lost per unit time which is the sum of those lost by radiation and those lost by quenching:

$$\eta(N_m) = (A_{mn} + k^q \cdot [M]) N_m \tag{9.30}$$

The emission rate per unit volume now becomes

$$\eta(\lambda) = N_m \cdot A_{mn} = \frac{\eta(N_m)}{A_{mn} + k^q \cdot [M]} \cdot A_{mn} \tag{9.31}$$

where λ is the wavelength corresponding to the energy transition $E_m - E_n$.

Now in the case of the $O(^1S)$ state there are two transition probabilities, $A_{3,2}$ which corresponds to $O(^1S)$–$O(^1D)$ at 5577 Å and $A_{3,1}$ which corresponds to $O(^1S)$–$O(^3P)$ at 2972 Å. Therefore the probability for radiation transition between the ^1S and ^1D states is just a fraction of the total transition probability

$$A_3 = A_{3,1} + A_{3,2} \tag{9.32}$$

from the ^1S state.

Furthermore, the $O(^1S)$ atom may go through different kinds of deactivating collisions such as

$$O(^1S) + O \to O(^3P) + O \tag{9.33}$$

with a rate coefficient k_1^q and

$$O(^1S) + O_2 \to O(^3P) + O_2 \tag{9.34}$$

with a rate coefficient k_2^q. Then the total loss of excited $O(^1S)$ atoms by quenching is given by

$$L(N_m) = (k_1^q[O] + k_2^q[O_2]) \tag{9.35}$$

and the 5577 Å volume emission rate is now

$$\eta(5577\text{ Å}) = \eta(O(^1S)) \cdot \frac{A_{3,2}}{A_{3,1} + A_{3,2} + k_1^q[O] + k_2^q[O_2]} \tag{9.36}$$

The effective lifetime of the $O(^1S)$ state then becomes

$$\tau_{eff} = [A_{3,2} + A_{3,1} + k_1^q[O] + k_2^q[O_2]]^{-1} \tag{9.37}$$

Concerning the 2972 Å emission

$$\eta(2972\text{ Å}) = \eta(5577\text{ Å}) \frac{A_{3,1}}{A_{3,2}} \tag{9.38}$$

For the 6300 Å emission, however, each emission of a 5577 Å photon results in a ^1D state.

The ^1D state is deactivated by collisions between N_2 and O_2 molecules according to the following schemes

$$O(^1D) + N_2 \rightarrow O(^3P) + N_2 \tag{9.39}$$

$$O(^1D) + O_2 \rightarrow O(^3P) + O_2 \tag{9.40}$$

with reaction rate coefficients k_3^q and k_4^q. We now obtain

$$\eta(6300\text{ Å}) = (\eta(O(^1D)) + \eta(5577\text{ Å})) \frac{A_{2,1}}{A_{2,1} + k_3^q \cdot [N_2] + k_4^q \cdot [O_2]} \tag{9.41}$$

where $\eta(O(^1D))$ is the direct excitation rate of the $O(^1D)$ state. For the $O(^1S)$ state the quenching is most important below 115 km while for the $O(^1D)$ quenching is important below 250–300 km. The latter will, however, depend heavily on the thermospheric temperature.

9.12 THE PROTON AURORA

The behaviour of the protons penetrating the atmosphere is fundamentally different from the behaviour of the penetrating electrons. Firstly, the probability of the deflection of protons when colliding with particles in the atmosphere, is almost negligible, and secondly, the possibility of a fast proton capturing an electron to form a fast neutral hydrogen atom is considerable. In a neutral state the hydrogen atom is free to move with respect to the magnetic field until a new collision occurs and the hydrogen atom is stripped off its electron. This proton–hydrogen interchange leads to a horizontal diffusion of the protons rendering the proton aurora less structured than the electron aurora. Fig. 9.30 illustrates the proton–hydrogen interchange and the diffusive action of this process.

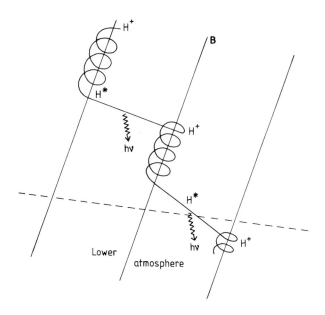

Fig. 9.30. A schematic illustration showing the charge loss for a proton in collision with a hydrogen atom and the charge capture of the latter when colliding with a proton. In the neutral state the hydrogen atom can be excited and emits either H$_\alpha$ (6563 Å) or H$_\beta$ (4861 Å). Due to the velocity of the H* atom along the field line, Doppler shifts will be observed in the emission lines.

After precipitation into the atmosphere the protons lose their kinetic energy through the following collisions with the atmospheric molecules to finally merge into the ambient density of neutral hydrogen:

1: Ionization

$$\mathrm{H}^+ + M \to \mathrm{H}^+ + M^+ + e \qquad (\Delta_{p1}) \qquad (9.42)$$

2: Excitation

$$\mathrm{H}^+ + M \to \mathrm{H}^+ + M^* \qquad (\Delta_{p2}) \qquad (9.43)$$

3: Charge capture

$$\mathrm{H}^+ + M \to \mathrm{H}^* + M^{*(+)} \qquad (\Delta_{p3}) \qquad (9.44)$$

The neutral hydrogen will then be involved in the following collision processes:

4: Charge loss or ionization

$$\mathrm{H} + M \to \mathrm{H}^+ + M^* + e \qquad (\Delta_{H1}) \qquad (9.45)$$

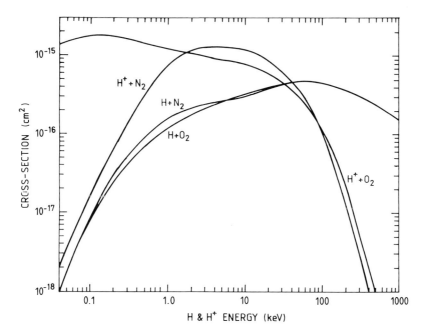

Fig. 9.31. Cross-sections for electron capture and electron stripping for H^+ and H in nitrogen and oxygen, respectively. (After Rees, 1989.)

5: Double ionization

$$H + M \rightarrow H^+ + M^{*(+)} + 2e \qquad (\Delta_{H2}) \qquad (9.46)$$

6: Excitation

$$H + M \rightarrow H^* + H \qquad (\Delta_{H3}) \qquad (9.47)$$

Here M^* and M^+ represent excited and ionized neutral atmospheric species, and the collision cross-sections Δ_{p1}, Δ_{p2}, Δ_{p3}, Δ_{H1}, Δ_{H2} and Δ_{H3} are indicated for each process, respectively. These cross-sections (Fig. 9.31) are typical between 10^{-23} and 10^{-19} m² in the energy range 0.1 keV to 1 MeV. Maximum cross-sections are found, however, around 10 keV.

Let us assume that at a given point in the beam of precipitating particles it contains $n(p)$ protons and $n(H)$ hydrogen atoms per unit volume (Fig. 9.32).

If Δ_p and Δ_H are the electron capture and loss cross-sections, respectively, then the variation in the number density of protons in the beam as it traverses at distance dz along the beam

$$dn(p) = \Delta_H \cdot dN \cdot n(H) - \Delta_p \, dN \cdot n(p) \qquad (9.48)$$

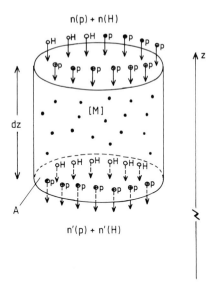

Fig. 9.32. A schematic drawing to illustrate a beam of protons and hydrogen atoms passing through a distance dz of the atmosphere. The sum of the densities of the protons and the hydrogen atoms must be the same at the top of the cylinder with area A as at the bottom.

where $dN = [M] \cdot dz$ is the number per unit area of neutral particles traversed by the beam. Let us consider a part of the beam in the form of a cylinder with cross-section A and height dz. After penetrating the height dz the densities of the proton and hydrogen atoms in the beam are $n'(p)$ and $n'(\mathrm{H})$, respectively, and since no particle is lost from the cylinder in Fig. 9.32

$$n(p) + n(\mathrm{H}) = n'(p) + n'(\mathrm{H}) = n_0 \qquad (9.49)$$

where n_0 is a constant number density. Let the relative number of protons be denoted by α; then

$$\alpha = \frac{n(p)}{n_0} = \frac{n(p)}{n(p) + n(\mathrm{H})} \qquad (9.50)$$

and

$$d\alpha = \frac{d(n(p))}{n_0} = \left(\Delta_{\mathrm{H}} \frac{n(\mathrm{H})}{n_0} - \Delta_p \frac{n(p)}{n_0} \right) dN \qquad (9.51)$$

Since $n(\mathrm{H}) = n_0(1 - \alpha)$, we have:

$$\frac{d\alpha}{dN} = -(\Delta_{\mathrm{H}} + \Delta_p)\alpha + \Delta_{\mathrm{H}} \qquad (9.52)$$

This is a linear differential equation with constant coefficients which can be solved analytically; and the solution is

$$\alpha = \frac{\Delta_H}{\Delta_H + \Delta_p} + \mathcal{C}\exp[-(\Delta_H + \Delta_p)N] \tag{9.53}$$

where $\mathcal{C}$ is a constant to be determined according to the boundary conditions. At the top of the column where $N = 0$, all particles in the beam are protons ($\alpha = 1$), and therefore

$$\alpha = \frac{\Delta_H}{\Delta_H + \Delta_p}\left(1 + \frac{\Delta_p}{\Delta_H}\exp[-(\Delta_H + \Delta_p)N]\right) \tag{9.54}$$

A steady-state value of α given by

$$\alpha_s = \frac{\Delta_H}{\Delta_H + \Delta_p} \tag{9.55}$$

is reached at a rate determined by

$$N_s = \frac{1}{\Delta_H + \Delta_p} \tag{9.56}$$

From Fig. 9.31 we notice that Δ_p and Δ_H are typically of the order of 5×10^{-20} m^2 in the energy range between 10 and 300 keV. The total number N of traversed neutral particles in the column must therefore be larger than 2×10^{19} m^{-2} before equilibrium can be reached. Since this occurs at a height close to 300 km in the Earth's atmosphere, the steady state is virtually completed in the beam below this height. Table 9.4 gives the ionization degree and the altitude for 90% of the ionization degree for different initial energies of the protons in keV.

The incident proton flux will dominate when the beam has a high energy, while neutral hydrogen will dominate at lower energies.

Table 9.4

Proton energy ε (keV)	Ionization degree $\Delta_p/(\Delta_p + \Delta_H)$	Height of 90% of ionization degree (km)
3	0.11	310
10	0.26	305
30	0.46	295
100	0.80	268
300	0.99	252

The Norwegian physicist Vegard observed in the years 1939–41 in Tromsø emissions from the hydrogen lines H$_\alpha$ (6563 Å) and H$_\beta$ (4861 Å) aurora showing clear frequency shifts toward shorter wavelengths when observed along the magnetic field line (Fig. 9.33a). Observed perpendicularly to the field line the H_α and H_β do not show a Doppler shift but rather a Doppler broadening (Fig. 9.33b). These observations do indicate that the emissions are due to the hydrogen atoms moving toward the observer with a relative speed. The average Doppler shift corresponds to a velocity of 3–400 km/s or a kinetic energy of 0.5–1 keV. This finding, however, does not represent the initial energy of the protons but rather the intermediate energy after the protons have been slowed down to kinetic energies in the neighbourhood of the maximum cross-section for hydrogen excitation. The wing of the emission line in Fig. 9.33b does correspond to a speed of 2000 km/s, however. Some of the incident protons must therefore have energies of the order of 20 keV or more.

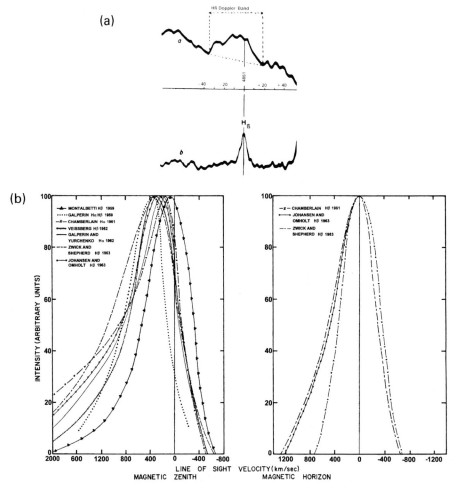

Fig. 9.33. (a) Original observations done by Vegard, showing the Doppler shift of the H_β line and also a broadening of the line due to the velocity distribution of the H* atoms along the magnetic field line. (From Vegard, 1952.) (b) Observed hydrogen line profiles. (From Eather, 1967.)

Vegard interpreted his measurements as evidence for protons penetrating the polar atmosphere together with electrons during an aurora and that the emissions observed are due to excited hydrogen atoms moving with high downward velocities due to the charge exchange with the protons, as explained. This represented an important support to the idea that the auroral particles originated in the solar atmosphere.

9.13 THE AURORAL SUBSTORM

It has become customary to divide a major auroral event into 4 phases, more or less clearly discernible from each other. The complete sequence of these phases, however, is often called the auroral substorm. The auroral substorm constitutes only a small event in a much larger and radical change of the magnetosphere, the magnetospheric substorm, but it is only through the aurora that nature exposes its inner turmoil of the magnetosphere to the naked eye.

The four phases of this display are the quiet phase the growth phase, the expansion phase and the recovery phase.

9.13.1 The quiet phase

Usually, when the interplanetary magnetic field is northward, the magnetosphere is in a stationary state, and momentum is transported from the solar wind to the magnetospheric tail plasma through plasma waves and frictional forces of some kind. This transfer of momentum induces a convection motion in the magnetosphere as already described in Chapter 7. As we have seen, there is a westward cross-tail or dawn–dusk electric field, which is transferred to the ionosphere along the magnetic field lines which eventually set up a twin cell current system, S_q. The auroral activity is generally quiet, and the electron aurora appears poleward of the oval; sometimes cross-polar-cap auroras might be seen (Fig. 9.34). The auroral motion is slow and directed toward the equator around midnight, and eastward and westward in the morning and evening hours, respectively. These motions are replicas of the general convective motion of the magnetosphere as just described.

9.13.2 The growth phase

The phase lasts for up to 2 hours before the onset of the dynamic process. Usually the onset of the substorm is associated with a turning of the interplanetary field from northwards to southwards. This implies the formation of an eastward current in front of the magnetopause on the dayside, which leads to a reconfiguration of the dayside magnetic field system. Magnetic field lines are probably swept toward the nightside carrying plasma across the polar cap from the dayside to the nightside. This enhances the convection across the polar cap and also the S_q current cells. Due to this enhanced convection on the afternoon side where the electron density is relatively high, a positive magnetic bay is often observed in the afternoon at high latitudes during a growth phase, in agreement with an increase in the eastward current.

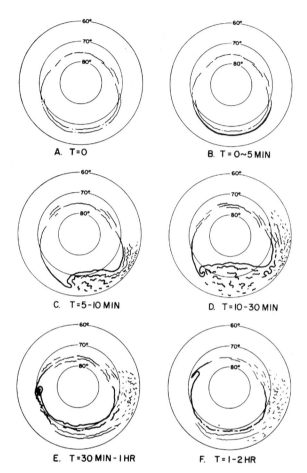

Fig. 9.34. The auroral substorm phases visualized by the expansion and contraction of the auroral oval as well as the occurrence of different auroral forms. (a) $T = 0$ min. Weak auroral arcs occur in the midnight region at high geomagnetic latitude ($\lambda \approx 75°$). (b) $T = 0 \sim 5$ min. The southernmost arc brightens and moves southwards. The onset phase. (c) $T = 5$–10 min. The bright arc folds and forms a bulge which expands northward and moves westward at a high speed; this is called the westward travelling surge. South of the bulge patchy forms are formed. Expansion phase starts. (d) $T = 10$–30 min. The bulge continues to expand westward and northward, and patchy and pulsating aurora occurs on the morning side. Expansion phase maximum. (e) $T = 30$–60 min. Several weak arcs are formed which start to contract toward the north. The pulsating aurora continues at lower latitudes in the morning hours. The recovery phase starts. (f) $T = 1$–2 hrs. The disturbance weakens and the situation retreats to the pre-onset condition except the pulsating patches in the morning hours may continue for some time. A new substorm may occur after 2 to 3 hours. Recovery phase continues. (From Akasofu, 1968.)

The magnetospheric tail also becomes more strongly stretched due to the enhanced transfer of energy from the solar wind. The plasma sheet becomes thinner as plasma is forced toward the equatorial plane. The cross-tail current is enhanced and the dipolar field closer to Earth is weakened at the inner edge of the plasma sheet. The plasma sheet can move closer towards the Earth. The high-energy trapped particles will have their trapping region disturbed by the intrusion of the plasma sheet particles. Pitch-angle diffusion occurs, and some particles penetrate into the atmosphere at the equatorward edge of the auroral oval where they form brighter auroral arcs moving equatorward as the plasma sheet moves closer towards the Earth. This equatorward arcs are probably formed at the boundary between the closed and open field lines, and the particles have rather high energies (> 21 keV).

9.13.3 The expansion phase
This phase starts at the moment an abrupt change is observed in the auroral display and lasts for about 20 min. It often begins with a sudden brightening of the most equatorward auroral arc and is followed by a rapid poleward expansion of the auroral arcs, indicating that a large volume of the magnetosphere suddenly is becoming a source of auroral particles. The most dramatic effects are seen around the midnight sector at auroral latitudes as a bulk of auroras grows and rapidly moves poleward also accompanied by a strong vortex-like auroral form moving westward at high speed, often termed the westward travelling surge.

There are auroral motion on the morning side, too, at the lower latitude region of the auroral oval. The morning side auroral forms are more patchy and display quasiperiodic intensity variations, auroral pulsations, often associated with X-ray pulsations, indicating fairly high-energy precipitating particles.

The horizontal magnetic field component decreases very strongly at stations in the midnight sector due to an enhancement of the westward auroral electrojet. Stations on the afternoon side may still observe positive enhancements in the horizontal component while stations on the evening side closer to the main centre of the substorm see a gradual change from positive to often strongly negative magnetic bays. Close to the substorm centre the enhancement of the negative bay may become very abrupt.

During the expansion phase the cross-tail current in the near-Earth plasma sheet collapses, and the current is probably routed down along the magnetic field lines and short-circuited through the ionosphere forming the auroral electrojet.

9.13.4 The recovery phase
The poleward expansion of the auroral bulk comes to a halt, and the auroral emission starts gradually to retreat equatorward and finally returns to its equatorward edge of the oval. This process may take from 30 min to 2 hours. While most of the evening aurora fades away, the morning side aurora becomes even fainter and more patchy. The poleward auroral arcs and bands remain quite

stable and can be observed for a considerable length of time. The magnetosphere returns back to a stronger dipolar shape, and will remain more dipole-like until a new event is triggered for some unknown reason.

— ooo —

The scenario outlined above can be said to be typical, although one auroral substorm is as different from another as most individual natural phenomena. The substorm, however, can be regarded as a process in which electromagnetic energy supplied from the solar wind is stored in the magnetosphere until the energy reaches such a magnitude that the magnetosphere no longer can sustain its thrust. The situation is unstable, and return to a new stability can only happen by a sweeping collapse of the magnetosphere. This can be envisaged by a short-circuiting of the cross-tail current through the ionosphere, creating the most gigantic discharge displaying itself in the Earth's atmosphere.

This summary is not intended to be the ultimate explanation of this phenomenon, as this will continue to be a matter of conflicting opinions for years to come. It is, however, a phenomenon in the framework of plasma physics and of central interest to similar astrophysical phenomena of different kinds, such as formation of solar flares and similar emissions from other celestial bodies. The auroral substorm and its magnetospheric cradle represent the closest and most familiar natural display of a fundamental plasma discharge of astrophysical scope. Therefore, through intensive studies of the auroral substorm, from the ground as well as from space, we can obtain new knowledge within plasma physics which can expand our understanding of the whole universe. If we are not able to understand the processes behind these phenomena which we can probe with a multitude of instruments, how can we imagine we can understand similar phenomena light years away.

9.14 REFERENCES

Akasofu, S.-I. (1968) *Polar and Magnetospheric Substorms,* D. Reidel, Dordrecht, The Netherlands.

Akasofu, S.-I. and Chapman, S. (1972) *Solar–Terrestrial Physics,* Oxford, at the Clarendon Press, London.

Berger, M. J., Seltzer, S. M. and Maeda, K. (1970) *J. Atmos. Terr. Phys.,* **32**, 1015–1045.

Birkeland, K. (1913) *The Norwegian Aurora Polaris Expedition 1902–03,* Vols. I and II, Aschehoug, Christiania, Norway.

Brekke, A. and Egeland, A. (1983) *The Northern Light. From Mythology to Space Research,* Springer-Verlag, Berlin, Heidelberg, New York, Tokyo.

Brekke, A., Hall, C. and Hansen, T. L. (1989) *Ann. Geophysicae,* **7**, 269–280.

Eather, R. H. (1967) *Rev. Geophys.,* **5**, 207–285.

Frank, L. A. and Craven, J. D. (1988) *Rev. Geophys.,* **26**, 249–283.

Fritz, H. (1881) *Das Polarlicht*, F.A. Brockhaus, Leipzig.
Hällström, G. G. (1847) *Acta. Soc. Scient. Fenn.*, **2**, 363–376.
Hansteen, C. (1827) *Philos. Mag. Ann. Philos. New Ser.*, **2**, 333–334.
Henriksen, K. (1974) Thesis, The Auroral Observatory, Tromsø, Norway.
Kamide, Y. (1988) *Electrodynamic Processes in the Earth's Ionosphere and Magnetosphere*, Kyoto Sangyo University Press, Kyoto, Japan.
Kamiyama, H. (1966) *Rep. Ionosph. Space Res. Japan*, **20**, 171–187.
Luhmann, J. G. (1995) in *Introduction to Space Physics*, Kievelson, M. G. and Russell, C. T. (Eds.), p. 568, Cambridge University Press, Cambridge.
Nevanlinna, H. (1995) *J. Geomag. Geoelectr.*, **47**, 953–960.
Nordenskiöld, A. E. (1880–81) *Vegas Färd Kring Asien och Europa*, F. & G. Beijer, Stockholm.
Rees, M. H. (1989) *Physics and Chemistry of the Upper Atmosphere*, Cambridge Atmospheric and Space Science Series, Cambridge University Press, Cambridge.
Réthly, A. and Berkes, Z. (1963) *Nordlichtbeobachtungen in Ungarn (1523–1960)*, Akadémiai Kiadó, Verlag der Ungarischen Akademie der Wissenschaften, Budapest.
Rubenson, R. (1879, 1882) *Kungl. Svenska Vetensk. Acad. Handl.*, Part **1**, 15 (5); Part **2**, 18 (1).
Sharp, R. D. and Johnson, R. G. (1974) in *The Radiating Atmosphere*, McCormack, B. M. (Ed.), D. Reidel, Dordrecht, The Netherlands, p. 239.
Starkov, G. V. and Feldstein, Ya. I. (1967) *Geomagnetism and Aeronomy*, **7**, 48–54.
Størmer, C. (1955) *The Polar Aurora*, Oxford University Press, London.
Tohmatsu, T. (1990) *Compendium of Aeronomy*. Translated and revised by T. Ogawa, Terra Sci. Publ. Comp., Tokyo.
Vallance Jones, A. (1974) *The Aurora*, D. Reidel, Dordrecht, The Netherlands.
Vegard, L. (1952) *Geofys. Publ.*, **18**, No. 5.
Winningham, J. D. and Heikkila, W. J. (1974) *J. Geophys. Res.*, **79**, 949–957.

9.15 EXERCISES

1. What is the most dominant emission of the aurora?

2. When an aurora is moving, what colour will you observe in the front of the aurora, and what colour will the trailing edge have?

3. Assume that the $O(^1S)$ also is produced by an indirect chemical reaction such as the one indicated by (9.18). Let the time-dependent density of the $N_2(A^2\Sigma)$ state be indicated $N_2(t)$. We can then write the continuity equation for $I_O(t)$ as follows:

$$\frac{dI_O(t)}{dt} = k_1 I_N(t) + k_2 N_2(t) - \frac{I_O(t)}{\tau_1} \qquad (9.57)$$

We also assume that the production of the $N_2(A^2\Sigma)$ state is proportional and in phase with the ionization and excitation of the N_2^+ ions. The continuity equation for $N_2(A^2\Sigma)$ can then be written as:

$$\frac{dN_2(t)}{dt} = k_2 I_N(t) - \frac{N_2(t)}{\tau_2} \qquad (9.58)$$

Find now the phase–angle relationship between $I_O(t)$ and $I_N(t)$.

Symbols

UNIVERSAL CONSTANTS

c	– the speed of light ($= 3 \times 10^8$ m/s)
σ	– the Stephan–Boltzmann constant ($= 5.67 \times 10^{-8}$ W m^{-2} K^{-4})
m_e	– electron mass ($= 9.1 \times 10^{-31}$ kg)
m_p	– proton mass ($= 1.672 \times 10^{-27}$ kg)
ε_0	– the permittivity constant ($= 8.854 \times 10^{-12}$ F/m)
μ_0	– the permeability constant ($= 4\pi \times 10^{-7}$ H/m)
m_0	– one atomic unit (1 amu $= 1.660 \times 10^{-27}$ kg)
N_A	– Avogadro's number ($= 6.02 \times 10^{23}$ molecules/mole)
R_0	– the universal gas constant ($= 8.3$ J/mole K)
κ	– the Boltzmann constant ($= 1.38 \times 10^{-23}$ J/K)
h	– Planck's constant ($= 6.63 \times 10^{-34}$ J/s)
e	– electronic charge ($= 1.60 \times 10^{-19}$ C)
G	– the constant of gravity ($= 6.67 \times 10^{-11}$ N m^2 kg^{-2})

PLANETARY CONSTANTS USED IN THE TEXT

R_e	– the mean radius of the Earth ($= 6.37 \times 10^6$ m)
$R_\odot$	– the mean solar radius ($= 6.96 \times 10^8$ m)
M_e	– the Earth's mass ($= 5.98 \times 10^{24}$ kg)
$M_\odot$	– the solar mass ($= 1.99 \times 10^{30}$ kg)
E_e	– the solar constant at 1 AU ($= 1380$ W/m^2)
1 AU	– one astronomical unit ($\approx 1.5 \times 10^{11}$ m)
H_0	– the mean magnetic field strength at the Equator ($= 3.0319 \times 10^{-5}$ tesla)
M_0	– magnetic dipole moment of the Earth ($= 7.91 \times 10^{22}$ A m^2) ($H_0 R_e^3 = 7.91 \times 10^{15}$ Wb m)
$g_\odot$	– acceleration of gravity at the solar surface ($= 2.74 \times 10^2$ m/s^2)
H	– mean scale height of the Earth's atmosphere ($= 8.43$ km at the surface)
α^*	– the adiabatic lapse rate ($= -9.8$ K/km at the Earth's surface)

Symbols

- v_{esc} – escape velocity ($= 11.2$ km at the Earth's surface)
- $v_{\odot e}$ – the escape velocity at the solar surface ($= 6.2 \times 10^5$ m/s)
- r_a – aphelion (1.46×10^{11} m)
- r_p – perihelion (1.52×10^{11} m)
- $T_\odot$ – mean solar radiation temperature ($= 6000$ K)
- T_e – mean temperature of the Earth ($= 288$ K)
- γ – adiabatic constant of the atmospheric gas ($= 1.4$ at the Earth's surface)
- M' – molecular mass number ($= 28.8$ of the Earth's atmosphere close to the surface)
- δ – angle between the Earth's magnetic and rotation axis ($= 11.2°$)
- ϕ_p – geographic longitude of the north magnetic dipole pole ($= 289.1°$ E)
- λ_p – geographic latitude of the north magnetic dipole pole ($= 78.8°$ N)
- c_v – the heat capacity at constant volume ($= 712$ J/K kg for air at the Earth's surface)
- c_p – the heat capacity at constant pressure ($= 996$ J/K kg for air at the Earth's surface)
- $\bar{V}_p$ – mean ionization threshold in air ($= 15$ eV)
- $\bar{\varepsilon}$ – mean energy for an ion-pair production in air ($= 34$ eV)
- Ω – rotation frequency of the Earth ($= 7.27 \times 10^{-5}$ s^{-1})
- g – acceleration of gravity at the Earth's surface ($= 9.81$ m/s^2)

Index

aa index, 24
acoustic mode, 60
acoustic waves, 311
adiabatic constant (γ), 35f, 258f
adiabatic invariant, 173ff
adiabatic lapse rate, 89–91
aerosol, 120
Africa, 152
agonic line, 147
albedo (α), 69f
Alfvén velocity, 58
Alfvén waves, 60–61
ambipolar diffusion, 221ff
Anchorage, 193
Antarctic, 105f, 131
antipodal, 24
aphelion, 12
A_p index, 297, 343
Appleton, E., 191
Archimedean spiral, 51
Atlantic anomaly, 131, 132
Atlantic Ocean, 131, 132
atmosphere of the Earth, 65ff, 258ff
atomic unit (1 a.m.u.), 82
attachment, 217f
aurora, 150, 192, 405, 435ff
aurora australis, 405
aurora borealis, 405
auroral arc, 362f, 424f
auroral oval 149, 151, 190, 446

auroral particles, 452ff
auroral spectrum, 450f
auroral substorm, 477ff
averaged speed, 81
Avogadro's number (N_A), 80

ballistic coefficient, (C_D) 77
Balmer lines, 8
Barnett, 191
barosphere, 66
Bartels, J., 2
Bernoulli, 37, 38
Birkeland, K., 347, 446
Birkeland current, 352f, 428f
Birkeland's terrella, 446
black body, 9, 10, 68f
Bode's law, 155
Boltzmann constant (κ), 10, 35, 80
Boström, R., 378, 382
Brahe, T., 435, 436
Breit, G., 191
bremsstrahlung effect, 465
brightness (B_ν), 9–14
Brunches epoch, 152
Brunt–Väisälä frequency, 303f
butterfly diagram, 26

Canada, 145, 147, 444
carbon cycle, 1
C^{14} concentration, 22

Celsius, A., 438, 440
CFC gas, 118–121
Chapman ionization profile, 209ff
Chapman, S., 99
characteristic potential (V_p), 206
charge exchange, 216f
Chasmata, 435
Chatanika Alaska Radar, 195, 284f, 346
China, 150, 435
chromosphere, 3, 4, 8, 9, 10
coefficient of viscosity (η), 407
collision frequencies
    electron collision frequency, 315f
    ion–neutral collision frequency, 314f
collision rate, 315f
conduction, 92
conductivity, 35, 44f, 316f, 325f, 407f
constant of gravity (G), 39–43, 49–50, 77
convection pattern, 426ff
convection zone, 3, 4
Copenhagen, 438
core field, 153
Coriolis force, 264f, 306
corona, 3, 4
corotation, 247f
cosmic noise absorption, 29
cosmic rays, 233f
Cowling conductance, 347
C-region, 191
critical frequency, 194
cross-section for collisions (σ), 87
crustal field, 153
current density, 44
current sheet, 393f
currents in the ionosphere, 313ff
curvature drift, 170
cycloid, 164
cyclotron motion, 158f

dawn–dusk electric field, 415f
declination, 134f
Definite Geomagnetic Reference Field (DGRF), 137
de Mairan, J. J., 441
DE–1 satellite, 146, 148
detachment, 217f

diffusion, 96ff
diffusion coefficient, 96f, 222f, 317f
dipole, 137f, 410
dipole field, 137f, 153, 391f
dipole latitude, 133ff
displacement current, 45
diurnal bulge, 282
D-layer, 191
Dobson units, 105–123
Dominion Astronomical Observatory, Ottawa, 14
Doppler shift, 286
drag force (F_D), 77, 279f
D-region, 191ff
drift motion, 158f

Earth, 1, 21, 24
Earth radius, (R_e) 68, 130
Earth's angular velocity, 262ff
Earth's atmosphere, 65ff
Earth's magnetic field, 24, 127ff, 389f
Earth's magnetic moment (M), 130ff, 389f
Earth's rotation axis, 130
Eddy, J. A., 22
EISCAT, 293f, 316, 332f, 358, 375
E-layer, 191
El Chichon, 120
electron density (n_e), 14, 16, 30
electronic charge (e), 14
electron mass (m_e), 14
electron–neutral collision frequency, 315f
electron temperature, 33ff
electrostatic potential, 166f, 363f
energy deposition, 458
energy state, 469f
equatorial anomaly, 193
equatorial fountain effect, 248
equatorial plane, 140f
equinox, 75
equivalent currents, 347f
E-region, 191ff, 290f
Europe, 152
EUV radiation, 8–29, 206f, 282f
excitation rate, 458f, 466f

Index

exobase, 67
exosphere, 86–88
exospheric temperature (T_∞), 76
expansion phase, 79
Explorer III, 78, 79

Fe, 8
Feldstein, 446
Ferraro's theorem, 243f
Ferrel cell, 270f
Fick's law, 96
field-aligned currents, 427f
first-order drift, 168ff
F-layer, 191
forbidden region, 182f
Fort Churchill, 272, 274
fountain effect, 194
$F_{10.7}$ radio flux, 14ff, 23
F-region, 191ff, 291f
Fritz, H., 443
frozen-in magnetic field, 44f, 405f

Galilei, G., 18
gamma ray, 1
garden hose, 50
gas constant, 80f, 257f
Gassendi, P., 438
Gauss epoch, 152
Gauss, G. F., 130, 445
Gaussian coefficients, 136f
Geneva, 438
geographic coordinates, 132ff
geomagnetic coordinates, 132ff
geomagnetic index (A_p), 76
geomagnetic pole, 131f
geomagnetism, 127ff
geostrophic wind, 264ff
Germany, 435
Gilbert, W., 127f
gravitational constant, 77
gravity waves, 310
Greece, 435
greenhouse effect, 68–72
Greenland, 131, 150, 437, 444, 445

group velocity, 59
growth phase, 477f
Guiding Centre System, 159f
gyrofrequency, 316f
gyro period, 159f
gyroradius, 159f

H_α, 42, 406, 472f
H_β, 472f
Hadley cell, 270f
Hall conductivity, 316f
Halley, E., 127, 129, 437
Halley Bay, 108–109
Hansteen, C., 445
Harang, L., 348–349
Harang discontinuity, 352
Hawaii, 152
He, 1, 2
He^+, 8
heating efficiency, 92
heliocentric latitude, 2
heliosphere, 36
helium line, 8
Herzberg continuum, 208
heterosphere, 65
Hiorter, O., 440
homosphere, 65ff
Hough function, 302
Huancayo, 193
hydrated ions, 235
hydrogen emissions, 471ff

Iceland, 444
ideal gas, 80f, 247f
inclination, 134f
Industrial Revolution, 123
infrared emission from Sun, 9, 17
infrared radiation from Earth, 70
internal waves, 306f
International Geomagnetic Reference Field (IGRF), 137, 138
International Geophysical Year (IGY), 361, 446
interplanetary magnetic field (IMF), 33f,

350f, 412f
interplanetary medium, 1
interplanetary space, 5, 56
invariant latitude (λ_m), 141
ionization, 5, 191ff, 461
ionization cross-section, 202ff
ionosphere, 10, 24, 44, 191ff
ionospheric currents, 24
Italy, 435

Japan, 435
Joule heating, 287f, 329f
Jupiter, 21

Kamide, Y., 365
Kamide–Richmond–Matsushita (KRM) method, 365f
Kokubun, S., 370
K_p index, 251–252
Krakatau, 119f

Laplace equation, 139
lapse rate, 88
Larmor radius, 159f
Laval's nozzle, 36
Legendre's functions, 302
London, 127f, 147
longitudinal mode, 59
Longyearbyen, Svalbard, 287
Lorentz force, 165f
loss rate (L_T), 91–93
L-value, 141f
Ly$_\alpha$, 8, 29, 206f
Ly$_\beta$, 8

Madrid, 443, 449
magnetic components, 134f
magnetic dip, 195
magnetic equator, 195
magnetic field, 33ff, 127ff
magnetic field merging, 402f
magnetic fluctuation, 344f
magnetic force, 403f
magnetic mirror, 175

magnetic moment, 130f
magnetic moment of
  78 Virginis, 155
  Earth, 155, 389f
  Hercules X-1, 155
  Jupiter, 155
  Mercury, 155
  Moon, 155
  Saturn, 155
  Sun, 155
  Venus, 155
magnetic polarity, 27
magnetic potential, 134f
magnetic reconnection, 403
magnetic tail, 399f
magnetometer, 136
magnetopause, 389, 400
magnetosheath, 400f
magnetoshock, 400
magnetosphere, 44, 389ff
magnetospheric convection, 244
Mars, 21
mass of Earth (M_e), 77
Matuyama epoch, 152
Maunder minimum, 22
Maxwell–Boltzmann distribution, 80ff
Maxwell's equations, 45ff, 171, 245, 313, 407f
mean molecular mass, 73, 80
Mercury, 21
Merry Dancers, 435
mesopause, 66, 67, 123
mesosphere, 8, 66, 67, 121–122
metastable state, 449
methane (CH_4), 121–122
Mg, 8
micropulsation, 62
mirror reflection motion, 158f
mobility coefficient, 316f
Moen, J., 358
molecular mass, 80ff
molecular mass number, 80ff
Moon, 21
most probable speed, 81

Index

M-regions, 2
Mt. Pinatubo, 120
Muncke, 443

N, 8
Nagata, T., 370
Neptune, 21
neutral wind, 271ff
neutrino, 1, 2
noctilucent clouds, 121f
Nordenskiöld, A., 445
Norse mythology, 435
North America, 152
Norway, 145, 150
Novaya Zemlja, 444

O, 8
Ohm's law, 44, 363f, 407f
Oslo, 147
oxygen atmosphere, 99–103
ozone (O_3), 69, 72, 99ff, 297f
ozone layer, 103ff, 241

Paris, 147, 438
partial ring current, 254, 423
particle energy, 457f
particle precipitation, 457
particle spectra, 452f
penumbra, 25
Pedersen conductivity, 316f
perihelion, 12
permeability in vacuum (μ_0), 45
permitted region, 182f
permittivity of a vacuum (ε_0), 14
photoabsorption, 92
photodissociation, 206
photoelectron, 209
photoionization, 92, 206
photons, 3
photosphere, 3, 4, 8, 30
pitch angle, 160f
Planck's constant (h), 10
Planck's radiation law, 10, 12
plasma, 35

plasma frequency, 14, 30
plasmahorn, 254
plasma sheath, 254
plasmasphere, 241ff
polar warming, 66
positron, 1, 2
potential function, 263
potential temperature, 259
Poynting vector, 411
production rate (q_T), 91–93, 202ff
proton aurora, 471f
proton–proton chain, 1
protons, 34, 186f
proton temperature, 34

quenching, 469
quiet Sun, 33

radiation belt, 186f
radiation flux, 14, 15
radiation intensity, 202ff
radio bursts, 14, 29, 30, 32
radio wave absorption, 66
recombination, 213ff
recombination coefficient, 213ff
recovery phase, 479
Reynold's number, 49
rigidity, 186
Rome, 443
root mean square velocity, 81
Rubenson, R., 441–443

Sabine, 440
satellite, 76f, 389
satellite orbit, 77
Saturn, 21
scale height, 30, 50, 66ff, 204ff, 260f, 304f
Scandinavia, 150
Schuman region, 7
Schuster, A., 191
Schwabe, H., 18, 440
Schøning, G., 436
scintillation, 5
Scotland, 435

second-order drift, 172f
secular variations, 145
shock wave, 389, 390
Si, 8
Siberia, 150, 444
sidereal day, 32
Silsbee, 349
solar
    acceleration, 1
    activity, 10
    arcs, 7
    atmosphere, 3
    bursts, 29
    constant, 12, 69
    convection zone, 3
    core, 3
    corona, 3ff
    cycle, 9ff, 66f
    day, 32
    disc, 17
    eclipse, 5, 9, 42
    electromagnetic radiation, 28
    escape velocity, 1
    fans, 7
    flare, 29, 405, 406
    magnetic field, 7, 16ff, 42
    mass, 1, 39, 49
    mass density, 1
    maximum, 6, 16, 66f
    minimum, 6, 16, 27, 66f
    plages, 14
    plumes, 7
    radiation zone, 3
    radio emission, 12ff, 72f, 80
    radius, 1, 12, 33, 40–42
    rotation, 2
    spectral irradiance, 9, 10, 12
    spectrum, 7, 11
    temperature, 1
    wind, 33, 33ff, 389f, 416
solstice, 74, 75
South America, 131, 132
specific heat for constant pressure (c_p), 89f, 258f

specific heat for constant volume (c_v), 89f, 258f
spectral brightness (B_ν), 10, 12
speed of sound (c_s), 37–42
spherical harmonics, 136f
spiculae, 3
spontaneous transition, 469
Sputnik III, 78, 79
Spörer minimum, 22
S_q current, 360f
Starfish, 188
Stephan–Boltzmann constant (σ), 11, 68
Stephan–Boltzmann law, 68
Stewart, B., 191
St. Helens, 119
stopping altitude, 461
stopping cross-section, 92–95
St. Petersburg, 438
stratopause, 65, 66
stratosphere, 65, 66
stratospheric warming, 66
stream function, 365
Størmer, C. F., 178ff, 439, 440
Størmer length, 180f
Størmer-tron, 186
subsonic, 37–42
Sun, 1ff
sunspot number, 19
sunspots, 2ff
supersonic, 37–42
synodic angular velocity, 2

tail lobes, 414
temperatures
    atmospheric, 65ff
    electron, 196ff
    proton, 196ff
terella, 128
the Bible, 435
the Einstein factor, 469f
The King's Mirror, 150, 437
the northern light, 150
thermopause, 282f
thermosphere, 8, 282ff

tidal oscillations, 297f
TIROS, 109–110
Tokyo, 147
TOVS, 109–110
transversal mode, 58
triangulation, 439, 440
Tromholt, S., 441, 442
tropopause, 65f, 123
troposphere, 65
turbosphere, 66
Tuve, M. A., 191

umbra, 25, 26
unipolar inductor, 157f, 249
universal constant (R_0), 80
Uppsala, Sweden, 438
Uraniborg, 435
Uranus, 21
USA, 147
UV-radiation, 7ff, 28f, 72

van Allen belt, 186
Vanguard, 78, 79
variometer, 136
Vega Expedition, 445
Venturi tube, 38
Vestine, E. H., 349
Viking satellite, 146, 421
visible radiation, 7–9, 17

Washington, 193
water vapour, 298
westward travelling surge, 478
Wien's displacement law, 10, 70
Wolf, R. A., 18, 440
Wolf sunspot number (R), 18, 22

X-ray, 1ff, 28f, 206

zenith angle, 202ff, 358, 359
zeroth-order drift, 163f

Ørsted, 440